PCR
PROTOCOLS

PCR PROTOCOLS

A GUIDE TO METHODS AND APPLICATIONS

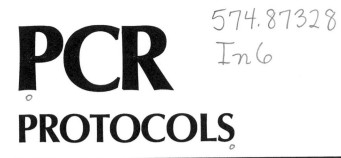

Edited by

Michael A. Innis,
David H. Gelfand, John J. Sninsky
Cetus Corporation, Emeryville, California

Thomas J. White
Hoffmann-La Roche, Inc., Emeryville, California

ACADEMIC PRESS, INC.
Harcourt Brace Jovanovich, Publishers

San Diego   New York   Berkeley   Boston   London   Sydney   Tokyo   Toronto

Academic Press, Inc., San Diego, California 92101

United Kingdom Edition published by
Academic Press Limited, 24–28 Oval Road, London NW1 7DX

Library of Congress Cataloging-in-Publication Data

PCR protocols.
    Includes index.
    1. Polymerase chain reaction. 2. Gene amplification.
I. Innis, Michael A.   [DNLM: 1. DNA Polymerases.
2. Gene Amplification--methods. 3. Genetic Engineering
--methods. 4. RNA Polymerases.   QH 442 P3479]
QP606.D46P36   1989   574.87'328   89-6938
ISBN   0-12-372180-6   (alk. paper)
ISBN   0-12-372181-4   (pbk. : alk. paper)

Printed in the United States of America
89 90 91 92   9 8 7 6 5 4 3 2 1

CONTENTS

Contributors   xi
Preface   xvii

Part One _____
BASIC METHODOLOGY

1. Optimization of PCRs   3
   Michael A. Innis and David H. Gelfand

2. Amplification of Genomic DNA   13
   Randall K. Saiki

3. Amplification of RNA   21
   Ernest S. Kawasaki

4. RACE: Rapid Amplification of cDNA Ends   28
   Michael A. Frohman

5. Degenerate Primers for DNA Amplification   39
   Teresa Compton

6. cDNA Cloning Using Degenerate Primers   46
   Cheng Chi Lee and C. Thomas Caskey

7. PCR with 7-Deaza-2'-Deoxyguanosine
   Triphosphate   54
   Michael A. Innis

8. Competitive PCR for Quantitation of mRNA   60
   Gary Gilliland, Steven Perrin, and H. Franklin Bunn

v

9. Quantitative PCR   70
   Alice M. Wang and David F. Mark

10. Production of Single-Stranded DNA by Asymmetric
    PCR   76
    Peter C. McCabe

11. Cloning with PCR   84
    Stephen J. Scharf

12. Oligonucleotide Ligation Assay   92
    Ulf Landegren, Robert Kaiser, and Leroy Hood

13. Nonisotopically Labeled Probes and Primers   99
    Corey Levenson and Chu-an Chang

14. Incorporation of Biotinylated dUTP   113
    Y.-M. Dennis Lo, Wajahat Z. Mehal, and Kenneth A. Fleming

15. Nonisotopic Detection of PCR Products   119
    Rhea Helmuth

16. Thermostable DNA Polymerases   129
    David H. Gelfand and Thomas J. White

17. Procedures to Minimize PCR-Product Carry-Over   142
    Shirley Kwok

18. Sample Preparation from Blood, Cells, and Other
    Fluids   146
    Ernest S. Kawasaki

19. Sample Preparation from Paraffin-Embedded
    Tissues   153
    Deann K. Wright and M. Michele Manos

20. Amplifying Ancient DNA   159
    Svante Pääbo

Part Two _____
RESEARCH APPLICATIONS

21. *In Vitro* Transcription of PCR Templates   169
    Michael J. Holland and Michael A. Innis

22. Recombinant PCR   177
    Russell Higuchi

23. DNase I Footprinting   184
    Barbara Krummel

24. Sequencing with *Taq* DNA Polymerase   189
    Mary Ann D. Brow

25. Direct Sequencing with the Aid of Phage
    Promoters   197
    *Steve S. Sommer, Gobinda Sarkar, Dwight D. Koeberl,
    Cynthia D. K. Bottema, Jean-Marie Buerstedde,
    David B. Schowalter, and Joslyn D. Cassady*

26. Identifying DNA Polymorphisms by Denaturing
    Gradient Gel Electrophoresis   206
    Val C. Sheffield, David R. Cox, and Richard M. Myers

27. Amplification of Flanking Sequences by Inverse
    PCR   219
    *Howard Ochman, Meetha M. Medhora, Dan Garza, and
    Daniel L. Hartl*

28. Detection of Homologous Recombinants   228
    Michael A. Frohman and Gail R. Martin

29. RNA Processing: Apo-B   237
    Lyn M. Powell

30. A Transcription-Based Amplification System   245
    *T. R. Gingeras, G. R. Davis, K. M. Whitfield, H. L. Chappelle,
    L. J. DiMichele, and D. Y. Kwoh*

31. Screening of λgt11 Libraries   253
    Kenneth D. Friedman, Nancy L. Rosen, Peter J. Newman, and
    Robert R. Montgomery

Part Three _____
GENETICS AND EVOLUTION

32. HLA DNA Typing   261
    Henry A. Erlich and Teodorica L. Bugawan

33. Multiplex PCR for the Diagnosis of Duchenne Muscular
    Dystrophy   272
    Jeffrey S. Chamberlain, Richard A. Gibbs, Joel E. Ranier, and
    C. Thomas Caskey

34. Isolation of DNA from Fungal Mycelia and Single
    Spores   282
    Steven B. Lee and John W. Taylor

35. Genetic Prediction of Hemophilia A   288
    Scott C. Kogan and Jane Gitschier

36. Haplotype Analysis from Single Sperm or Diploid
    Cells   300
    Ulf Gyllensten

37. Amplification of Ribosomal RNA Genes for Molecular
    Evolution Studies   307
    Mitchell L. Sogin

38. Amplification and Direct Sequencing of Fungal
    Ribosomal RNA Genes for Phylogenetics   315
    T. J. White, T. Bruns, S. Lee, and J. Taylor

Part Four _____
DIAGNOSTICS AND FORENSICS

39. Detection of Human T-Cell Lymphoma/Leukemia
    Viruses   325
    Garth D. Ehrlich, Steven Greenberg, and Mark A. Abbott

40. Detection of Human Immunodeficiency Virus   337
    David E. Kellogg and Shirley Kwok

41. Detection of Hepatitis B Virus   348
    I. Baginski, A. Ferrie, R. Watson, and D. Mack

42. Detection and Typing of Genital Human
    Papillomaviruses   356
    Yi Ting and M. Michele Manos

43. Detection of Human Cytomegalovirus   368
    Darryl Shibata

44. PCR Amplification of Enteroviruses   372
    Harley A. Rotbart

45. Novel Viruses   378
    David Mack, Oh-Sik Kwon, and Fred Faloona

46. Analysis of *ras* Gene Point Mutations by PCR and
    Oligonucleotide Hybridization   386
    John Lyons

47. B-Cell Lymphoma: t(14;18) Chromosome
    Rearrangement   392
    Marco Crescenzi

48. Detecting Bacterial Pathogens in Environmental Water
    Samples by Using PCR and Gene Probes   399
    Ronald M. Atlas and Asim K. Bej

49. PCR in the Diagnosis of Retinoblastoma   407
    Sang-Ho Park

50. Determination of Familial Relationships   416
    Cristián Orrego and Mary Claire King

Part Five _____
INSTRUMENTATION AND SUPPLIES

51. PCR in a Teacup: A Simple and Inexpensive Method for Thermocycling PCRs   429
Robert Watson

52. A Low-Cost Air-Driven Cycling Oven   435
Peter Denton and H. Reisner

53. Modification of a Histokinette for Use as an Automated PCR Machine   442
N. C. P. Cross, N. S. Foulkes, D. Chappel, J. McDonnell, and L. Luzzatto

54. Organizing a Laboratory for PCR Work   447
Cristián Orrego

55. Basic Equipment and Supplies   455
Roberta Madej and Stephen Scharf

Index   461

CONTRIBUTORS

Numbers in parentheses indicate the pages on which the authors' contributions begin.

Mark A. Abbott (325), Regional Oncology Center, State University of New York, Upstate Medical Center, Syracuse, New York 13210

Ronald M. Atlas (399), Department of Biology, University of Louisville, Louisville, Kentucky 40292

Isabele Baginski (348), INSERM, F69424 Lyons, Cedex 03, France

Asim K. Bej (399), Department of Biology, University of Louisville, Louisville, Kentucky 40292

Cynthia D. Bottema (197), Department of Biochemistry and Molecular Biology, Mayo Clinic/Foundation, Rochester, Minnesota 55905

Mary Ann D. Brow (189), University of Wisconsin, McArdle Laboratory for Cancer Research, Madison, Wisconsin 53706

Tom Bruns (315), Department of Plant Pathology, University of California, Berkeley, California 94720

Jean-Marie Buerstedde (197), Basel Institute of Immunology, Basel, Switzerland

Teodorica L. Bugawan (261), Cetus Corporation, Emeryville, California 94608

H. Franklin Bunn (60), Brigham and Women's Hospital, Harvard Medical School, Boston, Massachusetts 02115

C. Thomas Caskey (46, 272), Institute for Molecular Genetics, Howard Hughes Medical Institute, Baylor College of Medicine, Houston, Texas 77030

Joslyn Cassady (197), Department of Biochemistry and Molecular Biology, Mayo Clinic/Foundation, Rochester, Minnesota 55905

Jeffrey S. Chamberlain (272), Institute for Molecular Genetics, Howard Hughes Medical Institute, Baylor College of Medicine, Houston, Texas 77030

Chu-an Chang (99), Department of Nucleic Acid Chemistry, Chiron Corporation, Emeryville, California 94608

D. Chappel (442), Royal Postgraduate Medical School, Hammersmith Hospital, London W12 ONN, United Kingdom

H. L. Chappelle (245), The Salk Institute Biotechnology/Industrial Associates, Inc., San Diego, California 92138, and SISKA Diagnostics, Inc., P.O. Box 85453, San Diego, California 92138-9216

Teresa Compton (39), Research Institute of Scripps Clinic, Department of Immunology, La Jolla, California 92037

David R. Cox (205), Departments of Psychiatry, Biochemistry, and Biophysics, The University of California, San Francisco, California 94143

Marco Crescenzi (391), Laboratory of Cellular and Molecular Biology, National Cancer Institute, Bethesda, Maryland 20892

N. C. P. Cross (442), Royal Postgraduate Medical School, Hammersmith Hospital, London W12 ONN, United Kingdom

G. R. Davis (245), The Salk Institute Biotechnology/Industrial Associates, Inc., San Diego, California 92138, and SISKA Diagnostics, Inc., San Diego, California 92138-9216

Peter Denton (435), Department of Pathology, University of North Carolina, Chapel Hill, North Carolina 27599

L. J. DiMichele (245), The Salk Institute Biotechnology/Industrial Associates, Inc., San Diego, California 92138, and SISKA Diagnostics, Inc., San Diego, California 92138-9216

Garth D. Ehrlich (325), Regional Oncology Center, State University of New York, Upstate Medical Center, Syracuse, New York 13210

Henry A. Erlich (261), Cetus Corporation, Emeryville, California 94608

Fred Faloona (378), Cetus Corporation, Emeryville, California 94608

Andrew Ferrie (348), Digeme, University of Maryland, College Park, Maryland 20742

Kenneth A. Fleming (113), University of Oxford, Nuffield Department of Pathology, John Radcliffe Hospital, Oxford OX3 9DU, England

N. S. Foulkes (442), Royal Postgraduate Medical School, Hammersmith Hospital, London W12 ONN, United Kingdom

Kenneth D. Friedman (253), The Blood Center of Southeastern Wisconsin, Milwaukee, Wisconsin 53233

Michael A. Frohman (28, 227), Department of Anatomy, University of California, San Francisco, California 94143

Dan Garza (219), Department of Genetics, Washington University School of Medicine, St. Louis, Missouri 63110

David H. Gelfand (3, 129), Cetus Corporation, Emeryville, California 94608

Richard A. Gibbs (272), Institute for Molecular Genetics, Howard Hughes Medical Institute, Baylor College of Medicine, Houston, Texas 77030

Gary Gilliland (60), Brigham and Women's Hospital, Harvard Medical School, Boston, Massachusetts 02115

Thomas R. Gingeras (245), The Salk Institute Biotechnology/Industrial Associates, Inc., San Diego, California 92138, and SISKA Diagnostics, Inc., San Diego, California 92138-9216

Jane Gitschier (288), Howard Hughes Medical Institute, University of California, San Francisco, California 94143

Steven Greenburg (325), Regional Oncology Center, State University of New York Upstate Medical Center, Syracuse, New York 13210

Ulf Gyllensten (300), Department of Medical Genetics, Biomedical Center 23, Uppsala, Sweden

Daniel L. Hartl (219), Department of Genetics, Washington University School of Medicine, St. Louis, Missouri 63110

Rhea Helmuth (119), Cetus Corporation, Emeryville, California 94608

Russell Higuchi (177), Cetus Corporation, Emeryville, California 94608

Michael J. Holland (169), MSIA, Department of Biological Chemistry, University of California, Davis, California 95616

Leroy Hood (92), California Institute of Technology, Division of Biology, Pasadena, California 91125

Michael A. Innis (3, 54, 169), Cetus Corporation, Emeryville, California 94608

Robert Kaiser (92), California Institute of Technology, Division of Biology, Pasadena, California 91125

Ernest Kawasaki (21, 146), Cetus Corporation, Emeryville, California 94608

David E. Kellogg (337), Cetus Corporation, Emeryville, California 94608

Mary Claire King (416), School of Public Health, University of California, Berkeley, California 94720

Dwight D. Koeberl (197), Department of Biochemistry and Molecular Biology, Mayo Clinic/Foundation, Rochester, Minnesota 55905

Scott C. Kogan (288), Howard Hughes Medical Institute, University of California, San Francisco, California 94143

Barbara Krummel (184), Department of Biochemistry, University of California, Berkeley, California 94720

D. Y. Kwoh (245), The Salk Institute Biotechnology/Industrial Associates, Inc., San Diego, California 92138, and SISKA Diagnostics, Inc., San Diego, California 92138-9216

Shirley Kwok (142, 337), Cetus Corporation, Emeryville, California 94608

Oh-Sik Kwon (378), Cetus Corporation, Emeryville, California 94608

Ulf Landegren (92), Department of Medical Genetics, Biomedical Center, S751 #23 Uppsala, Sweden

Cheng Chi Lee (46), Institute for Molecular Genetics, Baylor College of Medicine, Houston, Texas 77030

Steven B. Lee (282, 315), Department of Plant Biology, University of California, Berkeley, California 94720

Corey Levenson (99), Cetus Corporation, Emeryville, California 94608

Y.-M. Dennis Lo (113), University of Oxford, Nuffield Department of Pathology, John Radcliffe Hospital, Oxford OX3 9DU, England

L. Luzzatto (442), Royal Postgraduate Medical School, Hammersmith Hospital, London W12 0NN, United Kingdom

John Lyons (386), Cetus Corporation, Emeryville, California 94608

David Mack (348, 378), Department of Molecular Genetics and Cellular Biology, University of Chicago, Cummings Life Science Center, Chicago, Illinois 60637

Roberta Madej (455), Cetus Corporation, Emeryville, California 94608

Michele Manos (153, 356), Cetus Corporation, Emeryville, California 94608

David F. Mark (70), Department of Microbial Chemotherapeutics and Molecular Biology, Merck Sharp & Dohme Research Laboratories, Rahway, New Jersey 07065-0900

Gail R. Martin (228), Department of Anatomy, University of California, San Francisco, California 94143

Peter C. McCabe (76), Cetus Corporation, Emeryville, California 94608

J. McDonnell (442), Royal Postgraduate Medical School, Hammersmith Hospital, London W12 0NN, United Kingdom

Meetha M. Medhora (219), Department of Genetics, Washington University School of Medicine, St. Louis, Missouri 63110

Wajahat Z. Mehal (113), University of Oxford, Nuffield Department of Pathology, John Radcliffe Hospital, Oxford OX3 9DU, England

Robert R. Montgomery (253), The Blood Center of Southeastern Wisconsin, Milwaukee, Wisconsin 53233

Richard M. Myers (206), Departments of Physiology and Biochemistry and Biophysics, University of California, San Francisco, California 94143

Peter J. Newman (253), The Blood Center of Southeastern Wisconsin, Milwaukee, Wisconsin 53233

Howard Ochman (219), Department of Genetics, Washington University School of Medicine, St. Louis, Missouri 63110

Christián Orrego[1] (416, 447) Department of Biochemistry, University of California, Berkeley, Califonia 94720.

Svante Pääbo (159), Department of Molecular, Cellular and Developmental Biology, University of California, Berkeley, California 94720

Sang-Ho Park (407), Whitehead Institute for Biomedical Research, 9 Cambridge Center, Cambridge, Massachusetts 02142, and Department of Biology, Massachusetts Institute of Technology, Cambridge, Massachusetts 01239

Steven Perrin (60), Brigham and Women's Hospital, Harvard Medical School, Boston, Massachusetts 02115

Lyn M. Powell (237), Cardiovascular Research, Genentech, South San Francisco, California 94080

Joel E. Ranier (272), Institute for Molecular Genetics, Howard Hughes Medical Institute, Baylor College of Medicine, Houston, Texas 77030

H. Reisner (442), Department of Pathology, University of North Carolina, Chapel Hill, North Carolina 27599

Nancy L. Rosen (253), The Blood Center of Southeastern Wisconsin, Milwaukee, Wisconsin 53233

Harley A. Rotbart (372), University of Colorado Health Sciences Center, Denver, Colorado 80262

Randall K. Saiki (13), Cetus Corporation, Emeryville, California 94608

Gobinda Sarkar (197), Department of Biochemistry and Molecular Biology, Mayo Clinic/Foundation, Rochester, Minnesota 55905

Stephen J. Scharf (84, 455), Cetus Corporation, Emeryville, California 94608

David B. Schowalter (197), Department of Biochemistry and Molecular Biology, Mayo Clinic/Foundation, Rochester, Minnesota 55905

Val C. Sheffield (206), Department of Pediatrics, The University of California, San Francisco, California 94143

Darryl Shibata (368), Los Angeles County–USC Medical Center, Los Angeles, California 90033

Mitchell L. Sogin (307), Division of Molecular and Cellular Biology, Department of Pediatrics, National Jewish Center for Immunology and Respiratory Medicine, 1400 Jackson Street, Denver, Colorado 80206

[1]Present address: Evolutionary Genetics Laboratory, Museum of Vertebrate Zoology, University of California, Berkely, California 94720.

Steve S. Sommer (197), Department of Biochemistry and Molecular Biology, Mayo Clinic/Foundation, Rochester, Minnesota 55905

J. Taylor (315, 282), Department of Plant Biology, University of California, Berkeley, California 94720

Yi Ting (355), Cetus Corporation, Emeryville, California 94608

Alice M. Wang (70), Cetus Corporation, Emeryville, California 94608

Robert Watson (348, 429), Cetus Corporation, Emeryville, California 94608

Thomas J. White (129, 315), Hoffmann-La Roche, Emeryville, California 94608

K. M. Whitfield (245), The Salk Institute Biotechnology/Industrial Associates, Inc., San Diego, California 92138, and SISKA Diagnostics, Inc., San Diego, California 92138-9216

Deann K. Wright (153), Cetus Corporation, Emeryville, California 94608

PREFACE

Since the unveiling of the polymerase chain reaction (PCR) method of DNA amplification at the American Society of Human Genetics Conference in October 1985, more than 600 publications involving the use of PCR have appeared in the scientific literature. Numerous modifications, improvements, and novel applications of PCR have been devised, yet there has been no source to which scientists could turn for basic instruction in the PCR method that is most suitable for the experimental problem at hand. Furthermore, there is no single set of instructions that works in every situation, even though some authors have drawn definitive conclusions from a single system about the importance or dispensability of certain parameters. This book is a collection of protocols for basic PCR methods which have been repeatedly tested in the authors' laboratories. It is intended to serve as an introduction to PCR for molecular biologists at the graduate level and beyond who are using the method for the first time, and to serve as a resource on novel variations and applications of PCR for scientists who may have considerable experience with the basic method. We have also included chapters for scientists in those fields (e.g., zoology, botany, and ecology) in which there may be little or no familiarity with molecular biological techniques. Our intent is to encourage scientists in these fields to utilize the *in vitro* PCR method to complement or circumvent more complex recombinant DNA methods.

The book has five sections. The first section, on basic methodology, contains chapters that provide protocols for many variations of PCR, a brief theoretical basis for each procedure, a comparison to other techniques, and helpful or cautionary notes on optimizing the procedure and avoiding pitfalls. These chapters contain the latest improvements to PCR and have been extensively tested for general applicability. The chapters in the subsequent sections were selected because they describe specific research and/or diagnostic applications of PCR and have been shown to be reliable procedures in the authors' laboratories. In some instances, the latter chapters may contain a procedure that the editors regard as suboptimal or in conflict with current information from more detailed studies; in these instances, cross-references to other chapters or editorial notes have been provided. The editors feel that this approach is preferable to altering the protocol, since untested revisions might cause the PCR to fail in the author's specific system.

The second section of the book addresses particular applications of PCR in basic research (sequencing, mutagenesis, etc.) and contains protocols and variations that complement and extend those of the first section. The third section addresses procedures that are useful for genetic analyses, diagnosis of inherited disorders and susceptibility to disease, and evolutionary analyses. The fourth section covers applications of PCR to specific diagnostic tests for infectious diseases and cancer and to forensic tests. Our intent is to provide medical scientists with procedures that can be useful for research on the epidemiology of infectious diseases and cancer as well as on methods for individual identification. The final section gives basic information on the equipment and reagents needed to perform the polymerase chain reaction and includes plans for several inexpensive devices for thermal cycling.

The editors extend their thanks and appreciation for the invaluable and patient efforts of Judy Davis, who formatted and copyedited the chapters for the publisher. They also thank Cetus Corporation, Hoffmann-La Roche, and Academic Press for their encouragement and support of the effort required to produce this book.

Thomas J. White (for the editors)

Part One

BASIC METHODOLOGY

OPTIMIZATION OF PCRs

Michael A. Innis and David H. Gelfand

Polymerase Chain Reaction (PCR) is an ingenious new tool for molecular biology that has had an effect on research similar to that of the discovery of restriction enzymes and the Southern blot. PCR is so sensitive that a single DNA molecule has been amplified, and single-copy genes are routinely extracted out of complex mixtures of genomic sequences and visualized as distinct bands on agarose gels. PCR can also be utilized for rapid screening and/or sequencing of inserts directly from aliquots of individual phage plaques or bacterial colonies. Enhancements, such as the use of thermostable DNA polymerases and automation, of the method invented by Kary Mullis (K. B. Mullis, U.S. patent 4,683,195, July 1987; U.S. patent 4,683,202, July 1987) (Saiki *et al.* 1985; Mullis *et al.* 1986; Mullis and Faloona 1987) have fostered the development of numerous and diverse PCR applications throughout the research community. Unquestionably, no single protocol will be appropriate to all situations. Consequently, each new PCR application is likely to require optimization. Some often encountered problems include: no detectable product or a low yield of the desired product; the presence of nonspecific background bands due to mispriming or misextension of the primers; the formation of "primer–dimers" that compete for amplification with the desired product; and mutations or heterogeneity due to misincorporation. The objective of this chapter is to expedite the optimization process by discussing parameters that influence the specificity, fidelity, and yield of the desired product. These recommendations derive from our practical experience using native or recombinant *Taq* DNA polymerases obtained from Perkin-Elmer Cetus Instruments.

Standard PCR Amplification Protocol

While the standard conditions will amplify most target sequences, they are presented here principally to provide starting conditions for designing new PCR applications. It can be highly advantageous to optimize the PCR for a given application, especially repetitive diagnostic or analytical procedures in which optimal performance is necessary.

---

1. Set up a 100-μl reaction in a 0.5-ml microfuge tube, mix, and overlay with 75 μl of mineral oil:
    Template DNA (10^5 to 10^6 target molecules*)
    20 pmol each primer ($T_m > 55°C$ preferred)
    20 mM Tris–HCl (pH 8.3) (20°C)
    1.5 mM MgCl$_2$
    25 mM KCl
    0.05% Tween 20
    100 μg/ml of autoclaved gelatin or nuclease-free bovine serum albumin
    50 μM each dNTP
    2 units of *Taq* DNA polymerase
    * 1 μg of human single-copy genomic DNA equals 3×10^5 targets; 10 ng of yeast DNA equals 3×10^5 targets; 1 ng of *Escherichia coli* DNA equals 3×10^5 targets; 1% of an M13 plaque equals 10^6 targets.
2. Perform 25 to 35 cycles of PCR using the following temperature profile:

Denaturation	96°C, 15 seconds (a longer initial time is usually desirable)
Primer Annealing	55°C, 30 seconds
Primer Extension	72°C, 1.5 minutes

3. Cycling should conclude with a final extension at 72°C for 5 minutes. Reactions are stopped by chilling to 4°C and/or by addition of EDTA to 10 mM.

Enzyme Concentration

A recommended concentration range for *Taq* DNA polymerase (Perkin-Elmer Cetus) is between 1 and 2.5 units (SA = 20 units/pmol) (Lawyer *et al.* 1989) per 100-μl reaction when other parameters are optimum. However, enzyme requirements may vary with respect to individual target templates or primers. When optimizing a PCR, we recommend testing enzyme concentrations ranging from 0.5 to 5 units/100 μl and assaying the results by gel electrophoresis. If the enzyme concentration is too high, nonspecific background products may accumulate, and if too low, an insufficient amount of desired product is made.

Note: *Taq* DNA polymerase from different suppliers may behave differently because of different formulations, assay conditions, and/or unit definitions.

Deoxynucleotide Triphosphates

Stock dNTP solutions should be neutralized to pH 7.0, and their concentrations should be determined spectrophotometrically. Primary stocks are diluted to 10 mM, aliquoted, and stored at -20°C. A working stock containing 1 mM each dNTP is recommended. The stability of the dNTPs during repeated cycles of PCR is such that approximately 50% remains as dNTP after 50 cycles (Corey Levenson, personal communication).

Deoxynucleotide concentrations between 20 and 200 μM each result in the optimal balance among yield, specificity, and fidelity. The four dNTPs should be used at equivalent concentrations to minimize misincorporation errors. Both the specificity and the fidelity of PCR are increased by using lower dNTP concentrations than those originally recommended for Klenow-mediated PCR (1.5 mM each).

Low dNTP concentrations minimize mispriming at nontarget sites and reduce the likelihood of extending misincorporated nucleotides (Innis *et al.* 1988). One should decide on the lowest dNTP concentration appropriate for the length and composition of the target sequence; e.g., 20 μM each dNTP in a 100-μl reaction is theoretically sufficient to synthesize 2.6 μg of DNA or 10 pmol of a 400-bp sequence. Recently, the use of low, uniform dNTP concentrations (2 μM each) enabled highly sensitive (1/10^7), allele-specific amplification of *ras* point mutations (Ehlen and Dubeau 1989).

Magnesium Concentration

It is beneficial to optimize the magnesium ion concentration. The magnesium concentration may affect all of the following: primer annealing, strand dissociation temperatures of both template and PCR product, product specificity, formation of primer–dimer artifacts, and enzyme activity and fidelity. *Taq* DNA polymerase requires free magnesium on top of that bound by template DNA, primers, and dNTPs. Accordingly, PCRs should contain 0.5 to 2.5 mM magnesium over the total dNTP concentration. Note that the presence of EDTA or other chelators in the primer stocks or template DNA may disturb the apparent magnesium optimum.

Other Reaction Components

A recommended buffer for PCR is 10 to 50 mM Tris–HCl (between pH 8.3 and 8.8) when measured at 20°C; however, an extensive survey of other buffers has not been performed. Tris is a dipolar ionic buffer having a pK_a of 8.3 at 20°C, and a Δ pK_a of −0.021/°C. Thus, the true pH of 20 mM Tris (pH 8.3) at 20°C varies between 7.8 and 6.8 during typical thermal cycling conditions.

Up to 50 mM KCl can be included in the reaction mixture to facilitate primer annealing. NaCl at 50 mM, or KCl above 50 mM, inhibits *Taq* DNA polymerase activity (Innis *et al.* 1988).

While DMSO is useful in PCRs performed with the Klenow fragment of *E. coli* DNA polymerase I, 10% DMSO inhibits the activity of *Taq* DNA polymerase by 50% (see Chapter 16) and its use is not recommended for most applications [one exception is the protocol of

Chamberlain *et al.* (1988) that describes amplifying multiple sequences in the same reaction; also see Chapter 33].

Gelatin or bovine serum albumin (100 μg/ml) and nonionic detergents such as Tween 20 or Laureth 12 (0.05 to 0.1%; Mazer Chemicals, Gurnee, Illinois) are included to help stabilize the enzyme, although many protocols work well without added protein.

Primer Annealing

The temperature and length of time required for primer annealing depend upon the base composition, length, and concentration of the amplification primers. An applicable annealing temperature is 5°C below the true T_m of the amplification primers. Because *Taq* DNA polymerase is active over a broad range of temperatures, primer extension will occur at low temperatures, including the annealing step (Innis *et al.* 1988). The range of enzyme activity varies by two orders of magnitude between 20 and 85°C. Annealing temperatures in the range of 55 to 72°C generally yield the best results. At typical primer concentrations (0.2 μM), annealing will require only a few seconds.

Increasing the annealing temperature enhances discrimination against incorrectly annealed primers and reduces misextension of incorrect nucleotides at the 3' end of primers. Therefore, stringent annealing temperatures, especially during the first several cycles, will help to increase specificity. For maximum specificity in the initial cycle, *Taq* DNA polymerase can be added after the first denaturation step during primer annealing. Low extension temperature together with high dNTP concentrations favors misextension of primers and extension of misincorporated nucleotides. For these reasons, some investigators have argued that PCRs should perform better using longer primers and only two temperatures; e.g., from 55 to 75°C for annealing and extension, and 94 to 97°C for denaturation and strand separation (Kim and Smithies 1988; Will Bloch, personal communication).

Primer Extension

Extension time depends upon the length and concentration of the target sequence and upon temperature. Primer extensions are traditionally performed at 72°C because this temperature was near optimal for extending primers on an M13-based model template (D. Gelfand, unpublished). Estimates for the rate of nucleotide incorporation at 72°C vary from 35 to 100 nucleotides second[-1] depending upon the buffer, pH, salt concentration, and the nature of the DNA template (Innis *et al.* 1988; Saiki and Gelfand 1989). An extension time of one minute at 72°C is considered sufficient for products up to 2 kb in length. However, longer extension times may be helpful in early cycles if the substrate concentration is very low, and at late cycles when product concentration exceeds enzyme concentration (approximately 1 nM) (Will Bloch, personal communication).

Denaturation Time and Temperature

The most likely cause for failure of a PCR is incomplete denaturation of the target template and/or the PCR product. Typical denaturation conditions are 95°C for 30 seconds, or 97°C for 15 seconds; however, higher temperatures may be appropriate, especially for G+C-rich targets. It only takes a few seconds to denature DNA at its strand-separation temperature (T_{ss}); however, there may be lag time involved in reaching T_{ss} inside the reaction tube. It is a good idea to monitor the temperature inside one reaction tube with a low-mass thermocouple probe (see Chapter 51). Incomplete denaturation allows the DNA strands to "snap back" and, thus, reduces product yield. In contrast, denaturation steps that are too high and/or too long lead to unnecessary loss of enzyme activity. The half-life of *Taq* DNA polymerase activity is >2 hours, 40 minutes, and 5 minutes at 92.5, 95, and 97.5°C, respectively (See Chapter 16).

Cycle Number

The optimum number of cycles will depend mainly upon the starting concentration of target DNA when other parameters are optimized. A common mistake is to execute too many cycles. To quote Kary Mullis, "If you have to go more than 40 cycles to amplify a single-copy gene, there is something seriously wrong with your PCR." Too many cycles can increase the amount and complexity of nonspecific background products (see Plateau Effect). Of course, too few cycles give low product yield. Some guidelines for number of cycles versus starting target concentration are provided:

Number of target molecules	Number of cycles
3×10^5	25 to 30
1.5×10^4	30 to 35
1×10^3	35 to 40
50	40 to 45

Primers

Primer concentrations between 0.1 and 0.5 μM are generally optimal. Higher primer concentrations may promote mispriming and accumulation of nonspecific product and may increase the probability of generating a template-independent artifact termed a primer–dimer. Nonspecific products and primer–dimer artifacts are themselves substrates for PCR and compete with the desired product for enzyme, dNTPs, and primers, resulting in a lower yield of the desired product.

Some simple rules aid in the design of efficient primers. Typical primers are 18 to 28 nucleotides in length having 50 to 60% G + C composition. The calculated T_ms for a given primer pair should be balanced. For this purpose, one can use the rule-of-thumb calculation of 2°C for A or T and 4°C for G or C (Thein and Wallace 1986). Depending on the application, T_ms between 55°C and 80°C are desired. One should avoid complementarity at the 3' ends of primer pairs as this promotes the formation of primer–dimer artifacts and reduces the yield of the desired product. Also, runs (three or more) of C's or G's at the 3' ends of primers may promote mispriming at G+C-rich sequences and should be avoided, when possible, as should palindromic sequences within primers. If all else fails, it usually helps to try a different primer pair. A less obvious reason for some primers failing to work is the presence of secondary structure in the template DNA. In this case, substitution of 7-deaza-2'-deoxyGTP for dGTP has been very useful (see Chapter 7).

The design of special-purpose primers is discussed in other chapters. Briefly, primers may contain 5' extensions or mismatches for incorporating restriction enzyme sites, an ATG start codon, or promoter sequences into the target sequence (see Chapter 11). Mismatched bases can be placed internally for mutagenesis (see Chapter 22). Degenerate primers can be used to isolate novel genes on the basis of similarity and/or amino acid sequence (see Chapters 5 and 6). Some authors have suggested using inosine in primers instead of using degenerate primers (Knoth *et al.* 1988). When using degenerate primers, it helps to avoid degeneracy at the 3' ends, because mismatched bases are inefficiently extended.

Plateau Effect

The term "plateau effect" is used to describe the attenuation in the exponential rate of product accumulation that occurs during late PCR cycles concomitantly with the accumulation of 0.3 to 1 pmol of the intended product. Depending on reaction conditions and thermal cycling, one or more of the following may influence plateau: (1) utilization of substrates (dNPTs or primers); (2) stability of reactants (dNTPs or enzyme); (3) end-product inhibition (pyrophosphate, duplex DNA); (4) competition for reactants by nonspecific products or primer–dimer; (5) reannealing of specific product at concentrations above 10^{-8} M (may decrease the extension rate or processivity of *Taq* DNA polymerase or cause branch-migration of product strands and displacement of primers); and (6) incomplete denaturation/strand separation of product at high product concentration.

An important consequence of reaching plateau is that an initially low concentration of nonspecific products resulting from mispriming events may continue to amplify preferentially. Optimizing the number of PCR cycles is the best way to avoid amplifying background products.

Fidelity Considerations

Conditions that promote misincorporation include when deoxynucleotide concentrations are well below the K_m (i.e., <1 μM) or when the concentration of one dNTP is low relative to the other three. We demonstrated that a respectable "chain-termination" sequencing ladder can be generated without using dideoxynucleotides by limiting one dNTP in each of four separate sequencing reactions (Innis *et al.* 1988). In contrast, we did not observe misincorporation bands (background on sequencing gels) when the concentration of the four dNTPs was >10 μM each and balanced. We recommend using balanced concentrations of dNTPs to diminish misincorporation errors. Because misincorporated bases cannot be proofread (*Taq* lacks a 3' to 5' exonuclease activity) and mismatched bases are inefficiently extended, misincorporation errors that do occur during PCR promote chain termination. Chain termination restricts the amplification of defective molecules and helps to maintain fidelity.

What determines whether a mismatch is extended? Petruska *et al.* (1988) showed (for *Drosophila* DNA polymerase α) that enzymatic discrimination against elongating mismatched termini is based mainly on K_m differences: a matched A–T terminus was found to be extended 200 times faster than a G–T mismatch was and 1400 and 2500 times faster than C–T and T–T mismatches were, respectively. The same is likely to be true for *Taq* DNA polymerase; therefore, the concentration of dNTPs in the reaction is predicted to have a substantial effect on the fidelity of PCR (at high dNTP concentrations, i.e., >1 mM, mismatches will be extended more efficiently). In combination, high-temperature annealing/extension (>55°C) and low dNTP concentrations (10 to 50 μM each) give the highest fidelity in the final PCR product.

Literature Cited

Chamberlain, J. S., R. A. Gibbs, J. E. Ranier, P. N. Nguyen, and C. T. Caskey. 1988. Deletion screening of the Duchenne muscular dystrophy locus via multiplex DNA amplification. *Nucleic Acids Res.* **16**:11141–11156.

Ehlen, T., and L. Dubeau. 1989. Detection of *ras* point mutations by polymerase chain reaction using mutation-specific, inosine-containing oligonucleotide primers. *Biochem. Biophys. Res. Commun.* **160**:441–447.

Innis, M. A., K. B. Myambo, D. H. Gelfand, and M. A. D. Brow. 1988. DNA sequencing with *Thermus aquaticus* DNA polymerase and direct sequencing of polymerase chain reaction-amplified DNA. *Proc. Natl. Acad. Sci. USA* **85**:9436–9440.

Kim, H.-S., and O. Smithies. 1988. Recombinant fragment assay for gene targeting based on the polymerase chain reaction. *Nucleic Acids Res.* **16**:8887–8903.

Knoth, K., S. Roberds, C. Poteet, and M. Tamkun. 1988. Highly degenerate, inosine-containing primers specifically amplify rare cDNA using the polymerase chain reaction. *Nucleic Acids Res.* **16**:10932.

Lawyer, F. C., S. Stoffel, R. K. Saiki, K. Myambo, R. Drummond, and D. H. Gelfand. 1989. Isolation, characterization, and expression in *Escherichia coli* of the DNA polymerase gene from *Thermus aquaticus*. *J. Biol. Chem.* **264**:6427–6437.

Mullis, K. B., and F. A. Faloona. 1987. Specific synthesis of DNA *in vitro* via a polymerase-catalyzed chain reaction. *Methods Enzymol.* **155**:335–350.

Mullis, K., F. Faloona, S. Scharf, R. Saiki, G. Horn, and H. Erlich. 1986. Specific enzymatic amplification of DNA *in vitro*: the polymerase chain reaction. *Cold
Spring Harbor Symp. Quant. Biol.* **51**:263–273.

Petruska, J., M. F. Goodman, M. S. Boosalis, L. C. Sowers, C. Chaejoon, and I. Tinoco,
Jr. 1988. Comparison between DNA melting thermodynamics and DNA polymerase fidelity. *Proc. Natl. Acad. Sci. USA* **85**:6252–6256.

Saiki, R. K., and D. H. Gelfand. 1989. Introducing AmpliTaq DNA polymerase. *Amplifications* **1**:4–6.

Saiki, R. K., S. Scharf, F. Faloona, K. B. Mullis, G. T. Horn, H. A. Erlich, and N.
Arnheim. 1985. Enzymatic amplification of β-globin genomic sequences and restriction site analysis for diagnosis of sickle cell anemia. *Science* **230**:1350–1354.

Thein, S. L., and R. B. Wallace. 1986. The use of synthetic oligonucleotides as specific
hybridization probes in the diagnosis of genetic disorders. In *Human genetic diseases: a practical approach* (ed. K. E. Davis), p. 33–50. IRL Press, Herndon,
Virginia.

2

AMPLIFICATION OF GENOMIC DNA

Randall K. Saiki

PCR is an *in vitro* method of nucleic acid synthesis by which a particular segment of DNA can be specifically replicated. It involves two oligonucleotide primers that flank the DNA fragment to be amplified and repeated cycles of heat denaturation of the DNA, annealing of the primers to their complementary sequences, and extension of the annealed primers with DNA polymerase. These primers hybridize to opposite strands of the target sequence and are oriented so that DNA synthesis by the polymerase proceeds across the region between the primers. Since the extension products themselves are also complementary to and capable of binding primers, successive cycles of amplification essentially double the amount of the target DNA synthesized in the previous cycle. The result is an exponential accumulation of the specific target fragment, approximately 2^n, where n is the number of cycles of amplification performed.

The human β-globin gene was one of the first DNA sequences to be amplified by PCR (Saiki *et al.* 1985; Mullis *et al.* 1986; Mullis and Faloona 1987), and the specific amplification of single-copy sequences from complex genomic samples continues to be one of the most common applications of this technique. The high specificity and yield of PCR with the thermostable *Taq* DNA polymerase make it an ideal method for the isolation of a particular genomic fragment. This chapter will cover some of the salient features of amplifying segments from eukaryotic genomes. In principle, one might expect the amplification of a single-copy genetic locus from a high-complexity DNA sample to be substantially more difficult than, say, the amplification of a fragment from a plasmid or phage. In practice, however, genomic amplifications are usually quite straightforward and are not necessarily more complicated than those of simpler systems. This equivalence can be attributed to the high specificity of the priming step, which, in turn, is due to the elevated temperatures at which PCR with *Taq* polymerase can be performed (Saiki *et al.* 1988).

Protocols

Sample Preparation

One of the most appealing features of PCR is that the quantity and quality of the DNA sample to be subjected to amplification do not need to be high. A single cell, or crude lysates prepared by simply boiling cells in water, or specimens with an average molecular length of only a few hundred base pairs are usually adequate for successful amplification (Saiki *et al.* 1986; Kogan *et al.* 1987; Higuchi *et al.* 1988; Li *et al.* 1988; Bugawan *et al.* 1988). The essential criteria are that the sample contain at least one intact DNA strand encompassing the region to be amplified and that impurities are sufficiently dilute so as not to inhibit polymerization.

Lysing cells (see procedure below) in a hypotonic solution at high temperature (i.e., boiling in water) is a quick and effective method of preparing DNA for PCR. The main limitation is that only a few cells, usually 10^4 or less, can be used; otherwise, the accumulation of cellular debris will begin to inhibit the reaction. In addition, cells in complex biological fluids (e.g., blood) or cells resistant to lysis (e.g., sperm) require additional manipulations (see Chapters 18, 19, and 20).

1. Pellet 10^2-10^4 cells in a 15-ml conical tube in a benchtop centrifuge at $1200-1500 \times g$ for 10 minutes.

2. Resuspend the cells in 5 ml of phosphate-buffered saline and repellet. (If necessary, repeat this step to remove residual amounts of the original suspension buffer.)
3. Resuspend gently in 25–50 µl distilled water and transfer to a 0.5- or 1.5-ml microcentrifuge tube.
4. Incubate in a 95°C heat block or in boiling water for 3–5 minutes. Spin briefly to remove condensate.
5. Optional: Pellet the cellular debris in a microcentrifuge for 3 minutes and transfer the cleared lysate to a new tube.
6. Add an equal volume of a 2× amplification mix (salts, primers, dNTPs, and enzyme) and amplify.

When DNA of known concentration is available, amounts of 0.05 to 1.0 µg are typically used for amplifications of single-copy loci. Correspondingly less DNA can be used for amplification of multi-copy genes (e.g., since nuclear ribosomal RNA genes are repeated about 200 to 500 times in eukaryotes, one can start with 0.5 to 2 ng of DNA).

Primer Selection

The approach to the selection of efficient and specific primers remains somewhat empirical; there are no discrete rules that will guarantee the choice of an effective primer pair. Fortunately, the majority of primers can be made to work, and the following guidelines will help in their design.

1. Where possible, select primers with an average G+C content of around 50% and a random base distribution. Try to avoid primers with stretches of polypurines, polypyrimidines, or other unusual sequences.
2. Avoid sequences with significant secondary structure. Computer programs such as Squiggles or Circles, available from the University of Wisconsin, are very useful for revealing these structures.
3. Check the primers against each other for complementarity. In particular, avoiding primers with 3' overlaps will reduce the incidence of "primer–dimer" artifacts (see Chapter 1).

Most primers will be between 20 and 30 bases in length, and the optimal amount to use in an amplification will vary. In general, concentrations ranging from 0.1 to 0.5 μM of each primer should prove acceptable.

Other Reaction Components

The standard buffer containing 50 mM KCl, 10 mM Tris–HCl (pH 8.4), 1.5 mM MgCl$_2$, and 100 μg/ml of gelatin will be adequate for the majority of genomic PCRs. In some circumstances (depending on the particular primer pair being used), altering the MgCl$_2$ concentration from 1 to 10 mM can have dramatic effects on the specificity and yield of an amplification. Further, the reduction or elimination of KCl and gelatin may also improve the performance of the PCR (Innis *et al*. 1988).

Deoxynucleotide triphosphates (dATP, dCTP, dGTP, and dTTP) are usually present at 50 to 200 μM of each and should be balanced (i.e., all four at the same concentration). At these concentrations, there is sufficient precursor to synthesize about 6.5 to 25 μg of DNA.

The optimal concentration for *Taq* DNA polymerase is about 2 units/100-μl reaction. As a general rule, concentrations in excess of 4 units tend to result in the accumulation of nonspecific amplification products, whereas amounts less than 1 unit usually reduce the yield of the desired product. Below we outline the procedure by which a PCR can be set up to produce a 536-bp fragment of the human β-globin gene.

Reagents

DNA sample	Prepare a DNA sample as either a 50-μl crude lysate of 10^4 cells from a human cell line (see the procedure in "Sample Preparation") or use 0.05 μg of purified human DNA in 50 μl of water.
10× PCR buffer	500 mM KCl, 100 mM Tris–HCl (pH 8.4 at 20°C), 15 mM MgCl$_2$, and 1 mg/ml of gelatin
10× dNTP stock	2 mM dATP, 2 mM dCTP, 2mM dGTP, and 2mM dTTP (neutralized to pH 7 with NaOH)
10× Primer 1	2.5 μM in TE (10 mM Tris, 0.1 mM EDTA, pH 8) 5'-GGTTGGCCAATCTACTCCCAGG-3' (KM29)
10× Primer 2	2.5 μM in TE 5'-GCTCACTCAGTGTGGCAAAG-3' (RS42)
Taq polymerase	5000 units/ml (Perkin-Elmer Cetus Instruments)

Procedure

1. Combine 10μl each of 10× PCR buffer, 10× dNTP stock, 10× Primer 1, 10× Primer 2, and distilled water to get 50 μl of a 2× PCR mix.
2. Add 2.5 units (0.5 μl) of *Taq* DNA polymerase. For multiple DNA samples, one should prepare a master 2× mix that includes polymerase.
3. Transfer the 2× mix to the tube containing the DNA sample. Mix and overlay with a few drops of mineral oil.
4. Amplify.

Cycling Parameters

PCR is performed by incubating the samples at three temperatures corresponding to the three steps (denaturation, annealing, and extension) in a cycle of amplification. This cycling can be accomplished either manually with preset water baths or automatically with thermal cyclers available from several manufacturers.

In a typical reaction, the double-stranded DNA is denatured by briefly heating the sample to 90 to 95°C. The primers are allowed to anneal to their complementary sequences by briefly cooling to 40 to 60°C, followed by heating to 70 to 75°C to extend the annealed primers with the *Taq* polymerase. It is oftentimes helpful to precede the first cycle with an initial denaturation step of 3 minutes at 93°C. Specificity can be improved by adding the *Taq* polymerase at an elevated temperature rather than having it present in the reaction prior to the first denaturation step. The time of incubation at 70° to 75°C varies according to the length of target being amplified; allowing one minute for each kilobase of sequence is probably excessive, but it is a good place to begin. Shorter times should be tried once the other amplification conditions have been settled. (The extension step can be eliminated altogether if the target sequence is approximately 150 bases or less. During the thermal transition from annealing to denaturation, the sample will be within the 70 to 75° range for the few seconds required to completely extend the annealed primers.)

Below we describe the manual and automated cycling procedures that have been used to amplify genomic sequences ranging from 100 to 3000 bp.

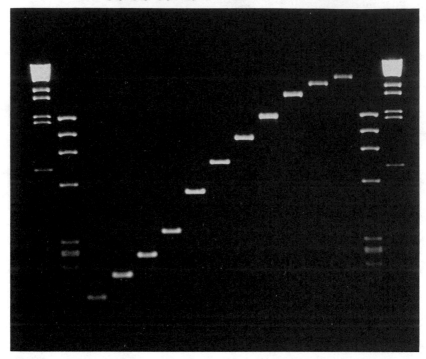

Figure 1. Amplification of β-globin fragments ranging from 150 to 2951 bp. Samples of 100 μl containing standard buffer, 200 μM each dNTP, and 250 nM each primer, 100 ng of human genomic DNA, and 2.5 units of *Taq* polymerase were subjected to 30 cycles of amplification. Each sample (2 μl) was resolved on a 1.6% agarose gel and visualized by ethidium bromide fluorescence. The lengths of the products are indicated at the top of each lane. The markers are *Bst*EII-digested phage λ (500 ng) and *Hae*III-digested φx174-RF (250 ng).

Cycling Procedure for Manual PCR

1. Set three water baths at 94, 55, and 72°C. (Covering the 94°C bath may be necessary to maintain its temperature.)
2. Prepare PCRs in 0.5- or 1.5-ml microcentrifuge tubes.
3. Place the reaction tubes in the 94°C water bath for 1 minute (denature).
4. Transfer the tubes to the 55°C water bath for 1 minute (anneal).

5. Transfer the tubes to the 72°C water bath for 1 minute (extension).
6. Repeat steps 3–5 until 35 cycles have been completed.
7. After the 35th cycle, leave the reaction in the 72°C bath for an additional 5 minutes to ensure that all the amplified DNA is fully double stranded.
8. Optional: Extract the mineral oil with an equal volume of chloroform. (The aqueous phase will be the upper phase.)
9. Analyze the sample for specificity and yield by gel electrophoresis (Fig. 1)

Cycling Program for Automated PCR

This procedure is written for the DNA Thermal Cycler available from Perkin-Elmer Cetus. Details for programming the device are described in the owner's manual.

1. Place the sample to be amplified in the block.
2. Preheat the block to 72°C. (Use the "Step-Cycle" program to heat to 72°C for 20 seconds for 1 cycle then link to the next file.)
3. Perform the cycling reaction. (Use the "Step-Cycle" program to heat to 94°C for 20 seconds [denature], cool to 55°C for 20 seconds [anneal], and heat to 72°C for 30 seconds [extend]. Repeat for 35 cycles and then link to the next file.)
4. After the 35th cycle, extend for an additional 5 minutes to ensure that the amplified DNA is double stranded. (Use the "Thermal Delay" program to incubate at 72°C for 5 minutes and then link to the next file.)
5. Cool down and hold at 15°C until the sample is retrieved. (Use the "Soak" program to hold the block at 15°C for an indefinite period.)
6. Analyze the sample (Manual PCR Procedure, Steps 8 and 9).

Literature Cited

Bugawan, T. L., R. K. Saiki, C. H. Levenson, R. W. Watson, and H. A. Erlich. 1988. The use of non-radioactive oligonucleotide probes to analyze enzymatically amplified DNA for prenatal diagnosis and forensic HLA typing. *Bio/Technology* **6**:943–947.

Higuchi, R., C. H. von Beroldingen, G. F. Sensabaugh, and H. A. Erlich. 1988. DNA typing from single hairs. *Nature (London)* **332**:543–546.

Innis, M. A., K. B. Myambo, D. H. Gelfand, and M. A. D. Brow. 1988. DNA sequencing with *Thermus aquaticus* DNA polymerase and direct sequencing of polymerase chain reaction-amplified DNA. *Proc. Natl. Acad. Sci. USA* **85**:9436–9440.

Kogan, S. C., M. Doherty, and J. Gitschier. 1987. An improved method for prenatal diagnosis of genetic diseases by analysis of amplified DNA sequences. Application to hemophilia A. *N. Engl. J. Med.* **317**:985–990.

Li, H. H., U. B. Gyllensten, X. F. Cui, R. K. Saiki, H. A. Erlich, and N. Arnheim. 1988. Amplification and analysis of DNA sequences in single human sperm and diploid cells. *Nature (London)* **335**:414–417.

Mullis, K. B., F. A. Faloona, S. J. Scharf, R. K. Saiki, G. T. Horn, and H. A. Erlich. 1986. Specific enzymatic amplification of DNA *in vitro:* the polymerase chain reaction. *Cold Spring Harbor Symp. Quant. Biol.* **51**:263–273.

Mullis, K. B., and F. A. Faloona. 1987. Specific synthesis of DNA *in vitro* via a polymerase-catalyzed chain reaction. *Methods Enzymol.* **155**:335–350.

Saiki, R., S. Scharf, F. Faloona, K. B. Mullis, G. T. Horn, H. A. Erlich, and N. Arnheim. 1985. Enzymatic amplification of β-globin genomic sequences and restriction site analysis for diagnosis of sickle cell anemia. *Science* **230**:1350–1354.

Saiki, R. K., T. L. Bugawan, G. T. Horn, K. B. Mullis, and H. A. Erlich. 1986. Analysis of enzymatically amplified β-globin and HLA- DQα DNA with allele-specific oligonucleotide probes. *Nature (London)* **324**:163–166.

Saiki, R. K., D. H. Gelfand, S. Stoffel, S. J. Scharf, R. Higuchi, G. T. Horn, K. B. Mullis, and H. A. Erlich. 1988. Primer-directed enzymatic amplification of DNA with a thermostable DNA polymerase. *Science* **239**:487–491.

3

AMPLIFICATION OF RNA

Ernest S. Kawasaki

Sensitive methods for the detection and analysis of RNA molecules are an important aspect of most cell/molecular biology studies. Methods commonly in use include *in situ* hybridization, Northern gels, dot or slot blots, S-1 nuclease assays, and RNase A protection studies. Detailed descriptions of these techniques can be found in several laboratory manuals (Davis *et al.* 1986; Ausubel *et al.* 1987; Berger and Kimmel 1987). The most sensitive of these methods is *in situ* hybridization, in which 10 to 100 molecules can be detected in a single cell. However, the *in situ* hybridization method can be technically difficult and does not lend itself to the processing of a large number of samples. With other techniques, the level of detection is about 0.1 to 1.0 pg of the target sequence. For an average-sized mRNA this translates to 10^5 to 10^6 target sequence molecules, and for most practical purposes the detection limit has been reached.

Adaptations of PCR technology have provided a breakthrough in the area. Seeburg and co-workers were the first to describe (P. Seeburg *et al.* 1986, UCLA Symposium, unpublished) the use of PCR to amplify mRNA sequences from cDNA. The first published description of the technique was in 1987 (Veres *et al.* 1987), when it was used to study point mutations in the mouse ornithine trans-carbamylase gene by using subclones derived from an amplified segment of the mRNA. This simple modification of the PCR method has not only increased the sensitivity of detection by several orders of magnitude (with the exception of *in situ* hybridization) but has also made possible the sequence analysis of RNA starting with extremely small amounts of template. In this chapter we describe how RNA sequences are amplified by the use of combined complementary DNA (cDNA) and PCR methodologies.

Protocols

Use autoclaved tubes and solutions wherever possible and wear gloves to minimize nuclease contamination from fingers.

Reagents and Supplies

$10\times$ PCR buffer: 500 mM KCl, 200 mM Tris–HCl (pH 8.4) at 23°C, 25 mM MgCl$_2$, and 1 mg/ml of nuclease-free bovine serum albumin.

Deoxynucleotide triphosphates: Neutralized 100 mM solutions from Pharmacia. The dNTPs are combined to make a stock solution (10 mM each dNTP) by using 10 mM Tris–HCl (pH 7.5) as diluent.

RNasin: RNase inhibitor from Promega Corporation at 20 to 40 units/μl.

Random hexamer oligonucleotides: 100 pmol/μl solution in TE [10 mM Tris–HCl, 1 mM EDTA (pH 8.0)]. Hexamers can be purchased from Pharmacia.

PCR primers: 18 to 22 bases in length, 10 to 100 pmol/μl in TE.

Reverse transcriptase: Moloney murine leukemia virus (MoMuLV) from Bethesda Research Laboratories (BRL) at 200 units/μl. Other reverse transcriptases can be used. See "Discussion and Helpful Hints."

Taq polymerase: 5 units/μl from Perkin-Elmer Cetus.

Light white mineral oil: Sigma

Chloroform: Any reagent grade saturated with TE.

NuSieve and ME agarose: Obtained from FMC Corporation.

TEA electrophoresis buffer: 40 mM Tris–HCl, 1 mM EDTA, and 5 mM sodium acetate (pH 7.5).

DNA markers: 123-bp or 1-kb ladder from BRL.

Microfuge tubes: Use only tubes specified for use in the Thermal Cycler. This is an important point, because ill-fitting tubes will result in inconsistent and inefficient amplifications.

Reverse Transcriptase Reaction

Assemble the following reagents in a final volume of 20 μl 1× PCR buffer: 1 mM each dNTP, 1 unit/μl of RNasin, 100 pmol random hexamer, 1 to 5 μl of RNA sample (amount will be variable), and 200 units of MoMuLV reverse transcriptase. Incubate 10 minutes at 23°C, and then 30 to 60 minutes at 42°C. Heat the reaction at 95°C for 5 to 10 minutes in a water bath and then quick-chill on ice. The heat treatment denatures the RNA–cDNA hybrid and inactivates the reverse transcriptase. It may be helpful to heat treat the RNA sample at 90°C for 5 minutes and quick chill on ice before addition to the reaction; presumably this breaks up aggregates and some secondary structures that may inhibit the cDNA priming step.

PCR Amplification

To the heat-treated reverse transcriptase reaction add 80 μl of 1× PCR buffer containing 10 to 50 pmol each of upstream and downstream primer and 1 to 2 units of *Taq* polymerase. To prevent evaporation of liquid during thermal cycling, layer 100 μl of mineral oil on top of the PCR solution. The number of PCR cycles required depends on the abundance of the target. Usually, somewhere between 20 and 50 cycles is used, but the optimal number should be determined in each case. A thermal cycle profile that works well for amplification of a target of more than 500 base pairs is (1) denaturing for 30 seconds at 95°C, (2) cooling over 1 minute to 55°C, (3) annealing primers for 30 seconds at 55°C, (4) heating over 30 seconds to 72°C, (5) extending the primers for 30 seconds at 72°C, (6) heating over 1 minute to 95°C, and so on. Variations in this thermal cycle profile will work just as well, but one should be careful about the times allotted for heating and cooling. It is essential that enough time be allowed for the solutions in the tubes to equilibrate to the correct temperatures; otherwise, the amplification efficiency may be very low.

Analysis of Amplification Products

After amplification, the mineral oil is removed by extraction with 200 to 300 μl of TE-saturated chloroform. The upper (aqueous) phase is saved, and 5 to 10 μl is used for analysis in a 3% NuSieve–1% ME agarose composite gel made in TEA buffer. An 8 or 10% polyacrylamide gel in TBE buffer is also suitable. Use the 123-bp or 1-kb ladder as a convenient marker for size estimates of the products. Stain the gel with ethidium bromide and photograph. Other analytical methods such as Southern gels or dot/slot blots can be done at this point. Detailed descriptions of these standard methodologies can be found in Davis *et al.* (1986), Ausubel *et al.* (1987), and Berger and Kimmel (1987). Further manipulations of the PCR product such as subcloning (Chapter 11) and direct sequencing (Chapter 24) are discussed elsewhere in this book and will not be described here.

Discussion and Helpful Hints

The same buffer is used for both the reverse transcription reaction and the PCR. We have found that the use of PCR buffer throughout does not seem to negatively affect the efficiency of amplification. Although this has simplified the protocol somewhat, one should be cautious if trying to make long cDNA products. We have not rigorously tested the PCR buffer for this purpose.

First-strand cDNA synthesis may be accomplished by extension with random hexamers, the downstream primer, or oligo(dT). In many cases, it does not seem to matter which priming method is used. With the downstream primer, we have successfully used from 5 to 100 pmol in the cDNA reaction. If oligo(dT) is used, 0.1 to 0.2 μg works well. Figure 1 shows the results of an experiment where

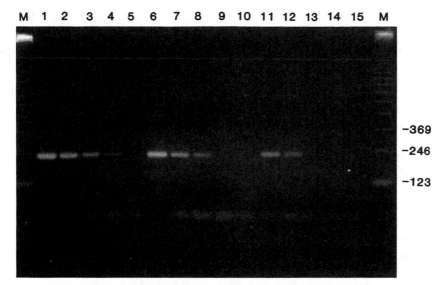

Figure 1 Comparison of random hexamers, downstream primer, and oligo(dT) in the cDNA PCRs. Samples in lanes 1–5 were amplified starting with random hexamers, lanes 6–10 with downstream primer, and lanes 11–15 with oligo (dT). Lanes 1–5 represent amplified cytoplasmic MoMuLV RNA from 50,000, 5000, 500, 50, and 0 cells, respectively. Lanes 6–10 and 11–15 represent the same dilution series but using the different first-strand primers.

the same RNA target (*Pol* gene of MoMuLV) was amplified after reverse transcription with random hexamers (lanes 1–5), downstream primer (lanes 6–10), or oligo(dT) (lanes 11–15). Lanes 1 through 5 represent amplifications starting with viral RNA from 50,000, 5000, 500, 50, and 0 cells, respectively. Lanes 6–10 and lanes 11–15 contain correspondingly the same amounts of starting RNA. As shown, the random hexamer reactions were the most efficient, containing the highest amount of final amplified product. In general, we and others (Veres *et al*. 1987; Noonan and Roninson 1988) have found that the random hexamer approach is the most consistent and results in the highest amplification of target sequence.

The source and type of reverse transcriptase do not seem to be of critical importance. We have used MuLV and AMV enzymes from BRL and Boehringer Mannheim Biochemicals with good results, so enzymes from any reputable source should do as well. For our own research, we have used only the *Taq* polymerase from Perkin-Elmer Cetus, so we cannot comment on the performance of other thermostable polymerases.

It may be useful to titrate the primer oligonucleotides in the PCR to find the smallest amount that can be used to give a "good" amplified product. We have found that a large excess of primer usually gives more extraneous amplified products, which hinders subsequent analysis. As little as 5 pmol of primer pairs has been used to give a very clean amplification. However, it is best to optimize the amount for each sequence you wish to amplify.

The concentration of each dNTP in the PCR should not exceed 0.2 mM. This is the reason for the fivefold dilution of the reverse transcriptase reaction; 1 mM of each dNTP is lowered to 0.2 mM. As little as 0.02 mM can work in PCR, but this should be checked for each case. Triphosphate concentrations higher than 0.2 mM increase the misincorporation rate or mutation frequency for *Taq* polymerase. This point is crucial if one wishes to subclone a cDNA sequence.

The magnesium concentration is also critical, so care should be taken that the addition of reagents does not lower the magnesium molarity; i.e., some nucleic acid buffers contain 1 mM EDTA, and this can chelate out much of the magnesium. In general, try to keep the free magnesium concentration at about 2 mM.

Use the smallest number of PCR cycles that gives you the "cleanest" result. More cycles often just give you more nonspecific amplification products. Also, with a large number of amplification cycles, you may run into contamination problems. You may start picking up false positives because of the presence of extremely small amounts of unwanted target sequences.

We have not discussed methods for RNA extraction because detailed descriptions of the latest protocols are described in several laboratory manuals (Davis *et al.* 1986; Ausubel *et al.* 1987; Berger and Kimmel 1987). Purified RNA usually gives the best results, but if the highest sensitivity is not required, protease- or diethylpyrocarbonate-treated samples will work in this system (see Chapters 18, 19, and 20).

Usually, 1 μg of cytoplasmic RNA is sufficient for amplification of rare mRNA sequences (1 to 10 copies per cell). Since a "typical" mammalian cell may contain ~10 pg RNA per cell cytoplasm, 1 μg represents the RNA from about 100,000 cells. Thus the number of target sequences in 1 μg is probably greater than 100,000 and should be easily amplifiable. Therefore, the isolation of poly(A)$^+$ RNA is not required in cases where one is analyzing a homogeneous cell population. In fact, the time-consuming step of mRNA purification should rarely, if ever, be necessary, since one can detect specific mRNA sequences from the equivalent of 1 to 1000 cells (Kawasaki *et al.* 1987; Rappolee *et al.* 1988).

Choose your PCR primers to be about 18 to 22 bases in length and not too high or low in G+C content. If you are studying eukaryotic mRNAs, try to use primers derived from separate exons; this will inhibit amplification of any contaminating genomic DNA sequences. When the genomic structure is not known, use primers separated by 300 to 400 bases in the 5' portion of the coding region. Exons larger than 300 bases in this area are fairly rare in vertebrates (Hawkins 1988), so the primers will have a good chance of residing in separate exons. If the gene in question has no introns, or you are studying bacterial mRNA, RNA viruses, RNA transcripts from viral integrates, etc., a thorough DNase treatment of the RNA will probably be necessary to obtain meaningful PCR results. It requires only a minuscule amount of contaminating genomic DNA to give a false-positive signal in this type of assay.

In summary, the RNA-PCR technique is a powerful new method for the analysis of RNA transcripts. Experiments that were previously extremely difficult or even impossible, such as mRNA quantitation from single cells, can now be easily performed by this modified PCR procedure.

Literature Cited

Ausubel, F. M., R. Brent, R. E. Kingston, D. D. Moore, J. G. Seidman, J. A. Smith, and K. Struhl (ed.). 1987. *Current protocols in molecular biology*. Greene Publishing Associates and Wiley-Interscience, New York.

Berger, S. L., and A. R. Kimmel. 1987. Guide to molecular cloning techniques. *Methods Enzymol.* **152**:215–304.

Davis, L. G., M. D. Dibner, and J. F. Battey. 1986. *Basic methods in molecular biology*. Elsevier Science Publishing Co., Inc., New York.

Hawkins, J. D. 1988. A survey on intron and exon lengths. *Nucleic Acids Res.* **16**:9893–9908.

Kawasaki, E. S., S. S. Clark, M. Y. Coyne, S. D. Smith, R. Champlin, O. N. Witte, and F. P. McCormick. 1987. Diagnosis of chronic myeloid and acute lymphocytic leukemias by detection of leukemia-specific mRNA sequences amplified *in vitro*. *Proc. Natl. Acad. Sci. USA* **85**:5698–5702.

Noonan, K. E., and I. B. Roninson. 1988. mRNA phenotyping by enzymatic amplification of randomly primed cDNA. *Nucleic Acids Res.* **16**:10366.

Rappolee, D. A., D. Mark, M. J. Banda, and Z. Werb. 1988. Wound macrophages express TGF-α and other growth factors *in vivo*: analysis by mRNA phenotyping. *Science* **241**:708–712.

Veres, G., R. A. Gibbs, S. E. Scherer, and C. T. Caskey. 1987. The molecular basis of the sparse fur mouse mutation. *Science* **237**:415–417.

4

RACE: RAPID AMPLIFICATION OF cDNA ENDS

Michael A. Frohman

Although the technology of cDNA cloning has been steadily growing more powerful (Gubler and Hoffman 1983; Heidecker and Messing 1987; Efstratiadis *et al.* 1977; Okayama *et al.* 1987; Coleclough 1988), the generation of full-length cDNA copies of mRNA transcripts is often challenging, particularly with respect to obtaining the 5′ ends of messages. The first step, reverse transcription of mRNA, is usually the limiting one, and many variations have been described, including the use of different types of reverse transcriptase, pretreatment of the mRNA to decrease secondary structure, alteration of reverse transcriptase conditions, and the use of primer-extended cDNA instead of oligo(dT)-primed reverse transcription. Such modifications can profoundly affect the likelihood of obtaining full-length cDNA clones, but optimal conditions for reverse transcription of a given mRNA are rarely established, since weeks to months are required to prepare and screen a single cDNA library and to isolate and analyze candidate cDNAs.

The method described here, known as the Rapid Amplification of cDNA Ends (RACE) protocol (Frohman *et al.* 1988), differs markedly in that the analysis of the reverse transcribed products can be performed within 1 to 2 days of the start of the experiment. The mRNA reverse transcription step can be modified and repeated until it is clear that the desired cDNA has been produced. Furthermore, information about alternate splicing and promoter use is obtained, and the RACE protocol can be used to separate cDNAs produced from different transcripts prior to cloning steps. Finally, RACE products can be efficiently cloned into standard high-copy plasmid vectors, allowing many independent isolates to be obtained from a single plate of transformants.

In essence, the RACE protocol generates cDNAs by using PCR to amplify copies of the region between a single point in the transcript and the 3' or 5' end. To use the RACE protocol, one must know a short stretch of sequence from an exon. From this region, primers oriented in the 3' and 5' directions are chosen that will produce overlapping cDNAs when fully extended. These primers provide specificity to the amplification step. Extension of the cDNAs from the ends of the messages to the specific primer sequences is accomplished by using primers that anneal to the natural (3' end) or a synthetic (5' end) poly(A) tail. Finally, the overlapping 3'- and 5'-end RACE products are combined to produce an intact full-length cDNA.

Figure 1 illustrates this strategy in more detail. In brief, for the 3' end, mRNA is reverse transcribed using a "hybrid" primer consisting of oligo(dT) (17 residues) linked to a unique 17-base oligonucleotide ("adapter") primer. Amplification is subsequently performed using the adapter primer, which binds to each cDNA at its 3' end, and a primer specific to the gene of interest. For the 5' end, reverse transcription is performed using a gene-specific primer. A homopolymer is then appended by using terminal transferase to tail the first-strand reaction products. Finally, amplification is accomplished using the hybrid primer previously described and a second gene-specific primer upstream of the first one. Additional factors considered in the development of the RACE protocol are described elsewhere (Frohman *et al.* 1988).

RACE Protocol

Materials

RNA: 1 μg of poly(A)⁺ RNA is ideal; one can use as little as 200 ng of total RNA without modifying the procedure.

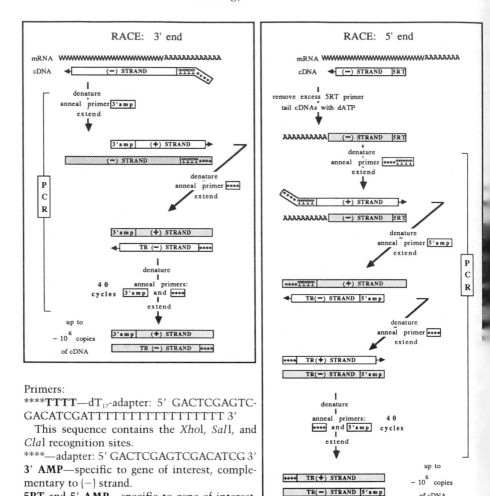

Primers:
****TTTT—dT$_{17}$-adapter: 5' GACTCGAGTC-
GACATCGATTTTTTTTTTTTTTTTTT 3'
    This sequence contains the *Xho*l, *Sal*1, and
*Cla*1 recognition sites.
****—adapter: 5' GACTCGAGTCGACATCG 3'
3' AMP—specific to gene of interest, complementary to (−) strand.
5RT and **5' AMP**—specific to gene of interest,
complementary to (+) strand.

Figure 1   Schematic representation of the RACE protocol. Open rectangles represent
DNA strands actively being synthesized; shaded rectangles represent DNA previously
synthesized. At each step the diagram is simplified to illustrate only how the new
product formed during the previous step is utilized. A (−) or (+) strand is designated
as "truncated" (TR) when it is shorter than the original (−) or (+) strand, respectively.

    10× RTC buffer: 500 m*M* Tris–HCl (pH 8.15) at 41°C, 60 m*M* MgCl$_2$,
400 m*M* KCl, 10 m*M* DTT, each dNTP at 10 m*M*
    Hybrid dT$_{17}$-adapter primer:
GACTCGAGTCGACATCGATTTTTTTTTTTTTTTTTT—stock (1 μg/ml)
    Adapter primer: GACTCGAGTCGACATCG
    Gene-specific primers: see Fig. 1

10× PCR buffer: 670 mM Tris–HCl (pH 8.8 at room temperature), 67 mM MgCl$_2$, 1.7 mg/ml of bovine serum albumin, 166 mM (NH$_4$)$_2$SO$_4$

TE: 10 mM Tris–HCl (pH 7.6), 1 mM ETDA.

Amplification of cDNA 3' Ends

Reverse Transcription

1. Assemble on ice: 2 μl 10× RTC buffer, 10 units of RNasin (Promega Biotech), 0.5 μg of dT$_{17}$-adapter primer, and 10 units of AMV reverse transcriptase (Life Sciences) in a total volume of 3.5 μl.
2. Heat RNA in 16.5 μl of H$_2$0 for 3 minutes at 65°C; quench on ice and add to above.
3. Incubate at 42°C for 1 hour, and then at 52°C for 30 minutes. Dilute to 1 ml with TE and store at 4°C ("cDNA pool").

Amplification

1. Prepare PCR cocktail   5 μl 10× PCR Buffer
        5 μl DMSO
        5 μl 10× dNTPs (15 mM each)
        30 μl H$_2$O
        1 μl adapter primer (25 pmol/μl)
        1 μl gene-specific primer (25 pmol/μl)
        1 to 5 μl cDNA pool
2. Denature at 95°C for 5 minutes; cool to 72°C.
   Add 2.5 units of *Taq* polymerase and overlay with 30 μl of mineral oil (heated to 72°C).
   Anneal at 55°C for 5 minutes; extend at 72°C for 40 minutes.
   Cycle: 95°C for 40 seconds / 55°C for 1 minute / 72°C for 3 minutes—for 40 cycles.
   Extend 72°C for 15 minutes.

Amplification of cDNA 5' Ends

1. Reverse transcribe RNA as previously described, but substitute 10 pmol gene-specific primer 1 for dT$_{17}$-adapter primer.

2. Remove excess primer by using a Centricon 100 spin filter (Amicon Corp.). Dilute reverse transcription mixture with 2 ml of $0.1 \times$ TE and centrifuge at $1000 \times g$ for 20 minutes. Repeat and collect retained liquid. Concentrate to 10 μl using Speed Vac centrifugation.
3. Add 4 μl of $5\times$ Tailing Buffer (Bethesda Research Laboratories, Inc.), 4 μl of 1 mM dATP, and 10 units of Terminal d Transferase (BRL). Incubate 5 minutes at 37°C, and then 5 minutes at 65°C. Dilute to 500 μl with TE.
4. Amplify 1 to 10 μl of cDNAs as previously described, adding dT_{17}-adapter primer (10 pmol) in addition to the adapter primer and gene-specific primer 2.

Analysis

Perform a Southern blot analysis of the amplification products. Test for single-stranded DNA by removing the PCR buffer (Centricon filter [any type], spun column, or ethanol precipitation) and incubating in mung bean nuclease buffer (BRL) and 1 unit of mung bean nuclease for 30 minutes at 30°C (Frohman et al. 1988). Compare to the samples incubated in buffer alone and in TE.

Cloning RACE Products

Blunt Cloning

Remove the PCR buffer and excess primer by three rounds of Centricon 100 filtration. Kinase the cDNAs (Maniatis et al. 1982) and isolate the fragments of interest after gel electrophoresis. Ligate to the blunt-cut and phosphatased vector and identify the recombinants with a colony lift procedure (Buluwela et al. 1989) by using a probe containing sequences within the amplified fragment.

Directional Cloning

Inactivate the residual Taq Polymerase with phenol–chloroform and ethanol precipitation. Cleave the amplified products at one of the adapter primer restriction sites (XhoI, SalI, or ClaI) and at a site in the gene-specific primer or one known to exist in the amplified product. Gel purify and subclone as usual. Although a very high proportion of the recombinant plasmids may contain the desired product, it is still helpful to use colony lifts and Southern blot hybridization (as previously described) whenever possible. It is also important to compare the size of amplified products before and after restriction to ensure that the endonucleases chosen are not cleaving within the uncharacterized portion of the cDNA. If this occurs, try other restriction enzymes or clone the cDNAs intact.

Troubleshooting and Comments on Protocol Steps

Artifactual and Truncated Products

The most obvious artifact of the RACE protocol is nonspecific amplification of cDNAs that share no significant homology with the gene of interest and that fail to hybridize to an appropriate probe under Southern blot analysis conditions. Several ways of minimizing this problem are discussed below.

It is also possible to generate specific products that are undesirable because they are truncated or because they are derived from contaminating genomic DNA.

Truncation of cDNA Ends During Reverse Transcription   Occasionally, truncated cDNA 3' ends are produced because of annealing of the dT_{17}-adapter primer to an A-rich region in the transcript upstream of the poly(A) tail. This can be minimized by decreasing the concentration of adapter primer to 15 ng per reverse transcription reaction.

For the 5' end, a primer that binds inappropriately to the message of interest within the region to be amplified will cause premature termination of the reverse transcription reaction at that site, resulting in truncated ends. Problems of this sort are minimized by (1) using short primers (12 to 16 nucleotides) rich in A+T, (2) increasing the reverse transcription temperature to 52°C, (3) decreasing the primer concentration to 15 ng or lower per reaction, and (4) trying different primers for reverse transcription. It should also be noted that primers rich in G+C produce more nonspecific reverse transcription and thus increase background amplification.

Truncation of cDNAs During Amplification   Truncation of cDNAs can also occur during amplification through mismatched annealing of any of the primers. Some of these products can be detected by leaving out one primer from the reaction mixture and determining whether specific products are still amplified. More precisely, examine the products of an amplification reaction in which only the gene-specific primer (for 3'-end amplifications) or the gene-specific primer and the adapter primer (for 5'-end amplifications) are present. Under these conditions, a specific product should not appear, or at least should be a different size compared to that of the amplification products generated by reactions containing all of the necessary primers. If such problems arise, mismatched annealing can often be reduced by decreasing the primer concentrations (to as little as 2.5 pmol in some cases).

Production of Specific Products by Amplification of Genomic DNA
It is often possible, through mismatched annealing of one of the RACE or gene-specific primers, to obtain amplification of a fragment of the gene of interest from genomic DNA. Although reverse-transcribed cDNAs should normally be present in vast excess over residual genomic DNA, technical problems with preparation of cDNA templates or with inefficient amplification through regions with secondary structure may result in preferential genomic amplification. Genomic DNA (50 ng) should be amplified under the same conditions as those for the cDNA pool; the results of this amplification will indicate whether such products can be efficiently created and will provide information about their size.

One way to minimize the problem is to use primers that span splice borders in the cDNA to force amplification of cDNAs only. A primer thus chosen should fail to anneal to the genomic DNA, since in that substrate the primer sequence is interrupted by an intron sequence.

More generally, it is useful to design the amplification scheme such that the final RACE products are composed of more than one exon. This ensures that the material recovered derives from cDNAs and not from the amplification of genomic DNA.

Reverse Transcription

There are $\sim 10^5$ copies of a rare transcript in 200 ng of total RNA. Thus, very little RNA is required to provide starting material sufficient for many amplifications. If it proves difficult to obtain fully extended cDNAs, try using Mo-MuLV reverse transcriptase, increasing reverse transcription temperature, or pretreating the RNA with methyl mercury. For the amplification of some genes, RNase H and RNase A treatment of the cDNA–RNA hybrids after reverse transcription appears to result in a higher yield of specific product.

Tailing

An obvious advantage of tailing the first-strand cDNAs with dAs is that the same dT_{17}-adapter primer used in 3'-end cDNA amplification reactions can be used as a primer for the second-strand synthesis of 5'-end cDNAs. In theory, any nucleotide can be used in the tailing reaction. In practice, however, it is probably best not to use either dC or dG, since this is more likely to lead to truncation of products due to nonspecific primer annealing during amplification (see "Truncation of cDNAs During Amplification") than tailing with dA or dT would. This is because homopolymers of C or G anneal with much higher affinity than do homopolymers of A or T. Therefore, much shorter stretches of Cs or Gs than As or Ts are required in a cDNA sequence for nonspecific binding of the complementary homopolymer-adapter primer and consequent truncation to occur during amplification.

Amplifications

Nonspecific amplification is quite temperature-dependent, and the relative abundance of specific product often increases dramatically with increased annealing temperatures. Optimize the annealing temperature by repeating the amplification step at increasing temperatures (by increments of 2°C) until specific products are no longer obtained.

Gene-Specific Primers for Amplification

Nonspecific amplification will also be minimized if all of the primers have similar melting temperatures. Thus, choose gene-specific primers that match the adapter primer, both in length and G+C content.

Controls

Always include the following controls: (1) no DNA; (2) genomic DNA (see "Production of Specific Products by Amplification of Genomic DNA"); and (3) cDNA pool, amplified with gene-specific primer only (see "Truncation of cDNAs During Amplification").

Optional controls that can be included are: (1) "sham cDNA" (no specific amplification product should be obtained if reverse transcriptase is left out when one is preparing the cDNA pool, demonstrating that the putative specific product is not generated by amplification of tailed contaminant or residual genomic DNA) and (2) "DNased" RNA (if genomic DNA proves to be responsible for a significant fraction of the gene-specific amplification, there are several RNase-free DNases available that can be used before reverse transcription to eliminate the genomic DNA).

Cloning

When nonspecific amplification is a problem, an increase in the frequency of the desired products can be obtained during the cloning step. First, as noted in the protocol, the specific products can be isolated from agarose gels after Southern blot analysis, and cloned. By using a blunt cloning method, one can clone the RACE products intact, eliminating the possibility of restriction endonuclease cleavage within the uncharacterized part of the specific product. On the other hand, blunt cloning requires fully extending and kinasing the PCR products and is inefficient.

A further increase in the frequency of the desired product can be obtained by directional cloning of the RACE products after cleavage with appropriate restriction enzymes. By cleaving the RACE products with two endonucleases, each specific to a different primer, the cloning of any nonspecific amplification product that has the same primer at both ends will be precluded. Directional cloning can be used to obtain an even higher frequency of the desired product, providing that a restriction enzyme that recognizes infrequent sites can be used to cleave the specific product at a site internal to the gene-specific primer. Most products arising from nonspecific amplification will not contain this recognition site and thus will not readily clone. Directional cloning also dictates the orientation of the insert; this can be useful for mapping and sequencing purposes. The major limitation of directional cloning is that cleavage of restriction sites within the uncharacterized portion of the RACE products can result in an unclonable product. This possibility can be minimized, as described in the "Protocol" section, but not eliminated.

5'-End Fidelity

Extended products often end with one-to-a-few extra bases, generally a combination of As and Gs. This is presumably because of the snapback feature of reverse transcriptase, rather than because of the PCR amplification itself. Confirm 5' ends with genomic sequence and primer extension/RNase protection analysis.

Construction of Full-Length cDNAs

Ultimately, a full-length cDNA is often sought. The overlapping 3'- and 5'-end RACE products can be linked by several approaches. First, any unique restriction site in the region of overlap can be used to join the two ends in a standard subcloning procedure. Second, the rapid mutagenesis procedure (Chapter 22) can be used at any position within the overlapping region to join the two ends. Finally, information obtained from the RACE protocol allows one to make primers that represent the sequences at the extreme 3' and 5' ends of the mRNA. These primers can then be used to amplify full-length cDNAs from the original cDNA pool. This approach has several advantages. First, it minimizes the overall number of amplification cycles, thus decreasing the number of polymerase-induced mutations. Second, this approach "strips" the cDNA of the homopolymeric sequences appended to the 5' end by the RACE protocol, which may in some cases inhibit translation. Finally, in the case of multiple 3' or 5' ends, successful production of a full-length cDNA confirms that the sequences used for amplification represent bona fide ends of an mRNA present in the original population.

Literature Cited

Buluwela, L., A. Forster, T. Boehm, and T. H. Rabbitts. 1989. A rapid procedure for colony screening using nylon filters. *Nucleic Acids Res.* **17**:452.

Coleclough, C. 1987. Use of primer-restriction end adapters in cDNA cloning. *Methods Enzymol.* **159**:64–83.

Efstratiadis, A., F. C. Kafatos, and T. Maniatis. 1977. The primary structure of rabbit β-globin mRNA as determined from cloned DNA. *Cell* **10**:571–586.

Frohman, M. A., M. K. Dush, and G. R. Martin. 1988. Rapid production of full-length cDNAs from rare transcripts: amplification using a single gene-specific oligonucleotide primer. *Proc. Natl. Acad. Sci. USA* **85**:8998–9002.

Gubler, U., and B. J. Hoffman. 1983. A simple and very efficient method for generating cDNA libraries. *Gene* **25**:263–269.

Heidecker, G., and J. Messing. 1987. A method for cloning full-length cDNA in plasmid vectors. *Methods Enzymol.* **159**:28–41.

Maniatis, T., E. F. Fritsch, and J. Sambrook. 1982. In *Molecular cloning: A laboratory manual*, p. 123. Cold Spring Harbor Laboratory, Cold Spring Harbor, New York.

Okayama, H., M. Kawaichi, M. Brownstein, F. Lee, T. Yokota, and K. Arai. 1987. High-efficiency cloning of full-length cDNA: construction and screening of cDNA expression libraries for mammalian cells. *Methods Enzymol.* **159**:3–27.

5

DEGENERATE PRIMERS FOR DNA AMPLIFICATION

Teresa Compton

A mixture of oligonucleotides varying in base sequence but with the same number of bases ("degenerate") can be substituted for defined sequence oligonucleotides as primers and successfully coupled with PCR, resulting in the desired gene-specific amplification products. There are many instances in which the use of degenerate primers may be necessary or desired. Mixed oligonucleotide primers derived from an amino acid sequence are useful when only a limited portion of a protein sequence is known for a sought-after gene. Such a strategy has been used to clone the urate oxidase gene (Lee *et al.* 1988) and the diabetes-associated peptide (Girgis *et al.* 1988). An extension of this strategy may be applied when the search is for new or uncharacterized sequences related to a known family of genes. Examples of this approach were the simultaneous detection of the mammalian and avian members of the hepadnaviruses by using degenerate primers based on conserved regions of the amino acid sequence of the viral reverse transcriptase (Mack and Sninsky 1988). Finally, one may wish to use degenerate primers that incorporate sequence variation when the goal is to amplify many known members of a gene family for subsequent typing or identification.

PCR Protocols: A Guide to Methods and Applications
Copyright © 1990 by Academic Press, Inc. All rights of reproduction in any form reserved.

Recommended Procedures

Primer Design

There are several considerations when designing degenerate primers for PCR. First, the degeneracy of the genetic code for the selected amino acids of the region targeted for amplification must be examined. Although methionine and tryptophan are encoded by a single codon, the other amino acids may be encoded by two to six different triplets (see Table 1). Selection of amino acids with minimal degeneracy is desired. While the system will tolerate a high degree of degeneracy (516-fold degenerate primers have been successfully used), obviously the lower this degeneracy, the higher the specificity. Therefore, given a choice, the use of peptide regions containing amino acids requiring four or six codons should be avoided. Second, if desired, the degeneracy of the chosen primer may be further restricted by considering codon bias for translation. The preferable use of a subset of possible codons is sometimes observed for a gene or organism or virus (Aota *et al.* 1988). Third, if the peptide sequence being examined is known to be at the extreme carboxyl terminus, the sequence for the translational stop codons could be used. Fourth, PCR primers may contain as few as 15 to 20 nucleotides; the same is true of mixed sequence primers, but degeneracy greater than 516-fold is not recommended. However, 15- to 20-mers require the identification of five to seven amino acids, which is not always possible. Fortunately, degenerate PCR primers based on as few as three consecutive amino acids can be successfully used by including 6 to 9 base 5' extensions that usually would include restriction enzyme sites. While not complementary to the template, these 5' extensions become incorporated into the amplified product at the second and

Table 1

Codon Degeneracy

Amino Acid	Codons
M, W	1
C, D, E, F, H, K, N, Q, Y	2
I	3
A, G, P, T, V	4
L, R, S	6

all subsequent cycles of amplification. The effect is a higher overall efficiency of amplification due to the increased stability of the priming duplex (Mack and Sninsky 1988). Finally, since a single mismatch may obviate extension, degeneracy on the 3' end of the primer is best avoided. For example, in designing the sense primer from an amino acid sequence, a degenerate terminal base should not be included. In the end, however, empirical testing of the designed primers is required and may need further alteration of the degree of degeneracy and size.

Thermal Cycling

The key to success in applying degenerate primers in a PCR is to find conditions representing the optimum complementation between efficiency and specificity. For the best results, certain parameters of the reaction mixtures must be empirically determined for each primer pair (see Chapter 1 on the optimization of PCRs). Thermal cycling conditions that have worked with a wide variety of degenerate PCR primers are diagrammed in Fig. 1. One begins the reactions by using a low, nonstringent annealing temperature of 37°C during the initial two to five cycles of amplification. The relaxed annealing conditions allow for hybridization of the template-targeted portion of the primer. After the second cycle of amplification, the unit-length product that includes the 5' extensions will serve as the template for subsequent rounds of amplification. Another parameter that was found to be uniformly critical during this phase of the amplification was a slow ramp time between the annealing tempera-

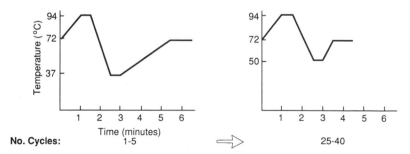

Figure 1 Schematic diagram of the thermal cycling parameters for a DNA Thermal Cycler (Perkin-Elmer Cetus) using short, highly degenerate oligonucleotide primers with 5' extensions.

ture and the extension temperature for the *Taq* polymerase (72°C). Once the number of cycles required at the 37°C annealing temperature is determined (I generally test each new primer pair over one, three, and five cycles), the reactions are shifted to a higher stringency of annealing based on the T_m of the entire primer duplex and subjected to an additional 25 to 35 cycles.

Examples of Use

An example of using degenerate primers to detect a new sequence related to a known family of genes is described below. Common to all the herpesviruses is a glycoprotein gene referred to as gB. To isolate the gB gene of the feline herpesvirus (FHV), the amino acid sequences of all known gBs were aligned, and regions of conservation were scored. A 23-amino-acid region was identified that contained a unique sequence flanked by conserved pentapeptides. An alignment of this region is shown in Table 2.

Although the first pentapeptide contains both a serine and an arginine (six codons each), there is a striking bias in the codon usage among the herpesviruses. In this case, therefore, primers were synthesized encompassing only the preferred codon usage of these genes. In addition, the downstream primer had degeneracy to detect both codons of tyrosine and glutamic acid. To include all the degeneracy, two primers were synthesized for each pentapeptide and pooled in equimolar amounts in the reactions. Both primers had restriction

Table 2

Herpesvirus Family gB Conserved Region

VZV:	CYSRP	LISIVSLNGSGTV	EGQLG
BHV:	CYSRP	VSFAFGNESEPV	EGQLG
CMV:	CYSRP	VVIFNFANSSYVQ	YGQLG
EBV:	CYSRP	LVSFSFINDTKTYE	EGQLG
PRV:	CYSRP	LVTFEHNGTGVI	EGQLG
HSV:	CYSRP	LVSFRYEDQGPLV	EGQLG
FHV:	CYSRP	LVSFRALNDSEYI	EGQLG

Abbreviations: CMV, cytomegalovirus; VZV, varicella zoster virus; BHV, bovine herpesvirus; EBV, Epstein Barr virus; PRV, pseudorabies virus; HSV, herpes simplex virus; FHV, feline herpesvirus.

enzyme recognition sites added to the 5' end. The primers and probe used are listed below. (In the oligonucleotide sequences Y = T + C and N = A + G + C + T.)

Sense primers based on "CYSRP"
    *Hin*dIII 22-mer (5') CA AAGCTT **TGY TAY TCN CGN CC**

Antisense primers based on "E/YGQLG"
    *Eco*RI 22-mer: (5') TC GAATTC **NCC YAA YTG NCC NT**

HSV internal probe
                    AGCTTTCGGTACGAA

Viral DNA was prepared from infected cells to serve as a template in the reactions. Herpes simplex viral (HSV) DNA was used as a positive control template. A product of 85 bp was predicted (69 bp from the coding region plus 16 bp from the 5' extensions). Primers (100 pmol each) were used in 100-μl reactions in standard *Taq* buffer [10 mM Tris (pH 8.5), 2.5 mM Mg^{2+}, 50 mM KCL, 100 μg/ml of BSA] plus 2.5 units of *Taq* polymerase (Perkin-Elmer Cetus), 200 μM dNTPs, and 1 ng of each viral template (see Chapter 1). Because of the somewhat high G + C content of this region, reactions were also included containing the nucleotide analog 7-deaza-2'-deoxyguanosine 5'-triphosphate at a ratio of 1 analog : 3 GTP (McConlogue *et al.* 1988; Chapter 7). The reactions were performed as described in Fig. 1, with three cycles of amplification including a 37°C annealing step linked to 30 cycles including a 55°C annealing step. One-tenth of the reaction mixtures were analyzed by agarose gel electrophoresis as shown in Fig. 2. As can be seen in panel A, a band of 85 bp was detected in both the HSV- and FHV-primed reactions. To determine the specificity of the reaction product, the gel shown in panel A was Southern blotted and probed with an HSV-specific internal probe (panel B). Although some nonspecific bands were present in the reaction products, only the 85-bp band in the HSV-primed sample hybridized with the probe. After purification of this 85-bp segment from the FHV-primed samples, the DNA was cloned by using the restriction enzyme sites at both termini, and the DNA sequence was determined. The translation of the FHV peptide as compared to the conserved region of the gB genes is shown at the bottom of Table 2.

This study demonstrated the ability to identify and clone a novel but related gene segment using degenerate primers in PCR. Such segments can then serve as probes to isolate the entire gene or procedures to obtain amplification of flanking regions have been de-

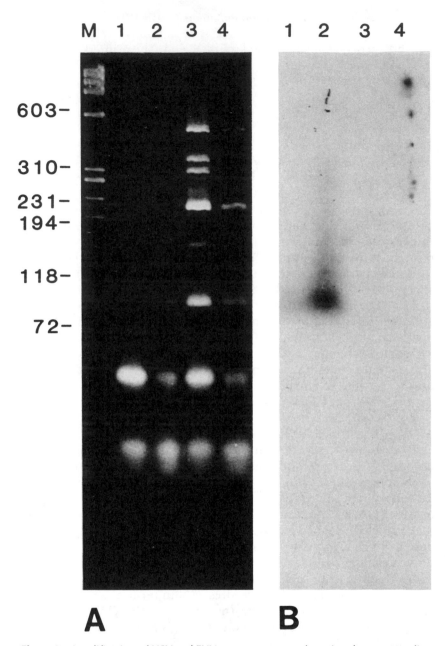

Figure 2 Amplification of HSV and FHV target sequences by using degenerate oligo-nucleotide primers. Purified HSV (1 ng) (lanes 1 and 2) and FHV DNA (1 ng) (lanes 3 and 4) were subjected to PCR as described in Fig. 1. Comparisons were made between 200 μM each dNTP (lanes 1 and 3) and 50 μM dc7GTP mixed with 150 μM dGTP (lanes 2 and 4). One-tenth of the reaction was examined by ethidium bromide/agarose gel electrophoresis (3% NuSieve agarose and 1% agarose) (Panel A). The gel was Southern blotted and hybridized with an internal HSV-specific probe (Panel B).

scribed (see Chapter 27). The use of degenerate primers coupled with PCR is a powerful and expeditious way to clone genes that are related to known gene families (Nunberg *et al.* 1989). This technique overcomes many of the limitations of using degenerate oligonucleotide probes in traditional hybridization, in which developing conditions to distinguish an authentic signal from background hybridization is difficult and time-consuming. Another exciting application of this technology is the detection of novel sequences. In the field of virology, for example, it is speculated that only a fraction of the viruses responsible for human disease have so far been identified. The use of degenerate primers and PCR may allow for the detection of viral genomes that are as yet unidentified members of existing virus families (Mack and Sninsky 1988; Chapter 45). In addition, other investigators have begun to explore the use of inosine-containing primers because of the ability of this base to pair with all bases to initiate amplification from partially known sequences. (Knoth *et al.* 1988).

Literature Cited

Aota, S., T. Gojobori, F. Ishibashi, T. Maruyama, and T. Ikemura. 1988. Codon usage tabulated from the GenBank Genetic Sequence Data. *Nucleic Acids Res.* **16** (supplement): r315–r402.

Girgis, S. I., M. Alevizaki, P. Denny, G. J. M. Ferrier, and S. Legon. 1988. Generation of DNA probes for peptides with highly degenerate codons using mixed primer PCR. *Nucleic Acids Res.* **16**: 10371.

Knoth, K., S. Roberds, C. Poteet, and M. Tamkun. 1988. Highly degenerate, inosine-containing primers specifically amplify rare cDNA using the polymerase chain reaction. *Nucleic Acids Res.* **16**: 10932.

Lee, C. C., X. Wu, R. A. Gibbs, R. G. Cook, D. M. Muzny, and C. T. Caskey. 1988. Generation of cDNA probes directed by amino acid sequence: cloning of urate oxidase. *Science* **239**: 1288–1291.

Mack, D., and J. J. Sninsky. 1988. A sensitive method for the identification of uncharacterized viruses related to known virus groups: hepadnavirus model system. *Proc. Natl. Acad. Sci. USA* **85**: 6977–6981.

McConlogue, L., M. A. D. Brow, and M. A. Innis. 1988. Structure-independent DNA amplification by PCR using 7-deaza-2'-deoxyguanosine. *Nucleic Acids Res.* **16**: 9869.

Nunberg, J. H., D. K. Wright, G. E. Cole, E. A. Petrovskis, L. E. Post, T. Compton, and J. H. Gilbert. 1989. Identification of the thymidine kinase gene of feline herpesvirus: use of degenerate oligonucleotides in the polymerase chain reaction to isolate herpesvirus gene homologs. *J. Virol.* **63**: 3240–3249.

6

cDNA CLONING USING DEGENERATE PRIMERS

Cheng Chi Lee and C. Thomas Caskey

A popular approach to complementary DNA (cDNA) cloning involves the synthesis of oligonucleotide probes to a known amino acid sequence. Predicting the codon usage when designing the oligonucleotide probes is difficult since the genetic code is degenerate (Crick 1968). The use of "guessmer" oligonucleotide probes that lack absolute complementarity to the target sequence generates spurious hybridization signals when used in cDNA library screening. Identifying hybridization conditions that distinguish authentic signals from spurious signals is difficult and frequently unproductive.

We have described a novel procedure for synthesizing authentic cDNA probes to a known amino acid sequence (Lee *et al.* 1988). The mixed oligonucleotides primed amplification of cDNA (MOPAC) procedure was developed to clone the porcine urate oxidase cDNA. Since our report, two other groups have successfully used this procedure to generate DNA probes (Girgis *et al.* 1988; Griffin *et al.* 1988). In this chapter we will summarize our MOPAC protocol and describe some recent improvements on the MOPAC protocol that we (Lee and Caskey, in press) and others have identified.

PCR Protocols: A Guide to Methods and Applications

Principle of the MOPAC Procedure

The MOPAC technique was conceived with the simplistic thought that if we could reverse translate an amino acid sequence to its authentic DNA sequence, we could improve the process of cDNA library screening. By synthesizing a perfectly matched cDNA probe, one could use stringent hybridization conditions, thereby eliminating spurious hybridization signals. A technology that could be adapted for such a purpose was PCR (Mullis and Faloona 1987). However, a major problem of selecting a complementary PCR primer set for the reaction had to be overcome. To resolve this problem we decided not to use a unique primer set but instead chose to synthesize every possible primer combination coding for the amino acid sequence. The fundamental assumption behind this approach was that the authentic sequence primer would selectively anneal to its target complementary sequence, out-competing the less complementary primers during the annealing process. In practice this strategy worked, but there were some surprises.

The most interesting of these surprises was the finding that a "perfect match" primer need not be present in the mixed oligonucleotide population for the procedure to be successful. Our studies indicate that there is tolerance of up to a 20% base-pair mismatch between the primer and the template during the MOPAC reaction. The amplified product characteristically possesses different primer sets, but the newly synthesized DNA sequence spanning the primers is unique. Mixed primers of up to 1024 combinations have been used to generate authentic cDNA probes. However, we have observed that an increase in primer complexity is also associated with an increase in nonspecific priming. The problem of nonspecific priming can be dealt with by two different approaches: (1) reducing the primer combinations by selecting the most frequently used codon sets [based upon tabulated codon usage frequencies (Lathe 1985)] in conjunction with the use of DNA polymerase I Klenow fragment as the polymerization enzyme and (2) selecting an annealing condition that would favor primers that are highly complementary to their target templates in conjunction with the use of *Taq* polymerase as the polymerization enzyme. In the former case the presence of primer mismatches to template will be tolerated by the lower annealing and polymerization temperature used in conjunction with Klenow enzyme (Lee and Caskey, in press). The latter approach with *Taq* polymerase may develop into the superior procedure once the annealing parameter is better clarified. Authentic products were obtained when primers were annealed at 48°C for a primer length corresponding to five amino acids (128 species). Annealing at 40°C did not generate any authentic product (Girgis *et al.* 1988). Authentic products were also generated when primers were annealed at 46°C when using a primer length corresponding to eight amino acids (256 and 1024 species). Annealing performed at 37°C, 56°C, and 65°C failed to generate authentic MOPAC products (Griffin *et al.* 1988). These studies would suggest that an annealing temperature of 46 to 48°C is suitable for complex primer mixtures when used in conjunction with *Taq* polymerase.

Further Advances in MOPAC Technology

Since our initial report we have made several advances in the MOPAC technique (Lee and Caskey, in press). We have asked the following questions: (1) Can the first-strand cDNA synthesis be specifically primed with the antisense primer rather than with the random or oligo(dT) primers? (2) Can total RNA be used in place of poly(A)$^+$ RNA as the initial template for cDNA synthesis? (3) Does removal of the mRNA template by alkaline hydrolysis improve MOPAC specificity? (4) Does the enzyme ribonuclease A (RNase A) reduce the level of amplification of nonspecific sequences that are due to nonspecific priming? and (5) Does the length of the priming oligonucleotide have any effect on the MOPAC procedure?

Our results show (see Figs. 1A and 1B) that: (1) specific priming by the antisense primer can replace random or oligo(dT) primers for the first-strand cDNA synthesis in the MOPAC procedure; (2) authentic MOPAC product can be generated from samples prepared from poly(A)$^+$ and total RNA, demonstrating that the MOPAC process can work equally well with total RNA; (3) similar MOPAC success was observed in the samples that had not been treated by alkaline hydrolysis, indicating that the denaturation of the DNA : RNA duplex occurs efficiently at the initial denaturation temperature and that the annealing primers will efficiently compete with the mRNA

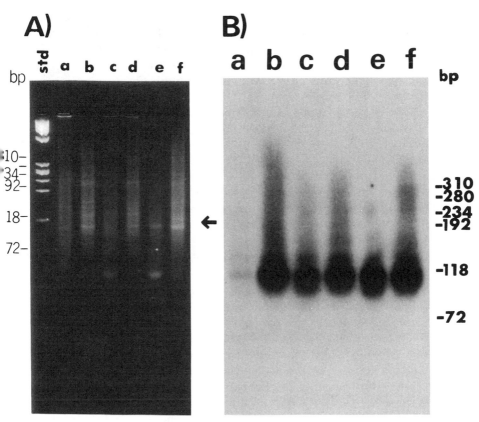

Figure 1 Analysis of the MOPAC reactions by agarose gel electrophoresis and by Southern analysis. (A) Analysis of MOPAC reactions by electrophoresis on a 4% NuSieve agarose gel electrophoresis. The MOPAC reactions (a), (b), (c), and (d) are from poly(A)$^+$ RNA whereas (e) and (f) are from total RNA. The reactions (a), (c), and (e) received RNase A treatment in contrast to (d) and (f), which are untreated. Reaction (b) was alkaline treated to remove the mRNA template. Reactions (b), (c), (d), (e), and (f) all used an antisense primer corresponding to seven amino acids, whereas reaction (a) used an antisense primer corresponding to five amino acids for the first-strand cDNA synthesis and for the MOPAC reaction in conjunction with the sense primer corresponding to six amino acids (see Lee and Caskey, in press). The arrow indicates the presence of the expected-size cDNA. (B) Autoradiograph of a Southern analysis of the above agarose gel probed by an internal 48-oligonucleotide guessmer probe. The guessmer oligonucleotide probe was end labeled in the presence of [^{32}P]ATP (3000 Ci/mmol) by T4 polynucleotide kinase. The probe hybridization was performed at 42°C in 6× SSC, 0.1% sodium pyrophosphate, 0.1% SDS, 0.1% Denhardt's, 50 mM Tris–HCl (pH 7.5), and 50 μg/ml of denatured herring sperm DNA. The blot was washed in 2× SSC, 0.1% SDS at 42°C.

for template annealing; (4) the addition of RNase A does produce a lower background, as supported by agarose gel and Southern analysis; however, it is not required for the overall success of the MOPAC reaction; (5) longer primers are superior in authentic MOPAC product generation compared to the shorter primers. On the basis of this study, we would suggest the use of a primer length corresponding to at least five amino acids and preferably six to seven amino acids.

Further Recommendations for the MOPAC Procedure

To rapidly determine whether the MOPAC reaction has been successful, we highly recommend the use of an internal oligonucleotide probe (Lee *et al.* 1988; Lee and Caskey, in press). To avoid the problem of amino acid substitutions that can occur between animal species, we suggest that the mRNA used in the MOPAC reaction be derived from the same animal species from which the protein sequence was derived. To facilitate a rapid rescue of the MOPAC product into a cloning vector, we suggest the inclusion of a restriction site when synthesizing the primers. We also recommend the fractionation of the MOPAC product by polyacrylamide gel electrophoresis to facilitate the rescue of the authentic product from the nonspecific products. In our experience, the introduction of a fractionation step will increase the percentage of authentic colonies by at least 10-fold, making a direct plasmid analysis of transformed colonies a feasible option in place of *in situ* colony hybridization.

Protocols

NOTE: When selecting bacterial colonies for miniplasmid analysis, it is important not to select bacterial colonies on the basis of color selection with Xgal/IPTG, since the small insert may not fully disrupt the β-galactosidase gene within the cloning vector.

First-Strand cDNA Synthesis

1. The first-strand cDNA synthesis is performed in a 50-μl reaction containing 50 mM Tris–HCl (pH 8.3), 75 mM KCl, 1 mM DTT, 15 mM MgCl$_2$, 1 mM of each dNTP, 400 units of MuLV reverse transcriptase (Bethesda Research Laboratories, Inc.), 20 μg of total RNA, and 400 ng of antisense primers. Alternatively, if the antisense primers are very complex, such that ribosomal RNA may be primed, replace the antisense primer with 400 ng of oligo(dT)$_{12-18}$. Heat the RNA at 65°C for 3 minutes and quickly cool on ice before use. The addition of ribonuclease inhibitor RNasin (50 units) is optional.

2. Incubate the cDNA synthesis reaction at 37°C for 1 hour, followed by ethanol precipitation of the mRNA : cDNA duplex in the presence of 200 mM ammonium acetate (pH 4.5) and two volumes of cold absolute ethanol.

3. Recover nucleic acid precipitate by centrifugation in a microcentrifuge for 10 minutes at maximum speed.

4. Wash the nucleic acid pellet in 70% ethanol and lyophilize. Dissolve lyophilized pellet in 50 μl of sterile distilled H$_2$O. The sample is now ready for use in the MOPAC reaction.

MOPAC with the Klenow Fragment

1. The MOPAC reaction is performed in a 100-μl reaction containing 10 mM Tris–HCl (pH 7.5), 50 mM NaCl, 10 mM MgCl$_2$, 1 mM DTT, 1.5 mM of each dNTP, 1 μg of RNase A (DNase free), synthesized cDNA, and 3 μM of each primer mixture.

2. Heat the reaction mixture to 95°C for 3 minutes.

3. Anneal and let the RNase A digestion proceed at 37°C for 15 minutes.

4. Add 5 units of Klenow fragment (USB), followed by chain polymerization at 37°C for 2 minutes.

5. Denature reaction mixture at 95°C for 1 minute.

6. Anneal at 37°C for 30 seconds.

7. Add 5 units of Klenow fragment, followed by a chain polymerization period of 2 minutes at 37°C.
8. After 29 cycles (steps 5–7), analyze a 10-μl aliquot in a 4%
   NuSieve agarose gel, and if the DNA fragment size is consistent
   with the predicted product, perform a Southern analysis on the
   gel using the internal guessmer probe.

MOPAC with *Taq* Polymerase

The protocol described has been summarized from Girgis *et al.*
(1988) and Griffin *et al.* (1988). MOPAC reactions using complex
primer mixtures ranging from 128 to 1024 species have been successful with *Taq* polymerase.

1. The MOPAC reaction is performed in the Cetus buffer containing 50 mM KCl, 10 mM Tris–HCl (pH 8.3), 1.5 mM MgCl$_2$, 0.01%
   (w/v) gelatin, 200 μM of each dNTP, 4 μM of each primer mixture, synthesized cDNA, and 4 units of *Taq* polymerase.
2. Denature reaction mixture at 95°C for 30 seconds.
3. Anneal at 48°C for 30 seconds.
4. Chain polymerization is performed at 70°C for 1 minute.
5. After 30 cycles (steps 2–4), analyze 10 μl of the reaction mixture.

Fractionation of the MOPAC Product

1. The fractionation of the MOPAC products is performed in a TBE
   buffer [89 mM Tris borate, 89 mM boric acid, and 0.2 mM EDTA
   (pH 8.0)] in an 8% polyacrylamide gel.
2. Electrophorese the polyacrylamide gel at 80 V (minigel) until the
   dye front reaches the bottom.
3. Soak the gel in TBE buffer containing 0.02% ethidium bromide
   for 20 minutes.
4. Visualize the fragments under UV and isolate the authentic band.
5. DNA from the gel can be eluted by soaking the gel slice in sterile H$_2$O overnight at 4°C and purifying by ethanol precipitation.

Note: Further DNA amplification can be performed using the isolated DNA fragments as the template.

6. The DNA fragments can now be subcloned into a cloning vector (pTZ 18/19R or pUC 18/19) by standard subcloning procedures.

Acknowledgments

This work has been supported in part by the Howard Hughes Medical Institute and PHS grants DK-31428 and GM 34438.

Literature Cited

Crick, F. H. C. 1968. The origin of the genetic code. *J. Mol. Biol.* **38**:367–379.

Girgis, S. I., M. Alevizaki, P. Denny, G. J. M. Ferrier, and S. Legon. 1988. Generation of DNA probes for peptides with highly degenerate codons using mixed primer PCR. *Nucleic Acids Res.* **16**:10371.

Griffin, L. D., G. R. MacGregor, D. M. Muzny, J. Harter, R. G. Cook, and E. R. B. McCabe. 1988. Synthesis of hexokinase I (HKI) cDNA probes by mixed oligonucleotide primed amplification of cDNA (MOPAC) using primer mixtures of high complexity. *Amer. J. Hum. Genet.* **43**:A185 and in *Biochem. Med. Met. Biol.* Manuscript in press.

Lathe, R. 1985. Synthetic oligonucleotide probes deduced from amino acid sequence data. Theoretical and practical considerations. *J. Mol. Biol.* **183**:1–12.

Lee, C. C., X. Wu, R. A. Gibbs, R. G. Cook, D. M. Muzny, and C. T. Caskey. 1988. Generation of cDNA probes directed by amino acid sequence: cloning of urate oxidase. *Science* **239**:1288–1291.

Lee, C. C., and C. T. Caskey. 1989. In *Genetic Engineering: Principles and Methods* (ed. J. K. Setlow), vol. 11, p. 159–170. Plenum Publishing Corp., New York. Manuscript in press.

Mullis, K. B., and F. A. Faloona. 1987. Specific synthesis of DNA *in vitro* via a polymerase-catalyzed chain reaction. *Methods Enzymol.* **155**:335–350.

PCR WITH 7-DEAZA-2'-DEOXYGUANOSINE TRIPHOSPHATE

Michael A. Innis

The presence of stable hairpin-loop structures, base compressions, or high G+C content in the target DNA template can present difficulties for *in vitro* DNA amplification by PCR. On occasion, either a given pair of primers yields either a high background of nonspecific products, including primer–dimer artifact and a low yield of the desired product, or there is no apparent amplification of the desired product. The problem of no amplification occurs infrequently, but unpredictably, and has even been observed with templates that contain neither obvious potential secondary structure nor a notably high G+C content. Usually, amplification problems can be overcome by choosing different primer pairs or optimizing the PCR conditions (e.g., using higher annealing and/or extension temperatures or varying the dNTP, Mg^{2+}, and/or *Taq* DNA polymerase concentrations [see Chapter 1]). However, optimization can be time-consuming: On one occasion it was necessary for an investigator to try four different primer pairs to find one pair that would efficiently amplify β-actin cDNA (E. Kawasaki, unpublished).

A likely cause for the difficulties in amplifying certain sequences is the presence of secondary structures in single-strand copies of the target DNA segment, which hinder annealing and/or extension of

PCR Protocols: A Guide to Methods and Applications

the primers. The use of 7-deaza-2'-deoxyguanosine (c⁷dGTP) in PCR incorporates this structure-destabilizing base analog into the amplified DNA. *Taq* DNA polymerase incorporates c⁷dGTP with kinetics similar to those of dGTP incorporation (Innis *et al.* 1988). Because the N-7 position of the guanine ring is replaced with a methine moiety (Barr *et al.* 1986), 7-deazaguanine precludes Hoogsteen bond formation (stacking) without affecting Watson-Crick base pairing. We found that the use of c⁷dGTP in PCR can significantly increase the specificity of the reaction with nucleic acid templates that contain stable secondary structures and/or have compressed regions (McConlogue *et al.* 1988). In the example given, a specific product was obtained only in the presence of c⁷dGTP. Furthermore, incorporation of c⁷dGTP does not affect the fidelity of PCR, and the 7-deaza-PCR product can be subcloned readily into *E. coli* and/or sequenced (McConlogue *et al.* 1988).

PCR Using a 3 : 1 c⁷dGTP : dGTP Mixture

PCRs (100 μl) are performed in 0.5-ml microcentrifuge tubes in a Thermal Cycler. Mineral oil overlays can usually be avoided if the PCR volume is $\geq$100 μl.

(variable)	Template DNA
20	pmol each primer
2.5	units of AmpliTaq
50	μM each dATP, dCTP, and dTTP
12.5	μM dGTP
37.5	μM c⁷dGTP
0.01%	autoclaved gelatin or nuclease-free bovine serum albumin (optional)
20	mM Tris–HCl (pH 8.8)
1.5	mM MgCl$_2$
50	mM KCl

These reaction components are standard except that a mixture (3 parts : 1 part) of c⁷dGTP and dGTP is recommended. PCR thermal cycling conditions depend upon the nature and concentration of starting DNA template, as discussed in Chapter 1.

A

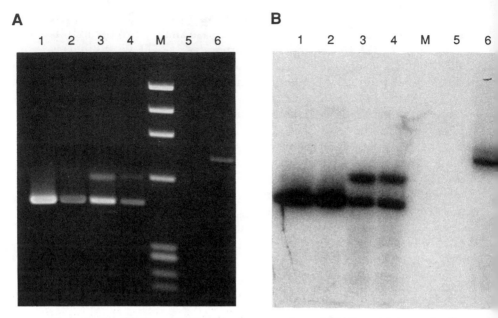

Figure 1   Quantitation of PCR products: comparison of ethidium staining and radio-activity. 100-μl amplifications were performed on phage plaques of two different inserts cloned into M13 mp18 with commercial primers (New England BioLabs, Inc., primer A: NEB #1233; primer B: NEB #1224) flanking the polylinker region (see Chapter 10). The target insert in lanes 1–4 was 350 bp and approximately 50% G + C. The target insert in lanes 5 and 6 was 600 bp and approximately 80% G + C. Standard reactions (lanes 1, 2, 5, and 6) used 20 pmol of each primer; asymmetric reactions (lanes 3 and 4) used 20 pmol of primer B and 1 pmol of primer A. In all reactions, primer B was kinase-labeled with ^{32}P (10^4 cpm/pmol). Reactions were 50 μM in each dNTP, with either 100% dGTP (lanes 1, 3 and 5) or 75% c^{7}dGTP : 25% dGTP (lanes 2, 4, and 6). Aliquots (5%) of each reaction were electrophoresed on a 1.5% TBE agarose gel and visualized by ethidium bromide staining (panel A). The stained gel was then dried and exposed to X-ray film (panel B). The upper band (lanes 3 and 4) is single-stranded DNA. Lane M in each panel is ϕX174 DNA digested with HaeIII.

Results and Discussion

In our initial publication describing the use of c^{7}dGTP for structure-independent PCR (McConlogue *et al.* 1988), we pointed out that PCRs with c^{7}dGTP (or mixtures of c^{7}dGTP and dGTP) appeared to be less efficient on most templates than were PCRs with dGTP alone. This has now been shown to be incorrect. We have discovered that

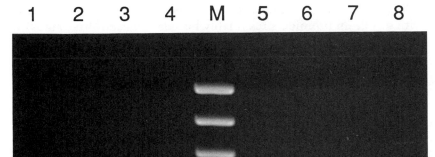

Figure 2   Restriction enzyme digestion of c⁷dGTP-containing DNA. The 350-bp M13 phage insert of Fig. 1 was amplified as described with either 100% dGTP (lanes 1–4) or 75% c⁷dGTP : 25% dGTP (lanes 5–8). Reactions were precipitated from 2 M NH₄OAc and 50% isopropanol, resuspended in water, and 20% of each reaction was digested with 20 units of EcoRI (lanes 2 and 6), PstI (lanes 3 and 7), or HindIII (lanes 4 and 8). Each enzyme cuts the amplified fragment once, near one end. Reaction products before (lanes 1 and 5) and after digestion were electrophoresed on a 1.5% TBE agarose gel, stained with ethidium bromide, and photographed. Lane M is ϕχ174 DNA digested with HaeIII.

PCR products containing c⁷dGTP simply do not stain efficiently with ethidium bromide, presumably because adjacent base stacking is diminished in the c⁷dGTP-containing DNA. In fact, PCR (including asymmetric PCR) with c⁷dGTP is as efficient as it is with dGTP for most templates, and, for difficult templates, is vastly superior to PCR with dGTP alone. This was demonstrated by using low-specific-activity (10^4 cpm/pmol) ^{32}P-labeled primers and analyzing the PCR products by both ethidium staining and autoradiography of the same gel (Fig. 1). Indeed, from these results it appears that the only reason to use a mixture of c⁷dGTP and dGTP is that incorporation of some dGTP is necessary for visualization of the product by ethidium staining.

We (McConlogue *et al.* 1988) also showed that DNA amplified in the presence of 100% c⁷dGTP was cleavable by *Taq* I restriction endonuclease (T'CGA). At that time, we had not tried digesting c⁷dGTP-amplified DNA with other restriction enzymes. In contrast, we show here that incorporation of c⁷dGTP during PCR can interfere with subsequent digestion by some enzymes (Fig. 2). Under conditions of approximately 20-fold overdigestion, about 95% of the dGTP-containing DNA was cleaved. In contrast, the c⁷dGTP-containing DNA (a 3 : 1 mix with dGTP) was cleaved only 10 to 20% by *Eco*RI (G'AATTC) and *Pst*I (CTGCA'G), and about 50% by *Hind*III (A'AGCTT). The inefficiency of cutting appears to reflect that portion of the product that contains c⁷dGTP within the restriction site. This raises the interesting possibility that one could use amplified fragments containing 100% c⁷dGTP for cloning (with linkers) regions that would otherwise contain multiple restriction sites. This is similar to the use of enzyme methylases for cloning fragments with multiple enzyme sites.

Acknowledgment

I thank Peter McCabe for assistance with Figs. 1 and 2.

Literature Cited

Barr, P. J., R. M. Thayer, P. Laybourn, R. C. Najarian, F. Sella, and D. R. Tolan. 1986. 7-Deaza-2'-deoxyguanosine-5'-triphosphate: enhanced resolution in M13 dideoxy sequencing. *BioTechniques* 4:428–432.

Innis, M. A., K. B. Myambo, D. H. Gelfand, and M. A. D. Brow. 1988. DNA sequencing with *Thermus aquaticus* DNA polymerase and direct sequencing of polymerase chain reaction-amplified DNA. *Proc. Natl. Acad. Sci. USA* **85**:9436–9440.

McConlogue, L., M. A. D. Brow, and M. A. Innis. 1988. Structure-independent DNA amplification by PCR using 7-deaza-2'-deoxyguanosine. *Nucleic Acids Res.* **16**:9869.

8

COMPETITIVE PCR FOR QUANTITATION OF mRNA

Gary Gilliland, Steven Perrin, and H. Franklin Bunn

PCR has proved useful in amplifying specific mRNAs, especially those present in low copy number. For example, it has been possible to amplify, subclone, and characterize low-abundance mRNA (Frohman *et al.* 1988) and to detect unique mRNA transcripts from abnormal cells in a background of normal cells (Chelly *et al.* 1988; Kawasaki *et al.* 1988; Lee *et al.* 1987).

Although it has been possible to detect and amplify large amounts of rare mRNA transcripts, it has been more difficult to quantitate the amount of mRNA present in the starting material. This has precluded, for example, the use of PCR in the analysis of induction of mRNA in response to exogenous stimuli. The main constraint in obtaining quantitative data is inherent in the amplification process. Because amplification is (at least initially) an exponential process, small differences in any of the variables that control the reaction rate will dramatically affect the yield of PCR product. Variables that influence the rate of the PCR include the concentrations of polymerase, dNTPs, Mg, DNA, and primers; annealing, extension, and denaturing temperatures; cycle length and cycle number; ramping times; rate of "primer–dimer" formation; and presence of contami-

PCR Protocols: A Guide to Methods and Applications

nating DNA (see Chapter 1). Even when these parameters are controlled precisely, there is sometimes a tube-to-tube variation that precludes accurate quantitation. For example, significant differences in yield occur in PCR samples that are prepared as a pool, aliquoted into separate tubes, and amplified in the same run. The basis for this variation is not certain; it may be related to events that occur during the first few cycles or small temperature variances across the thermal cycler block.

We describe here a technique that obviates these problems and allows the precise quantitation of specific mRNA species. The strategy involves co-amplification of a competitive template that uses the same primers as those of the target cDNA but can be distinguished from the target cDNA after amplification.

Ideally, the competitive template is a mutant cDNA containing a new restriction site. These mutants can be prepared in most cases by a single base-pair change and are easily synthesized by using PCR for site-directed mutagenesis (Higuchi *et al.* 1988; Chapter 22). The mutant template can be distinguished from the target cDNA by restriction enzyme digestion following PCR. Alternatively, genomic plasmid DNA can be used as a competitive template provided oligonucleotide primers are in separate exons and flank a small intron (100 to 200 bp). The amplified competitive template can then be distinguished from the target cDNA by size. A disadvantage in using genomic DNA is the possibility that it may not be amplified as efficiently as target cDNA is, either because of increased size or increased duplex melting temperature.

Target cDNA is co-amplified with a dilution series of competitor DNA of known concentration. Since a change in any of the variables previously listed will affect the yield of competitive template and target cDNA equally, relative ratios of the two should be preserved with amplification. The relative amounts of target cDNA versus competitor can be measured by direct scanning of ethidium-stained gels or by incorporation of radiolabeled dNTPs. Because the starting concentration of the competitive template is known, the initial concentration of the target cDNA can be determined.

This method can be used to accurately quantitate less than 1 pg of target cDNA from 1 ng of total mRNA and can distinguish twofold differences in mRNA concentration. The technique can be applied to quantitation of mRNA from as few as 10 cells and is thus useful in screening cultured colonies of cells or flow-sorted cells for specific mRNA production under various conditions.

Protocols

Reagents

1. *Choice of Primers.* Primers should be chosen according to the general guidelines outlined elsewhere in this book. We usually use 30-mers with noncomplementary 3' ends, 50 to 60% G + C content, and minimal overlap with known sequences by Gen-Bank analysis. Primers are chosen to flank an intron so that the amplified product is readily distinguished from contaminating genomic DNA that may be present. We generally choose primers that will give fragments between 200 and 600 bp.

2. *Preparation of Competitive Templates.* We prefer to use mutant competitive templates that are identical to the target cDNA sequence except that they either contain a single new restriction site or lack an existing restriction site. We have found the PCR-based method of Higuchi *et al.* (1988) (Chapter 22) to be a rapid and accurate means of generating site-specific mutants. Mutant template should be characterized by digestion with appropriate restriction enzymes and by accurate spectrophotometric determination of concentration. A precise dilution series should be prepared, ranging from 10 ng to 1 fg of template, in relatively large volumes (e.g. 1 ml) so that the same dilution series can be used for multiple concentration determinations.

3. dNTP stock, 2.5 mM of each nucleoside triphosphate. Concentrations of each nucleotide should be determined spectrophotometrically before mixing.

4. 10× PCR buffer [500 mM KCl, 100 mM Tris–HCl (pH 8.3), 15 mM MgCl$_2$, 0.1% w/v gelatin]

5. *Taq* polymerase, 5 units/μl

6. AMV or MuLV reverse transcriptase

7. RNase inhibitor

8. Dithiothreitol, 20 mM stock

Reverse Transcription

Several methods may be used to obtain mRNA suitable for reverse transcription. In general, highly purified mRNA gives the best results for quantitation (e.g., guanidinium isothiocyanate lysis followed by centrifugation over $CsCl_2$, with at least two precipitation steps). It is not necessary to use poly(A)-purified mRNA. However, in many circumstances it is desirable to use mRNA from very few cells, which effectively precludes standard techniques for mRNA purification. Adequate amounts of mRNA can be obtained from small numbers of cells by lysing in 0.5% NP-40, 10 mM Tris (pH 8.0), 10 mM NaCl, 3 mM $MgCl_2$ for 5 minutes on ice, microfuging for 2 minutes to remove nuclei and cell debris, and performing reverse transcription on the supernatant.

Reverse transcription may be primed by using specific antisense primer for the gene of interest, oligo(dT), or random hexanucleotide primers. Using oligo(dT) permits amplification of more than one gene from the same reverse transcriptase (RT) mixture, but the target sequence needs to be relatively close to the poly(A) tail. Random hexanucleotide primers will prime all species of RNA present and allow the amplification of any desired target sequence from the RT mix.

1. Prepare RT mixture containing
   mRNA to be amplified (10 to 100 ng of total RNA, or supernatant from 10 or more cells)
   Primer (20 to 50 pmol of specific antisense primer or oligo(dT), 100 pmol of hexanucleotide primer)
   dNTP (500 μM in each nucleotide)
   1× PCR buffer
   Reverse transcriptase (use number of units recommended by manufacturer)
   Dithiothreitol, 1 mM final concentration
   RNase inhibitor, 2 to 5 units
   Diethylpyrocarbonate-treated water to a final volume of 20 μl
2. Incubate at 37°C for 1 hour.

   NOTE: When multiple samples are to be analyzed for content of a given mRNA, a master mix may be prepared containing all the components except the mRNA source.

Polymerase Chain Reaction

The accuracy of this method is improved through the use of master mixes. We usually titrate against a broad range of dilutions in log increments for the first quantitation to obtain a rough estimate of the amount of cDNA present. We then perform a second titration over a narrower (100-fold) range for precise quantitation. In most cases, only the second series of titrations is necessary once the range is known for a given set of experiments. To quantitate the amount of PCR product, we usually add [α-³²P]dCTP to a final concentration of 50 μCi/ml. It is important not to adjust the concentration of individual dNTPs in the dNTP stock (e.g., lowering the dCTP concentration to increase specific activity), as this will interfere substantially with the amplification. Alternatively, a densitometer may be used to scan ethidium-stained gels to quantitate the amount of each PCR product present.

Prepare a master mix containing oligonucleotide primers (10 to 20 pmol each), dNTPs (200 μM in each as final concentration), 10× PCR buffer, Taq polymerase (0.5 units), and an appropriate amount of cDNA. (We generally add 10 μl of the RT mixture as the source of cDNA.) Add 90 μl of this mixture to 10 μl of previously prepared competitive template of known concentration in a dilution series. An example of a reaction mixture that we typically use follows. Because this technique is not dependent on variables previously noted, any conditions (annealing temperature, cycle length, cycle number, etc.) that give good amplification of template DNA can be used for quantitation.

After amplification, an aliquot of each sample is digested with an appropriate restriction enzyme. Triplicate cut and uncut samples are run in parallel on a NuSieve/agarose gel. When amplification is performed using a genomic intron-containing competing template, as in the example that follows, samples may be run directly on gels and cut out for counting without restriction digestion.

Example

The quantitation of granulocyte macrophage colony stimulating factor (GM-CSF) cDNA (cGM) using a genomic GM-CSF (gGM) competitive template is demonstrated in Figs. 1–3. Primers have been chosen from exon 1 and exon 2 that give a 197-bp fragment when cDNA is amplified and a 290-bp fragment when genomic DNA is amplified (Fig. 1).

Tube	gGM (competitive template)	
1.	10.0 ng/μl	10 μl
2.	1.0 ng/μl	"
3.	0.8 ng/μl	"
4.	0.6 ng/μl	"
5.	0.4 ng/μl	"
6.	0.2 ng/μl	"
7.	0.1 ng/μl	"
8.	0.08 ng/μl	"
9.	0.06 ng/μl	"
10.	0.04 ng/μl	"
11.	0.02 ng/μl	"
12.	0.01 ng/μl	"
13.	0.001 ng/μl	"
Master mix	GM-CSF sense primer (30-mer, 5 nmol/ml)	48 μl
	GM-CSF antisense primer	48 μl
	dNTP stock (2.5 mM in each)	96 μl
	cGM (known concentration of 0.6 ng/μl)	12 μl
	10× PCR buffer 120	μl
	Water 756	μl
	Taq (5.0 units/μl)	1.2 μl

Add 90 μl of master mix to each tube 1–11.

40 cycles (e.g., 94°C 1 minute / 62°C 1 minute / 72°C 2 minutes).

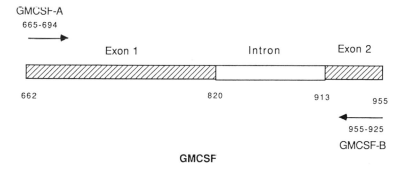

amplified genomic DNA = 290 bp

amplified cDNA = 197 bp

Figure 1   GM-CSF map with position of oligonucleotide primers.

When using mutant templates containing altered restriction sites, heat to 94°C for 4 minutes after completion of thermal cycling. (As discussed as follows, this will normalize for heteroduplex formation between mutant and wild-type strands.)

Run 20 μl/well in triplicate on a 2.5% NuSieve, 1% agarose gel containing ethidium bromide.

Analysis

Determine the amount of gGM and cGM present in each lane. The gel can be photographed and analyzed by densitometry (see Fig. 2), or, if bands contain [^{32}P]dCTP, the labeled bands can be cut out and counted. The amount of gGM is multiplied by the ratio of cGM bp per gGM bp to correct for increased label/ethidium staining per mole by the larger fragment. gGM per cGM can then be plotted as a function of the amount of known competitive gGM (see Fig. 3). The point of equivalence (i.e., where there is a 1 : 1 ratio) is where cGM equals gGM and represents the concentration of cDNA in the unknown.

When mutant template containing a new restriction site is used as a competitor, it is not necessary to correct for differences in molecular weight, since competitive and wild-type templates are identical in size. However, because wild-type and mutant fragments only differ by a single base pair, heteroduplex annealing can occur when the primer is not present in excess. Since heteroduplexes will not be cleaved by the restriction enzyme, the amount of wild-type template is overestimated. We have found this to be a minimal problem with runs of less than 40 cycles.

GM-CSF gDNA vs cDNA

1   **13**

290 bp
197 bp

Figure 2   GM-CSF gDNA versus cDNA. Lanes 1–13 contain various amounts of genomic GM-CSF (0.01 ng to 100 ng) competed against a fixed concentration of GM-CSF cDNA (0.6 ng) corresponding to tubes 1–13 in text.

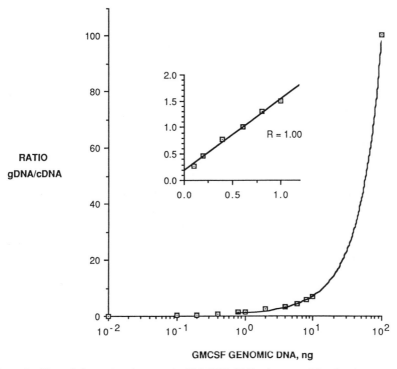

Figure 3 Plot of the ratio of genomic GM-CSF/cDNA after amplification versus genomic DNA added to the original mixture. Inset shows an expansion of the 0.1–1.0 ng range.

Discussion

Competitive PCR provides a rapid and reliable way to quantitate the amount of cDNA in a sample prepared from as few as 10 cells. The advantages of this technique are that (1) quantitation is independent of the many variables that affect amplification and (2) it is more sensitive than Northern blotting or ribonuclease protection assays are for quantitating specific mRNAs. For example, the technique may be readily applied to screening colonies of cells for specific mRNA. Several other methods have been described for quantitating cDNA species by using PCR, usually by co-amplifying a second, unrelated template (Rappolee *et al.* 1988). These methods are critically dependent on several variables, including cycle number and the amount of starting mRNA of each species. Even when these variables are adequately controlled, it is unlikely that the unrelated control template will be amplified at precisely the same rate as that of the unknown template. Small differences in the rate of amplification of the two templates are magnified during PCR and may grossly over- or underestimate the amount of the unknown template present.

As would be predicted, the ratio of competitive template per unknown template plotted against competitive template is a hyperbolic relationship that approaches an asymptote when one species is present in vast excess. For this reason, the most accurate results are obtained when competitive template and unknown template are amplified at nearly equivalent concentrations. As previously noted here, we generally perform an initial titration in log increments to determine the approximate concentration of the unknown cDNA. We then perform a finer titration (as in the example cited in this chapter) to obtain the most accurate results. Accordingly, twofold differences in cDNA concentrations can be accurately determined.

Competitive PCR should be readily applied to the assay of reporter gene mRNAs such as human growth hormone. This would provide a more accurate way for determining the activity of putative regulatory sequences, since, in contrast to the standard HGH reporter assay, it measures the transcription product directly and does not rely on synthesis and/or secretion of HGH protein by a transfected cell.

Competitive PCR has several limitations. First, it quantitates the amount of cDNA present in a given sample, but, if efficiency of reverse transcription is less than 100%, the method will underestimate the actual amount of mRNA present. To obtain an internal mRNA control, we have used random hexanucleotides to prime the RT reaction and have used competitive PCR to assay the amount of cDNA of a "housekeeping" gene, β-actin. Alternatively, a known amount of mutant cRNA template prepared by using T7 polymerase can be added to the RT mixture as an internal standard.

Literature Cited

Chelly, J., J.-C. Kaplan, P. Maire, S. Gautron, and A. Kahn. 1988. Transcription of the dystrophin gene in human muscle and non-muscle tissues. *Nature (London)* **333**: 858–860.

Frohman, M. A., M. K. Dush, and G. R. Martin. 1988. Rapid production of full-length cDNAs from rare transcripts: amplification using a single gene-specific oligonucleotide primer. *Proc. Natl. Acad. Sci. USA* **85**:8998–9002.

Higuchi, R., B. Krummel, and R. K. Saiki. 1988. A general method of *in vitro* preparation and specific mutagenesis of DNA fragments: study of protein and DNA interactions. *Nucleic Acids Res.* **16**:7351–7367.

Kawasaki, E. S., S. S. Clark, M. Y. Coyne, S. D. Smith, R. Champlin, O. N. Witte, and F. P. McCormick. 1988. Diagnosis of chronic myeloid and acute lymphocytic leukemias by detection of leukemia-specific mRNA sequences amplified *in vitro*. *Proc. Natl. Acad. Sci. USA* **85**:5698–5702.

Lee, M.-S., K.-S. Chang, F. Cabanillas, E. J. Freireich, J. M. Trujillo, and S. A. Stass. 1987. Detection of minimal residual cells carrying the t(14;18) by DNA sequence amplification. *Science* **237**:175–178.

Rappolee, D. A., D. Mark, M. J. Banda, and Z. Werb. 1988. Wound macrophages express TGF-α and other growth factors in vivo: analysis by mRNA phenotyping. *Science* **241**:708–712.

9

QUANTITATIVE PCR

Alice M. Wang and David F. Mark

PCR analysis has been used for RNA blot analysis, mRNA pheno-typing, and nuclease protection analysis for the study of short-lived, low-copy-number mRNA transcripts (Kawasaki *et al.* 1988; Rappolee *et al.* 1988; Harbarth and Vosberg 1988; Lee *et al.* 1988; Price *et al.* 1988; Dobrovic *et al.* 1988; Hermans *et al.* 1988). Quantitation of the PCR analysis would provide more information for these studies. Such an approach has been used to study the relative amount of dys-trophin transcript in different human tissues (Chelly *et al.* 1988). It was also used to quantitate the relative amounts of $apoB_{100}$ and $apoB_{48}$ message after thyroid hormone treatment to demonstrate that the apoB mRNA modification can be hormonally modulated (Davidson *et al.* 1988). A possible application of this approach includes the measurement of specific mRNA levels in drug-resistant and drug-sensitive human carcinoma cells. The early detection of specific mRNA patterns could lead to the development of more effective clinical protocols in cancer therapy (Kashani-Sabet *et al.* 1988).

It has been difficult to quantitate the amount of specific mRNA without an internal standard. We report here a technique that uses a synthetic AW106 cRNA as an internal standard for quantitating

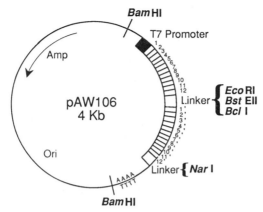

Figure 1 Structure of plasmid pAW106. Primers 1–12 are upstream primers of tumor necrosis factor (TNF), macrophage colony stimulating factor (M-CSF), platelet-derived growth factor A (PDGF-A), PDGF-B, apolipoprotein E (apo E), low-density lipoprotein receptor (LDL-R), 3-hydroxy-3-methylglutaryl coenzyme A reductase (HMG CoA-reductase), Interleukin-1α (IL-1α), IL-1β, IL-2, PDGF-R, and lipoprotein lipase (LPL), respectively, followed by the complementary sequences of their downstream primers 1′–12′ in the same order. Both upstream and downstream primers are followed by a restriction enzyme linker for easy insertion of primers for other genes.

the amount of specific mRNA by PCR. The AW106 cRNA was synthesized as a sense-strand from plasmid pAW106 by T7 polymerase (Fig. 1). This synthetic gene has twelve target genes' upstream primers connected in sequence followed by the complementary sequences of their downstream primers in the same order (Fig. 1). The PCR product synthesized from the cRNA standard by each primer set is around 300 bp and is designed not to overlap in size with the PCR products amplified by various primer pairs from the mRNA of these target genes. The AW106 cRNA contains polyadenylated sequences at the 3′ end, and it can be reverse transcribed and amplified together with the target mRNA in the same tube. The size difference in the PCR products permits easy separation of the cRNA product from the target mRNA product by gel electrophoresis. Since the same primer set is used in the PCR amplification on both templates, differences in primer efficiency are minimized. In the exponential phase of the amplification, the amount of target mRNA can be quantitated by extrapolating against the AW106 cRNA internal standard curve. In addition, the internal standard developed here contains the primer sequences for multiple genes so that the same standard can be used to quantitate a number of different mRNAs of interest.

Protocols

Reagents

1× PCR buffer   20 mM Tris–HCl (pH 8.3)
                           50 mM KCl
                           2.5 mM $MgCl_2$
                           100 μg/ml of bovine serum albumin

RNA Preparation

Use standard protocols to prepare total cellular RNA isolated from cell lines or tissues (see Chapter 18). Synthesize AW106 cRNA by the transcription system of Promega. Purify the resulting cRNA product through an oligo(dT) column and quantitate it by absorbance at 260 nm.

Reverse Transcriptase Reaction

To 10 μl of a reverse transcriptase reaction, add 1 μg of total cellular RNA, 1.77×10^2 to 1.77×10^7 molecules of AW106 cRNA, 1× PCR buffer, 1 mM DTT, 0.5 mM each dNTP, 10 units of RNasin, 0.1 μg of oligo(dT) primer, and 100 units of Moloney murine leukemia virus reverse transcriptase (Bethesda Research Laboratories, Inc.). Incubate 60 minutes at 37°C. Heat inactivate the enzyme at 95°C for 5 minutes, and then quick chill on ice.

Polymerase Chain Reaction

A threefold dilution series of the reverse transcriptase product is mixed with 1× PCR buffer, 50 μM of each dNTP, 5 pmol each of the upstream and downstream primers, 1×10^6 cpm of one ^{32}P-end-labeled primer, and 1 unit of Taq polymerase, in a 50-μl reaction volume. The mixture is overlaid with 100 μl of mineral oil to prevent evaporation and then amplified by PCR for 25 cycles. The amplification cycle profile is denaturation for 30 seconds at 95°C, cooling for 2 minutes to 55°C, annealing of primers for 30 seconds at 55°C, heating for 30 seconds to 72°C, and extension of primers for 1 minute at 72°C.

Gel Electrophoresis

A 5- to 10-µl portion of the PCR mixture is electrophoresed in 10% polyacrylamide gel in TBE buffer. Stain the gel with ethidium bromide and photograph. Appropriate bands are excised from the gel, and the radioactivity is determined by scintillation fluorography.

Quantitation

The amount of radioactivity recovered from the excised gel bands is plotted against the template concentrations. The fact that the amplification rates of internal standards and specific mRNA are identical within the exponential phase of the PCR allows one to construct a standard curve that can be used to quantitate the actual amount of a specific mRNA species. As shown in Fig. 2, 20 ng of PMA-induced

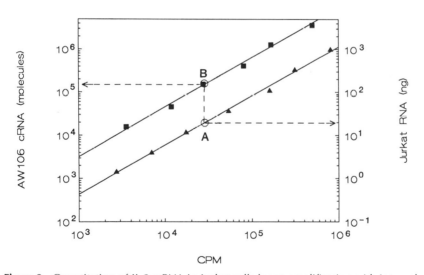

Figure 2 Quantitation of IL-2 mRNA in Jurkat cells by co-amplification with internal standard AW106 cRNA. The variable template concentrations of the internal standard AW106 cRNA and Jurkat RNA were plotted against the radioactivity of their PCR product. The PCR analysis was performed with IL-2 primers. (■) PCR product of AW106 cRNA. (▲) PCR product of Jurkat RNA.

Jurkat RNA yielded 30,000 cpm of interleukin-2 (IL-2)-specific PCR product, point A, which translated to point B on the standard curve as 1.5×10^5 molecules of AW106 cRNA. In other words, 20 ng of induced Jurkat total RNA contained 1.5×10^5 molecules of IL-2 mRNA.

Discussion

Larzul *et al.* (1988) demonstrated that the amount of amplified DNA fragment in a given sample has an influence on amplification efficiency. When a high template concentration is used, the primer concentration, *Taq* polymerase availability, amplified product snapback, etc., can be limiting factors for efficient amplification. To reliably quantitate the specific mRNA using an internal standard, the range of concentrations of both templates and the number of amplification cycles are chosen such that they stay within the exponential phase of the PCR. Therefore, it is necessary to first titrate the specific mRNA to find the range of concentrations that gives exponential amplification over a defined range of cycle numbers.

When a quantitative evaluation is required, the conditions have to be very strictly controlled in this extremely sensitive PCR technique. A negative control tube is necessary to monitor false positives and to give the amount of background counts.

Because of its high sensitivity, speed, and accuracy, the RNA/PCR quantitation method can be used to detect gene expression in a more extensive way. It can accurately quantitate the amount of specific mRNA. In many cases the detection of the differences in levels of expression of specific RNA molecules can provide useful information for the diagnosis of infectious disease or cancer.

Literature Cited

Chelly, J., J.-C. Kaplan, P. Maire, S. Gautron, and A. Kahn. 1988. Transcription of the dystrophin gene in human muscle and non-muscle tissues. *Nature (London)* **333**:858–860.

Davidson, N. O., L. M. Powell, S. C. Wallis, and J. Scott. 1988. Thyroid hormone modulates the introduction of a stop codon in rat liver apolipoprotein-B messenger RNA. *J. Biol. Chem.* **263**:13482–13485.

Dobrovic, A., K. J. Trainor, and A. A. Morley. 1988. Detection of the molecular abnormality in chronic myeloid leukemia by use of the polymerase chain reaction. *Blood* **72**: 2063–2065.

Harbarth, P., and H.-P. Vosberg. 1988. Enzymatic amplification of myosin heavy-chain mRNA sequences *in vitro*. *DNA* **7**: 297–306.

Hermans, A., L. Selleri, J. Gow, and G. C. Grosveld. 1988. Absence of alternative splicing in *bcr-abl* mRNA in chronic myeloid leukemia cell lines. *Blood* **72**: 2066–2069.

Kashani-Sabet, M., J. J. Rossi, Y. Lu, J. x. Ma, J. Chen, H. Miyachi, and K. J. Scanlon. 1988. Detection of drug resistance in human tumors by *in vitro* enzymatic amplification. *Cancer Res.* **48**: 5775–5778.

Kawasaki, E. S., S. S. Clark, M. Y. Coyne, S. D. Smith, R. Champlin, O. N. Witte, and F. P. McCormick. 1988. Diagnosis of chronic myeloid and acute lymphocytic leukemias by detection of leukemia-specific mRNA sequences amplified *in vitro*. *Proc. Natl. Acad. Sci. USA* **85**: 5698–5702.

Larzul, D., F. Guigue, J. J. Sninsky, D. H. Mack, C. Brechot, and J.-L. Guesdon. 1988. Detection of hepatitis B virus sequences in serum by using *in vitro* enzymatic amplification. *J. Virol. Meth.* **20**: 227–237.

Lee, M.-S., K.-S. Chang, E. J. Freireich, H. M. Kantarjian, M. Talpaz, J. M. Trujillo, and S. A. Stass. 1988. Detection of minimal residual *bcr/abl* transcripts by a modified polymerase chain reaction. *Blood* **72**: 893–897.

Price, C. M., F. Rassool, M. K. K. Shivji, J. Gow, C. J. Tew, C. Haworth, J. M. Goldman, and L. M. Wiedemann. 1988. Rearrangement of the breakpoint cluster region and expression of P210 *BCR-ABL* in a "masked" Philadelphia chromosome-positive acute myeloid leukemia. *Blood* **72**: 1829–1832.

Rappolee, D. A., D. Mark, M. J. Banda, and Z. Werb. 1988. Wound macrophages express TGF-α and other growth factors *in vivo*: analysis by mRNA phenotyping. *Science* **241**: 708–712.

PRODUCTION OF SINGLE-STRANDED DNA BY ASYMMETRIC PCR

Peter C. McCabe

The PCR technique provides direct access to defined segments of genomic DNA or messenger RNA for analysis. Frequently the analysis will include a determination of the DNA sequence of the amplified targets. For example, sequence variations within a population at a particular locus can be rapidly characterized by sequencing specific PCR products without the construction and screening of a genomic library for each individual. PCR can also provide a simple and rapid method of preparing DNA templates for sequencing from inserts in a variety of cloning vectors that can sometimes be difficult to sequence directly (i.e., lambda phage vectors). Cloned DNA segments can be amplified directly from individual bacterial colonies or phage plaques, thereby avoiding the time and effort spent on growth of cultures and DNA purification. PCR seems ideally suited for the automation of template preparation that will be necessary for large-scale DNA sequencing projects.

Directly sequencing products of the PCR without an additional cloning step is generally preferable to sequencing cloned products. In addition to the benefit of simplicity, this greatly reduces the potential for errors due to imperfect PCR fidelity, as any random misincorporations in an individual template molecule will not be

PCR Protocols: A Guide to Methods and Applications

detectable against the much greater signals of the "consensus" sequence. Although several reports have described direct sequencing of double-stranded PCR products (Wrischnik *et al.* 1987; Higuchi *et al.* 1988; Newton *et al.* 1988), the protocols available for preparing double-stranded template DNA for sequencing were developed for covalently closed circular plasmids. There are inherent difficulties in this method when applied to PCR-amplified fragments because of the rapid reassociation of the short linear template strands. This can be avoided by modifying the PCR in such a way that single-stranded DNA of a chosen strand is produced. This modified type of PCR utilizes an unequal, or asymmetric, concentration of the two amplification primers and has thus been termed asymmetric PCR (Gyllensten and Erlich 1988). During the initial 15 to 25 cycles, most of the product generated is double stranded and accumulates exponentially. As the low-concentration primer becomes depleted, further cycles generate an excess of one of the two strands, depending on which of the amplification primers was limited. This single-stranded DNA accumulates linearly and is complementary to the limiting primer. Typical primer ratios for asymmetric PCR are 50 : 1 to 100 : 1. The single-stranded template can be sequenced with either the limiting primer or a third, internal primer, which would provide an added degree of specificity.

Asymmetric PCR conditions for producing single-stranded DNA of cloned inserts directly from M13 phage plaques are described. Modifications of the basic protocol are discussed for lambda phage, bacterial plasmids, and genomic DNA targets.

Protocol

1. Pick a fresh phage plaque with the tip of a Pasteur pipette into 200 μl of TE [10 mM Tris–HCl (pH 8.0), 0.1 mM EDTA]; elute 30 minutes at 37°C.
2. Use 5 μl of phage eluate for a 100-μl PCR . Reactions contain:

    20 mM Tris (pH 8.5)

    1.5 mM MgCl$_2$

    50 mM KCl

    0.1% Tween 20

50 μM each dNTP

50 pmol excess primer

1 pmol limiting primer

2.5 units *Taq* DNA polymerase (Perkin-Elmer Cetus)

Primers are:

primer 1 (sense) AGCGGATAACAATTTCACACAGGA

primer 2 (antisense) CGCCAGGGTTTTCCCAGTCACGAC

These primers yielded the best results among those that were tested. They anneal about 20 bp away from each end of the polylinker region and are commercially available (e.g., New England BioLabs, Inc. #1233, 1224).

3. Run the PCR for 30 cycles in an automated thermocycler (e.g., Perkin-Elmer Cetus Instruments) with the following cycle profile:

95°C for 30 seconds

60°C for 30 seconds

72°C for 2 minutes

After cycling is complete, incubate an additional 5 minutes at 72°C to ensure completion of final extensions.

4. Reactions can be monitored for production of single-stranded DNA on a TBE agarose gel (1.4% for targets up to 1 kb; 12 to 15 V/cm). Single-stranded DNA product runs slower than the double-stranded DNA product does under these gel conditions and can usually be visualized by staining with ethidium bromide, although the fluorescence is much reduced relative to an equivalent amount of double-stranded DNA (see Fig. 1). Amplified inserts will contain an additional 100 to 150 bp because of the primers and flanking polylinker sequences. If accurate quantitation of yield is desired, a small amount (1 μCi) of ^{32}P-labeled dNTP can be added to a test reaction in addition to the normal amounts of unlabeled dNTPs. After 1% of the reaction is run on a gel, it is dried down and exposed to film.

5. Precipitate the reaction to concentrate it and remove residual dNTPs and primers by adding ammonium acetate and isopropanol to final concentrations of 2 M and 50%, respectively. Alternatively, spin dialysis can be used. The DNA is then ready for sequencing by the dideoxynucleotide chain-termination method, without further purification. If the recommended primers are employed for amplification, standard forward and re-

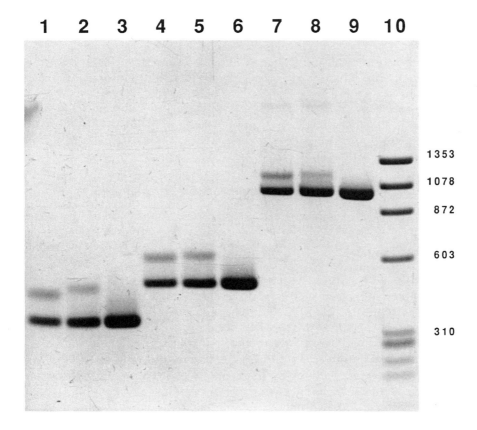

Figure 1 Standard and asymmetric PCR with M13 plaque samples. Reactions were performed as indicated in the protocol. Product sizes ranged from 350 to 1050 bp. Samples were resolved on a 1.4% TBE agarose gel as described. Lanes 1, 4, and 7: primer one > primer two. Lanes 2, 5, and 8: primer two > primer one. Lanes 3 , 6, and 9: primer one is equal to primer two. Lane 10: φX174/HaeIII digest. The upper bands in lanes 1, 2, 4, 5, 7, and 8 are single-stranded DNA.

verse sequencing primers can then be used as internal primers for subsequent DNA sequencing. The most consistent sequencing results with PCR-generated templates have been obtained using *Taq* DNA polymerase (Innis *et al*. 1988); also see Chapter 24.

Variations of the protocol described above can be used with other target sources and primers. Examples follow.

Lambda Phage Plaques

For inserts in lambda cloning vectors, pick plaques as previously described and use 10 μl of phage eluate (or 1 ng of purified DNA) for the reaction, which is run for 35 cycles. Standard and asymmetric PCR have recently been used in this manner to characterize and sequence cDNA clones in lambda gt11 (Trahey et al. 1988). Commercially available primers that anneal to sites directly flanking the EcoRI cloning site in lambda gt11 (New England BioLabs #1218, 1222) have been successfully used to asymmetrically amplify and sequence inserts up to 2 kb in length (unpublished results). An example of sequencing reactions performed on single-stranded templates generated from a lambda plaque is shown in Fig. 2.

Bacterial Colonies

For inserts in plasmid vectors, scrape fresh single colonies with a toothpick into 50 μl of 1% Triton X-100, 20 mM Tris (pH 8.5), 2 mM EDTA. Heat at 95°C for 10 minutes. Spin down bacterial debris and use 5 μl of the supernatant for the PCR. Alternatively, 0.1 ng of plasmid DNA can be used. More than 30 cycles may be required to achieve maximum levels of product. Inserts in plasmids with pUC-type lac sequences flanking the cloning site can be amplified with the primers described above for M13. The best results with colony samples have been obtained when reactions were extracted with phenol-chloroform before purification for sequencing.

Genomic DNA

For high-complexity genomic DNA samples, use up to 1 μg of purified DNA for single-copy genes and run the reaction for 35 to 40 cycles. An initial long denaturation step (5 minutes at 95°C) before the first cycle might be required to ensure sufficient denaturation of the DNA sample. Some investigators have increased the homogeneity of their double-stranded PCR products from genomic DNA by electrophoretic separation and reamplification of an aliquot of eluate from a selected gel slice (Higuchi et al. 1988). If the background from an asymmetric amplification of genomic DNA is problematic, this type of two-step protocol may be necessary.

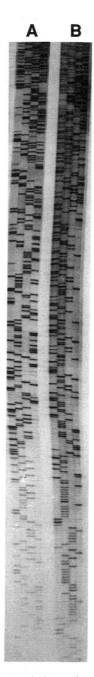

Figure 2 Sequencing of single-stranded templates produced from a λ gt11 plaque sample. An 850-bp insert was asymmetrically amplified with New England BioLabs primers #1218 and #1222, the reactions were purified by spin dialysis, and the same primers were subsequently used in DNA sequencing reactions with AmpliTaq DNA polymerase (Perkin-Elmer Cetus). PCRs contained (A) 50 pmol primer 1218 : 1 pmol primer 1222, (B) 50 pmol primer 1222 : 1 pmol primer 1218.

Discussion

Asymmetric amplification is somewhat less efficient than the equivalent standard PCR is, so more cycles are generally required to achieve a maximum yield of single-stranded DNA (Gyllensten and Erlich 1988). A total cycle number between 30 and 40 will give the best results with most targets. Variability in the single-stranded DNA yield and background level of the PCR has been observed with different primer pairs and ratios, and the reaction conditions for each amplification system will need to be adjusted for optimal results. The best ratio of primers will lie somewhere in a range between 0.5 pmol : 50 pmol and 5 pmol : 100 pmol. The amount of target DNA added to the reactions is also important, as it determines the proper number of cycles for maximal yield and minimal background. Product yield will be low if too little is included, while background levels can be excessive when too much target is present. The problem of low single-stranded DNA yield can also be addressed by doubling the amount of enzyme used in the reaction (to 5 units) and/or running 5 to 10 more cycles. The most consistent results have been obtained with target sizes not exceeding 1 kb in length.

Often the background observed when checking a reaction on an agarose gel does not interfere with the determination of the desired sequence, presumably because it is due to heterogeneity of the products at the 5' end, away from the sequencing primer. The amplification products should not be denatured before the sequencing primer is annealed, so that double-stranded DNA will not participate in the sequencing reactions.

The recommended concentration of dNTPs in the protocol is lower than what is typically used in standard PCR. This lower concentration is sufficient to yield maximal product levels, while greatly reducing the production of background products from certain targets (see Chapter 1). Protein additions, such as gelatin and bovine serum albumin, have not been included in the basic protocol in order to simplify the processing of samples for sequencing. Little benefit is gained from their inclusion when amplifying from phage plaques and bacterial colonies. Similarly, mineral oil has also been omitted, so that it would not be necessary to remove it by chloroform extraction. Evaporation does not seem to alter the resulting reaction products significantly when reaction volumes are 100 μl or more.

Acknowledgments

I thank Michael Innis and Mary Ann Brow for helpful discussions and advice.

Literature Cited

Gyllensten, U. B., and H. A. Erlich. 1988. Generation of single-stranded DNA by the polymerase chain reaction and its application to direct sequencing of the *HLA-DQA* locus. *Proc. Natl. Acad. Sci. USA* **85**:7652–7656.

Higuchi, R., C. H. von Beroldingen, G. F. Sensabaugh, and H. A. Erlich. 1988. DNA typing from single hairs. *Nature (London)* **332**:543–546.

Innis, M. A., K. B. Myambo, D. H. Gelfand, and M. A. D. Brow. 1988. DNA sequencing with *Thermus aquaticus* DNA polymerase and direct sequencing of polymerase chain reaction-amplified DNA. *Proc. Natl. Acad. Sci. USA* **85**:9436–9440.

Newton, C. R., N. Kalsheker, A. Graham, S. Powell, A. Gammack, J. Riley, and A. F. Markham. 1988. Diagnosis of α_1-antitrypsin deficiency by enzymatic amplification of human genomic DNA and direct sequencing of polymerase chain reaction products. *Nucleic Acids Res.* **16**:8233–8243.

Trahey, M., G. Wong, R. Halenbeck, B. Rubinfeld, G. A. Martin, M. Ladner, C. M. Long, W. J. Crosier, K. Watt, K. Koths, and F. McCormick. 1988. Molecular cloning of two types of GAP complementary DNA from human placenta. *Science* **242**:1697–1700.

Wrischnik, L. A., R. G. Higuchi, M. Stoneking, H. A. Erlich, N. Arnheim, and A. C. Wilson. 1987. Length mutations in human mitochondrial DNA: direct sequencing of enzymatically amplified DNA. *Nucleic Acids Res.* **15**:529–542.

11

CLONING WITH PCR

Stephen J. Scharf

The ability of PCR (Saiki *et al.* 1985; Mullis *et al.* 1986; Mullis and Faloona 1987) to generate microgram amounts of a specific DNA fragment can simplify the procedures necessary to clone single-copy gene fragments from genomic DNA. This amount of DNA allows direct cloning into a plasmid or M13 vector and obviates the need for the construction of phage or cosmid libraries, lengthy screening for recombinant clones, and restriction mapping and subcloning (Scharf *et al.* 1986; Scharf and Erlich 1988). By using PCR primers that incorporate sequences for creation of a restriction site, one can clone the PCR products into the desired vector after digestion of the amplified DNA. A limitation of PCR cloning compared to conventional genomic cloning is that some sequence information flanking the desired fragment is required. The advantages, however, include much faster isolation of the desired fragment with less time and effort required.

PCR Protocols: A Guide to Methods and Applications

Primer Design

It is possible to clone PCR products by using blunt-end cloning (see the blunt-end cloning protocol in this chapter). However, directed cloning using two enzymes that produce different termini (such as *Bam*HI and *Pst*I) is advantageous in that only one orientation is cloned, and it prevents the vector religating to itself. A decided advantage of PCR cloning is that one has flexibility in deciding which restriction enzyme sites to use for cloning, as the primers can be designed to contain sequences for creating any restriction site. Adding bases to the 5′ end of the primer is the simplest approach and has no effect on the PCR (Mullis *et al.* 1986; Scharf *et al.* 1986). When adding sequences to the 5′ end of a primer to create a restriction site, it is important to add a few bases (e.g., a "GG" or "CTC") to serve as a clamp to keep the 5′ ends from "breathing" during digestion, which would prevent the enzyme from cleaving the DNA. An example of a primer with bases added on to the 5′ end is shown:

5′-GG<u>GAATTC</u>AGCAGGTTAAACAT −3′

This primer has a "GG" clamp and an *Eco*RI site (underlined). Alternatively, the sequences added to create the restriction site can be incorporated "internally" in the primer. If one chooses to create an internal restriction site by modifying the existing sequence, it is advisable to choose a site that would result in the fewest possible changes when compared to the original sequence. Moreover, these changes should be made near the 5′ end of the primer so that the 3′ end matches the target sequence and is not mispaired; this is to prevent any possible mismatches at the 3′ end, which usually interferes with extension by the DNA polymerase. An example of a primer for amplifying β-globin with an internal restriction site is aligned with the actual β-globin sequence to show the changes used to create the restriction enzyme recognition sequence. The changes made in the primer compared to the native sequence are in lowercase letters, and the *Pst*I site is underlined:

β-globin primer:
　　　5′-CTT<u>CTGcag</u>CAACTGTGTTCACTAGC-3′
β-globin sequence:
TTACATTGCTTCTGACACAACTGTGTTCACTAGCAACCTC

Protocols

Reaction Conditions

The following conditions work well for the majority of genomic DNA targets.

Reaction Constituents

50 mM KCl

10 mM Tris–HCl (pH 8.8)

1.5 mM MgCl$_2$

100 μg/ml of gelatin

50 μM each dNTP

20 to 50 pmol upstream primer

20 to 50 pmol downstream primer

2.5 units of *Taq* DNA polymerase

100 ng to 1 μg of genomic DNA

dH$_2$O to a final volume of 100 μl

Overlay with 75 ml of mineral oil (prevents evaporation)

   With the Perkin-Elmer Cetus Instruments Thermal Cycler, the following temperature profiles work well for most applications: 15 seconds to 94°C, 30 seconds at 94°C, 15 seconds to 55°C, 30 seconds at 55°C, 15 seconds to 72°C, and 30 seconds at 72°C. Using a different temperature cycling machine may require quite different profiles to produce the best results. If one has to do the reaction manually, the times and temperatures listed above should work well for most applications. Twenty-six to twenty-eight cycles are generally sufficient to produce enough DNA for cloning. One should run the smallest number of cycles possible that will permit cloning of the target. This will minimize the possibility of cloning nonspecific products that accumulate after the desired target is no longer amplifying exponentially and has reached a plateau.

   One can often realize dramatic improvements in a particular amplification system by doing some optimization of the constituents and parameters of the reaction. Determining the optimal nucleoside triphosphate, primer, magnesium ion, and enzyme concentration for a particular set of primers can produce significant improvements in specificity and yield of the desired fragment. The cycling parameters can also affect the efficiency and/or specificity of the reaction, and one should experiment with different temperatures and dwell times for the denaturation, annealing, and extension steps. Amplification of a cloned template control (whenever possible) will help ensure that the target is being amplified by the primers. For cloning purposes, optimal specificity is more important than maximal efficiency, as this will minimize or eliminate cloning undesired products that are the result of nonspecific amplification. Confirm that the correct target was amplified by preparing a Southern blot of the reaction and probing with an oligonucleotide probe that hybridizes to a region that is internal to the region defined by the PCR primers.

Sticky-End Cloning

Restriction Endonuclease Digestion

1. Add 25 μl of PCR sample to a 1.5-ml Eppendorf microfuge tube.
2. Add 10 μl of 10× restriction buffer.
3. Add 5 to 10 units of desired restriction enzyme(s).
4. Dilute with sterile water to 100 μl and incubate 2 to 3 hours at optimal temperature for the restriction enzyme(s) used.
5. Inactivate enzyme by heating for 5 minutes at 70°C (or phenol extraction for heat-stable enzymes).

If one is using two different enzymes, as for a directed cloning, use a restriction buffer that is compatible with both restriction enzymes so as to allow simultaneous digestion of each restriction site. Enzyme preparations should be of a high enough concentration so that a minimal volume of enzyme is required; the glycerol concentration should be kept at 5% or less. While the quick-and-dirty approach of adding the restriction reaction directly to the ligation reaction can work, the digest should usually be purified and concentrated by either ammonium acetate/isopropanol precipitation (Treco 1988) or adsorption and elution with "glassmilk" (Geneclean kit, Bio 101, La Jolla, California). The dNTPs carried over from the PCR can inhibit ligation, and it is useful to concentrate the digested DNA so that the ligation can be in a smaller volume (which favors intermolecular ligation). Alternatively, one can ethanol-precipitate 20 to 25% of the PCR with ammonium acetate (Treco 1988) and digest it in a smaller volume. To the tube containing the precipitated and dried DNA pellet add:

1. 2 μl of appropriate 10× restriction buffer.
2. 1 to 2 units of restriction enzyme(s).
3. Sterile water to 20 μl final volume.

Digest at optimal temperature for enzyme(s) used for 2 to 3 hours. A portion of this digestion can be added directly to the ligation.

Ligation

10× Ligase Buffer:
    500 mM Tris–HCl (pH 7.4)
    100 mM MgCl₂
    10 mM spermidine
    2 mM ATP

Prepare a 10-μl ligation reaction by mixing the following:

1. 5 μl of eluted DNA to a 0.5-ml Eppendorf tube
2. 1 μl of 10× ligase buffer
3. 1 μl of 10 mM DTT
4. 1 μl of digested M13 vector (200 ng)
5. 1 μl of T4 DNA ligase (NEB-400 Weiss units)
6. 1 μl of sterile distilled water (to a final volume of 10 μl)

Ligate at 16°C. Transform into competent cells according to standard protocol for M13 cloning (Besmond 1988). Ligate for a minimum of 1 hour at 16°C; however, do not let the ligation proceed longer than 16 hours (overnight).

Blunt-End Cloning

It is possible to clone the double-stranded "short product" as a blunt-ended fragment if there are no available restriction sites or the DNA proves recalcitrant to digestion. For maximum cloning efficiency, repair any "ragged ends" of the amplified fragment by using

Klenow fragment to repair the 3′ termini (Tabor and Struhl 1988). This can greatly improve the efficiency of blunt-end cloning. Additionally, use samples that are fairly efficiently amplified, as only 5 to 12% of the PCR is added to the ligation reaction. For reactions that do not amplify particularly efficiently, it is possible to add a greater percentage of the reaction by ethanol-precipitating the reaction and resuspending the DNA in 14 μl of water.

10× Ligase buffer:

> 500 mM Tris–HCl (pH 7.4)
> 100 mM MgCl$_2$
> 10 mM spermidine
> 10 mM ATP

To a 0.5-ml Eppendorf tube add:

1. 1 μl of blunt-end-digested M13 vector DNA (200 ng)
2. 5 μl of PCR DNA
3. 2 μl of 10× ligase buffer
4. 2 μl of 10 mM DTT
5. 1 μl of T4 DNA ligase (NEB-400 Weiss units)
6. 9 μl of dH$_2$0 (to a final volume of 20 μl)

Ligate overnight at 16°C and transform the ligation as described for sticky-end cloning (Besmond 1988).

Discussion

Cloning a fragment amplified from a single-copy gene is generally as straightforward as subcloning a restriction fragment from a recombinant plasmid. For cloning, the amplified DNA should be purified from the PCR constituents, as the dNTPs are competitive inhibitors for ATP in the ligation reaction. This is less important for blunt-end cloning because high (1 mM) ATP concentrations are used. The selection of restriction enzymes (e.g., BamHI and PstI) to be used for sticky-end cloning should provide a directed ligation. If one desires to create a new restriction site by changing the native sequence (an "internal" restriction site), try to maintain primer specificity by se-lecting a sequence that requires as few changes as possible.

When cloning from multi-allelic gene families (such as the HLA Class II region) (Scharf *et al.* 1988a; Scharf *et al.* 1988b), one can ob-tain "shuffle clones." Such clones are mosaics that are created when the extension of one allele is incomplete and hybridizes to other al-leles during subsequent cycles (Saiki *et al.* 1988). Such amplification artifacts have been mistaken for examples of gene conversion. One should run as few cycles as possible so that the concentration of these incompletely extended products will not be so high that they compete with the primers for hybridizing to the target. The best ap-proach is to design PCR primers that are allele specific.

The real power of PCR cloning is that any restriction site, adapter-sequence, or promoter can be incorporated or added to the PCR primer(s). This avoids the constraint of using only the restriction sites that nature provides, or having to use site-directed mutagenesis at a later step to modify a sequence.

Literature Cited

Besmond, C. 1988. Preparing and using M13 vectors. In *Current protocols in molecu-lar biology* (ed. F. M. Ausubel, R. Brent, R. E. Kingston, D. D. Moore, J. G. Seid-man, J. A. Smith, and K. Struhl), vol. 1, p. 1.15.1. Greene Publishing Associates and Wiley-Interscience, New York.

Mullis, K., F. Faloona, S. Scharf, R. Saiki, G. Horn, and H. Erlich. 1986. Specific en-zymatic amplification of DNA in vitro: the polymerase chain reaction. *Cold Spring Harbor Symp. Quant. Biol.* **51**:263–273.

Mullis, K. B., and F. A. Faloona. 1987. Specific synthesis of DNA *in vitro* via a poly-merase-catalyzed chain reaction. *Methods Enzymol.* **155**:335–350.

Saiki, R. K., S. Scharf, F. Faloona, K. B. Mullis, G. T. Horn, H. A. Erlich, and N. Arnheim. 1985. Enzymatic amplification of β-globin genomic sequences and restriction site analysis for diagnosis of sickle cell anemia. *Science* **230**:1350–1354.

Saiki, R. K., D. H. Gelfand, S. Stoffel, S. J. Scharf, R. Higuchi, G. T. Horn, K. B. Mullis, and H. A. Erlich. 1988. Primer-directed enzymatic amplification of DNA with a thermostable DNA polymerase. *Science* **239**:487–491.

Scharf, S. J., G. T. Horn, and H. A. Erlich. 1986. Direct cloning and sequence analysis of enzymatically amplified genomic sequences. *Science* **233**:1076–1078.

Scharf, S., and H. A. Erlich. 1988. The polymerase chain reaction: in vitro enzymatic amplification of DNA. In *Current protocols in molecular biology* (ed. F. M. Aus-ubel, R. Brent, R. E. Kingston, D. D. Moore, J. G. Seidman, J. A. Smith, and K. Struhl), vol. 1, p. 3.17.1. Greene Publishing Associates and Wiley-Interscience, New York.

Scharf, S. J., C. M. Long, and H. A. Erlich. 1988a. Sequence analysis of the HLA-DRβ and HLA-DQβ loci from three *Pemphigus vulgaris* patients. *Hum. Immunology* **22**:61–69.

Scharf, S. J., A. Friedmann, C. Brautbar, F. Szafer, L. Steinman, G. Horn, U. Gyllen-
    sten, and H. A. Erlich. 1988b. HLA class II allelic variation and susceptibility to
    Pemphigus vulgaris. *Proc. Natl. Acad. Sci. USA* **85**:3504–3508.
Tabor, S., and K. Struhl. 1988. Klenow fragment of *Escherichia coli* DNA polymerase
    I. In *Current protocols in molecular biology* (ed. F. M. Ausubel, R. Brent, R. E.
    Kingston, D. D. Moore, J. G. Seidman, J. A. Smith, and K. Struhl), vol. 1, p. 3.5.7.
    Greene Publishing Associates and Wiley-Interscience, New York.
Treco, D. A. 1988. Removal of low-molecular-weight oligonucleotides and triphos-
    phates by ethanol precipitation. In *Current protocols in molecular biology* (ed. F.
    M. Ausubel, R. Brent, R. E. Kingston, D. D. Moore, J. G. Seidman, J. A. Smith, and
    K. Struhl), vol. 1, p. 2.1.4. Greene Publishing Associates and Wiley-Interscience,
    New York.

OLIGONUCLEOTIDE LIGATION ASSAY

Ulf Landegren, Robert Kaiser, and Leroy Hood

Assays for the detection of specific DNA sequences are being increasingly performed in both clinical and research settings. Recent developments will permit a reduction in cost and time expenditure for such analyses. We describe a method, the oligonucleotide ligation assay (OLA), that used alone or in conjunction with target amplification by PCR permits the rapid and standardized identification of closely related DNA sequences.

A majority of the sequence differences between homologous DNA segments in different individuals is limited to single nucleotide positions. Among 40 genes investigated for structural aberrations leading to human genetic disease, most of the sequence variants involved the substitution of one nucleotide for another (McKusick 1988). Likewise, a majority of sequence variations serving as markers for genetic linkage analysis are the consequence of point mutations, predominantly affecting noncoding areas of the genome (Kimura 1983). Therefore, an important requirement of a general gene detection technique is that differences in single-nucleotide positions may be readily detected. The distinction between closely related sequences may be achieved by using short synthetic oligonucleotide probes that are destabilized in their hybridization by even single nu-

PCR Protocols: A Guide to Methods and Applications

cleotide mismatches under very carefully controlled hybridization conditions. As an alternative, the substrate specificity of DNA-processing enzymes, including restriction endonucleases, polymerases, ligases, and ribonucleases, may be employed to identify sequence variants (Landegren *et al.* 1988b).

The most common technique used to identify DNA sequences is the Southern (DNA) blot technique. The variable presence of restriction enzyme recognition sequences gives rise to variable-sized fragments upon enzyme digestion of genomic DNA from different individuals. The fragments are separated by gel electrophoresis, transferred to a membrane, and identified by hybridization with a radiolabeled probe. Advantages to this technique include the ease with which any cloned DNA segment may be used, without further characterization, to serve as a probe, and the straightforward distinction between sequence variants. However, the technique has several limitations: (1) the turn-around time for the analysis is on the order of days; (2) the technique is relatively complex, rendering routine use and automation difficult; (3) the currently used radioactive detection presents safety and stability problems; (4) not all sequence variants affect the recognition sequence of the more than hundred different restriction enzymes now available; and (5) relatively large amounts of high-molecular-weight genomic DNA are required for an analysis.

We have developed a gene detection technique with the potential to circumvent each of the above problems (Landegren *et al.* 1988a). In the oligonucleotide ligation assay, two oligonucleotides are designed to hybridize in exact juxtaposition on the target DNA sequences, permitting their covalent joining by a DNA ligase. The ligation of oligonucleotides, mismatched at the junction, may be prevented by adjusting the concentration of ligase and NaCl. The successful ligation can be monitored by a variety of means, including visualizing the ligation product after size separation by gel electrophoresis or, as described here, immobilizing one of the two oligonucleotides on a support and, after washing, detecting the covalent joining to the other appropriately labeled oligonucleotide (Fig. 1).

The OLA technique is characterized by few false positives, since the chance that the two oligonucleotides will hybridize in immediate proximity in the absence of the target sequence is minimal. In addition, the bond formed by the ligase is covalent, permitting extensive washes to remove non-specifically-bound reagents. The test

A

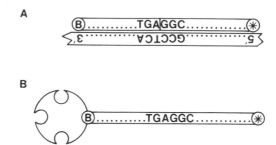

B

Figure 1 A diagrammatic representation of (A) the ligation of two oligonucleotides, labeled with biotin (B) and radioisotope (*). (B) After the assay the biotinylated oligonucleotides are collected on a solid support. The radiolabel remaining after a wash reflects the ligation of oligonucleotide pairs.

may be performed by simple, standard manipulations on samples of genomic DNA or, indeed, directly on nucleated cells after first treating them with detergent and a protease. The sensitivity of the assay may be greatly increased, however, by combining it with target amplification by PCR. This measure permits the analysis of minute DNA samples and will allow the use of alternative detectable moieties such as fluorophores, with concomitant gains in safety, stability of reagents, and expedience. Here we describe a standard protocol for ligase-mediated gene detection in conjunction with PCR. We will also briefly discuss some modifications of the strategy and some uses for this type of gene detection assay.

Protocol

Rationale

The target-dependent ligation of a biotin-labeled oligonucleotide to another oligonucleotide labeled with ^{32}P is measured by retrieving the ligation products on a streptavidin-coated support. After the support is washed, the signal is detected by autoradiography. Reagents specific for either of two alternate gene sequences are analyzed separately in parallel reactions. An example of a PCR-assisted OLA analysis of mutations in the β-globin gene is provided in Fig. 2.

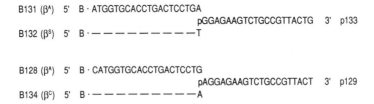

```
B131 (βᴬ)  5'  B · ATGGTGCACCTGACTCCTGA
                               pGGAGAAGTCTGCCGTTACTG  3'  p133
B132 (βˢ)  5'  B · — — — — — — — — —T


B128 (βᴬ)  5'  B · CATGGTGCACCTGACTCCTG
                               pAGGAGAAGTCTGCCGTTACT  3'  p129
B134 (βᶜ)  5'  B · — — — — — — — — —A
```

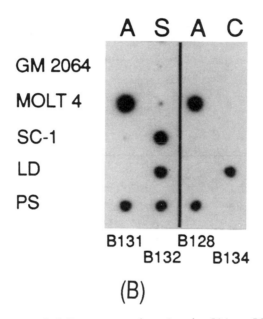

(B)

Figure 2 Distinction of allelic sequences by using the OLA on PCR products. (A) Primers designed to distinguish the β^A and β^S (B131 or B132, and P133) or the β^A and β^C (B128 or B134, and P129) alleles of the β globin gene were used to analyze the PCR products from various DNA samples. (B) Autoradiogram demonstrating the result of an analysis. The cell line GM 2064 has deleted the β globin locus. MOLT 4 and SC-1 are homozygous for the β^A and β^S globin alleles, respectively. Patient LD is a compound heterozygote ($\beta^{S/C}$) and PS is a carrier of the sickle cell allele β^S ($\beta^{S/A}$). Genomic DNA (1 μg) was amplified in 20 cycles as described by Saiki *et al.* (1985). A 3-μl sample of each reaction was assayed by the OLA as outlined in the text.

Design and Labeling of Oligonucleotides

The oligonucleotides used for the assay are designed as 20-mers, and their nucleotide sequence is selected such that any sequence differences to be distinguished are located immediately 5' to the junction of the two oligonucleotides. Two alternative oligonucleotides are designed so that each will have a 3' sequence specific for one of the DNA sequences to be distinguished. These oligonucleotides are equipped with a biotin group in the 5' end by reacting an aminothymidine residue, introduced during synthesis, with an N-hydroxysuccinimide ester of biotin. The products are purified by reverse-phase high pressure liquid chromatography (Landegren *et al.* 1988a). A third oligonucleotide is synthesized that can hybridize immediately 3' to either of the other two oligonucleotides. The ligation reaction requires that this oligonucleotide has a 5' phosphate group; this may conveniently be introduced as a radioisotope by using polynucleotide kinase and $[\gamma\text{-}^{32}P]ATP$ to provide a reporter group.

As an alternative, one can introduce a biotin group by using terminal deoxynucleotidyl transferase to add dUTP-biotin (Bethesda Research Laboratories, Inc.) to the 3' end of an oligonucleotide. A nonradioactive phosphate group is added to the 5' end of the oligonucleotide by using polynucleotide kinase to permit ligation. Enzymatically labeled oligonucleotides may be purified by using an affinity matrix such as Nensorb (Du Pont Biotechnology systems).

Reagents and Reactions of the OLA

1. Add to the wells of a flexible round-bottom microtiter plate in order:
    3 μl of a DNA sample amplified by PCR as described elsewhere in this book;
    1 μl of sheared salmon sperm DNA at 10 μg/ul;
    1 μl of 0.5 N NaOH (mix the droplets by a brief centrifugation at 400 rpm followed by a 10-minute incubation at room temperature);
    1 μl of 0.5 N HCl (mix with the other components before adding the subsequent reagents);
    1 μl containing 140 fmol of biotinylated oligonucleotide in water;
    1 μl containing 1.4 fmol of radiolabeled oligonucleotide in water;

2 μl of approximately 0.1 Weiss units T4 DNA ligase in 5× ligation buffer (mix and incubate for 1 hour at 37°C and 100%
humidity);
1 μl of 1.1 N NaOH (mix and incubate at room temperature for
10 minutes);
1 μl of 1.1 N HCl;
2 μl of 10% SDS;
3 μl of a 15% (v/v) slurry of streptavidin-coated agarose beads
(BRL) (incubate for 10 minutes at room temperature).
2. Transfer the contents of the wells to a dot blot manifold
(Schleicher & Schuell, Inc.) with a Whatman no. 4 filter paper
previously boiled in 0.5% dry milk, 1% SDS, and 100 μg/ml of
salmon sperm DNA. Wash the beads under suction with several
milliliters of 1% SDS followed by 0.1 N NaOH. The area on
which the beads are deposited may be reduced from 5 to 2 mm
in diameter by interposing, on the Whatman filter, a plastic disk
with conical holes.
3. Sandwich the membrane in plastic wrap and autoradiograph
with one enhancing screen overnight at −70°C to identify captured radiolabeled probes.

The composition of the 5× ligation buffer is 250 mM Tris–HCl (pH
7.5), 500 mM NaCl, 50 mM MgCl$_2$, 25 mM dithiothreitol, 5 mM ATP,
500 μg of bovine serum albumin per μl.

We find it convenient to use an electronic digital pipettor with a
repeat-dispense function (Rainin) to add multiple aliquots of the different reagents to the microtiter wells. Similarly, the washing of the
filter-trapped particles may be simplified by using a 96-tip dispenser
(Vaccu-pette/96, Culture Tek).

To establish optimal conditions to distinguish between closely
similar sequence variants, it is prudent to initially perform a titration
of the enzyme concentration used. In addition, the temperature of
the ligation reaction may be increased to 50°C to further enhance
discrimination between similar alleles.

Discussion

Several modifications of the described procedure will serve to simplify the analysis. It may be convenient to reserve one oligonucleotide to be employed both in the amplification and detection phase so

that only four different oligonucleotides will be required for the distinction of two allelic sequences. Direct detection of fluorescent groups incorporated into one of the oligonucleotides will offer advantages as a means of detection. Fluorophores are stable and safe moieties that may be detected instantaneously at the conclusion of an assay. Their major drawback is the limited detection sensitivity. We have demonstrated that combined with amplification by PCR, fluorophores afford a readily detectable signal in the analysis of single-copy genes (Landegren *et al.* 1988a). Gene detection techniques will attain their greatest importance if the procedure can be fully automated from the introduction of the DNA or tissue sample to the printout of the interpretation of the outcome. We are presently adapting a robotic workstation to perform the described analysis in large numbers and with a short turn-around time. Such an instrument, along with a large panel of ready-to-use detection reagents, could form the basis for a greatly simplified analysis of large numbers of DNA samples for the presence of gene variants that cause or predispose to disease, or that are associated with particular malignancies. The same technique also lends itself to the analysis of DNA sequences that vary among different individuals and facilitates the analysis of genetic linkage or the identification of individuals.

Acknowledgments

We thank Dr. Deborah Nickerson for critically reading the manuscript, Randall Saiki for providing genomic DNA from cell lines, and Dr. Tanaka for the patient samples. This work was supported by the Whittier Foundation and by a fellowship from the Knut and Alice Wallenberg Foundation to U.L.

Literature Cited

Kimura, M. 1983. *The neutral theory of molecular evolution.* Cambridge University Press, Cambridge.

Landegren, U., R. Kaiser, J. Sanders, and L. Hood. 1988a. A ligase-mediated gene detection technique. *Science* **241**: 1077–1080.

Landegren, U., R. Kaiser, C. T. Caskey, and L. Hood. 1988b. DNA diagnostics-molecular techniques and automation. *Science* **242**: 229–237.

McKusick, V. A. *molecular defects in mendelian disorders,* newsletter obtainable from Dr. McKusick, 25 March 1988.

Saiki, R. K., S. Scharf, F. Faloona, K. B. Mullis, G. T. Horn, H. A. Erlich, and N. Arnheim. 1985. Enzymatic amplification of β-globin genomic sequences and restriction site analysis for diagnosis of sickle cell anemia. *Science* **230**: 1350–1354.

13

NONISOTOPICALLY LABELED PROBES AND PRIMERS

Corey Levenson and Chu-an Chang

The PCR produces sufficient quantities of amplified DNA to enable the use of nonisotopic labels for the detection of target sequences. Labels may be incorporated directly into the PCR product via modified nucleoside triphosphates or primers that are labeled at their 5' termini. Alternatively, the PCR may be performed by using conventional oligonucleotide primers and triphosphates, and the product sequences can be detected by hybridization to an appropriately labeled nonradioactive probe. For primers, the most commonly employed 5'-terminal labels are biotin and a variety of fluorescent dyes (Chollet and Kawashima 1985; Smith *et al.* 1985; Smith *et al.* 1986; Adarichev *et al.* 1987; Smith *et al.* 1987; Brosalina and Grachev 1986; Ansorge *et al.* 1986; Ansorge *et al.* 1987). The most sensitive nonisotopic probes are conjugates between oligonucleotides and enzymes such as horseradish peroxidase (HRP) or alkaline phosphatase (Chu and Orgel 1988; Jablonski *et al.* 1986; Li *et al.* 1987; Urdea *et al.* 1988). The synthesis and purification of biotinylated, fluorescent-, and HRP-labeled oligomers are described below. The incorporation of labels via modified triphosphates will not be discussed in this chapter.

PCR Protocols: A Guide to Methods and Applications
Copyright © 1990 by Academic Press, Inc. All rights of reproduction in any form reserved.

Chemistry

A successful primer- or probe-labeling reaction depends upon the presence of a unique functional group at the terminus of the oligomer to be labeled. This functional group must be inherently more reactive toward labeling reagents than the other reactive groups (e.g., phosphates, exocyclic amino groups) are that are normally present in oligonucleotides. Generally, this functional group is a primary amine or thiol group that is introduced during the synthesis of the oligomer via an appropriately protected phosphoramidite. The incorporation of protected amines (Lin and Prusoff 1978; Smith *et al.* 1985; Coull *et al.* 1986; Wachter *et al.* 1986; Connolly 1987; Bischoff *et al.* 1987) and thiols (Connolly and Rider 1985) into oligonucleotides and their subsequent modification has been described by a number of groups. We have developed reagents that incorporate a tetraethylene glycol spacer between the amine (or thiol) and the 5′ hydroxyl of the oligomer; commercially available labeling reagents that incorporate linkers of six to twelve atoms have been found to be adequate. Once a primary amino group has been successfully intro-

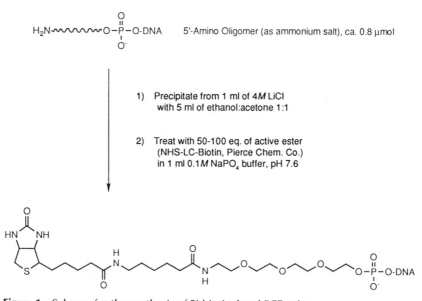

Figure 1   Scheme for the synthesis of 5′-biotinylated PCR primers.

duced, reaction with a variety of commercially available acylating reagents will yield the desired 5'-labeled oligomer. In the case of the 5'-thiolated oligomers, conjugates are produced via a reaction with an appropriately modified enzyme (i.e., maleimido labeled).

The synthesis of a biotinylated oligomer and the structure of our tetraethylene glycol-based linker are illustrated in Fig. 1. The reagents we use for the introduction of the 5'-amino or -thiol group are not currently commercially available; however, phosphoramidite reagents suitable for the incorporation of these functional groups are available from a variety of vendors (including Clonetech, Applied Biosystems, Glen Research, and ChemGenes). The protocol provided by the manufacturers should be adhered to during the coupling of the reagent to the oligonucleotide being synthesized. Once the addition of the modified amidite is complete, the oligomer should be deprotected (again following the protocol of the reagent supplier) and the crude oligomer should be treated as described in the protocol outlined as follows.

Protocols

Synthesis of Labeled PCR Primers from 5'-Amino Oligonucleotides

Conversion of the Ammonium Salt of the DNA to the Lithium Salt

1. Add 1 ml of 4 M lithium chloride to the dried residual deprotected oligomer generated from a 1-μmol-scale synthesis.
2. Sonicate (or vortex) briefly and transfer the suspension to a 1.5-ml Eppendorf tube and spin for about 2 minutes in a microfuge.
3. Filter the supernatant through 0.45-μm syringe filters (small diameter) into clean, labeled 12-ml silanized glass screw-cap tubes.
4. Add 5 ml of cold absolute ethanol : acetone (1 : 1) to each tube, mix, and chill for 1 hour in the freezer.
5. Centrifuge at about 4000 $\times$ g for 10 to 15 minutes.
6. Decant (discard) the supernatant.

Acylation of the Lithium Salt of the 5'-Amino-Labeled DNA

1. Using water-soluble acylating reagents (such as sulfo-N-hydroxy-succinimide esters), add to the pellet a solution of 50 to 100 μmol of labeling reagent (such as LC-NHS-biotin from Pierce Chemical Co.) dissolved in 1 ml of 0.1 M sodium phosphate buffer (pH 7.6), agitate to dissolve the pellet, and let the solution sit overnight.

2. Using water-insoluble acylating reagents (such as isothiocyanates or N-hydroxysuccinimide esters), dissolve the pellet in 0.9 ml of 0.1 M sodium phosphate buffer (pH 7.6), add 50 to 100 μmol of labeling reagent dissolved in 0.1 ml of DMSO or DMF, agitate, and let the solution sit overnight.

Removal of Excess Labeling Reagent

1. Pipette the reaction mixture onto the top of a short G-25 column (disposable NAP-10 columns from Pharmacia work well for this step) that has been pre-equilibrated with 20 ml of 0.1 M triethylammonium acetate [TEAA (pH about 7.4)]. Allow the mixture to enter the bed of the column.

2. Apply 1.5 ml of 0.1 M TEAA to the top of the column and collect the eluate.

Purification of the Labeled Primers

The labeled primers may be purified by HPLC or PAGE. It is impossible to provide a detailed protocol that is suitable for all labeled primers since the nature of the label and its linkage to the oligomer will alter the mobility of the oligomer in both chromatographic and electrophoretic systems.

In general, labeled primers will migrate more slowly on PAGE than unlabeled control oligomers do (unless, of course, the label is strongly anionic). Fluorescent primers can be easily differentiated on gels (unlike biotinylated primers) and will be much more strongly retained on reversed-phase HPLC columns than will either biotinylated or unmodified primers. When the labeling reagent exists as a mixture of isomers (i.e., N-hydroxysuccinimide esters of 5- and 6-carboxy fluorescein), the labeled primers resulting from each label isomer may often be resolved.

The linkers we have employed are based on tetraethylene glycol, and the following HPLC protocol works well with biotinylated primers.

HPLC Purification of Biotinylated Primers

Column: PRP-1 (7 by 305 mm, from The Hamilton Co.) with
             guard cartridge (from Alltech Associates, Inc.)
Flow Rate: 2 ml/minute
Solvent A: 0.1 M Triethylammonium Acetate [TEAA (pH about
             7.4)], containing 5% acetonitrile
Solvent B: Acetonitrile
Gradient: 30 minutes for equilibration
             5 to 10% B over 30 minutes
             10 to 50% B over 10 minutes
Detection: UV at 260 nm

The biotinylated oligomers should elute as large peaks at about 25 minutes (after the unmodified DNA peaks; see Fig. 2). Elution time may vary with the nature of the linker between the DNA and the biotin. The same column and solvent system have been used successfully with primers labeled with fluorescein, rhodamine, and Texas Red, although acetonitrile concentrations as high as 30% may be required to elute the labeled oligomers. Therefore, gradients need to be adjusted according to the label used.

Synthesis of HRP-Labeled Probes from 5'-Mercapto Oligomers

During solid-phase DNA synthesis (Fig. 3), sulfhydryl groups are introduced at the 5' termini of oligonucleotides as trityl mercaptans (Connolly and Rider 1985). Again, we have used phosphoramidite reagents that incorporate a tetraethylene glycol linkage between the tritylthiol and the 5' hydroxyl, although commercially available reagents that incorporate linkers of six to twelve carbon atoms work as well (e.g., ThioModifier from Clonetech, Palo Alto, California). The protocols provided by the manufacturer should be followed during the coupling of the reagent to the oligonucleotide being synthesized. The final detritylation step is omitted after addition of the modified

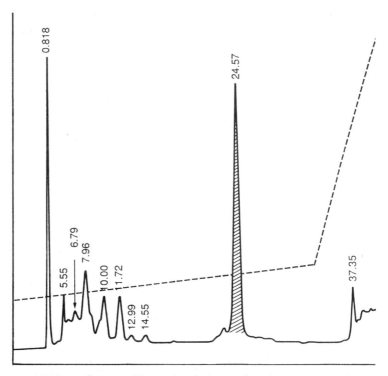

Figure 2 HPLC purification of biotinylated oligonucleotides. Vertical scale is absorbance at 260 nm. Retention times shown are in minutes. The dashed line shows the percentage of buffer B in the elution solvent. Details are given in the text.

amidite. The oligomer is deprotected with concentrated ammonia according to standard procedures and treated as described in the following procedure.

Purification of the 5'-Tritylmercapto Oligonucleotide

1. The ammoniacal solution of the crude oligomer is dried under a stream of air, and 1 ml of glass-distilled water is added to the residual crude oligomer. Vortex or sonicate the tube if necessary to suspend the residue and incubate for 5 minutes.
2. Filter the supernatant through a 0.45-μm syringe filter (Nylon Acrodisc from Gelman Sciences, Inc.). Rinse the filter with 0.8 ml of water and combine the washings with the supernatants.

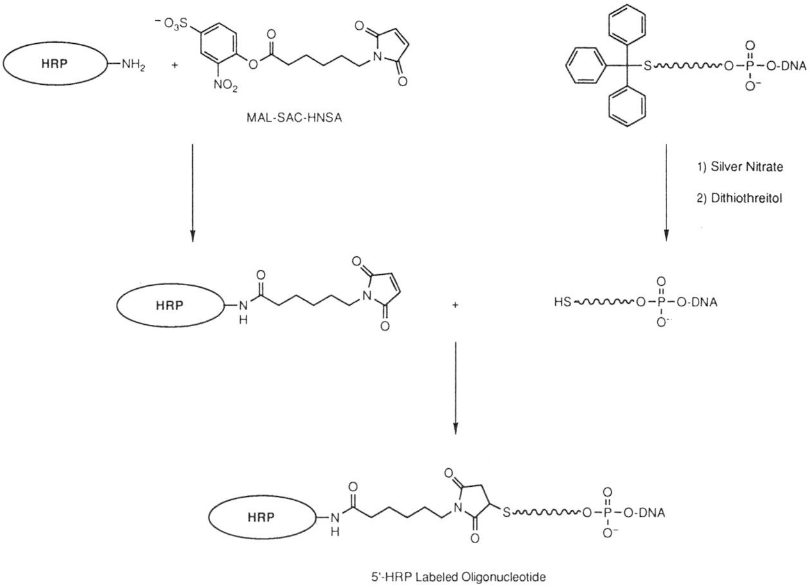

Figure 3 Scheme for the preparation of 5'-horseradish peroxidase (HRP)-labeled probes.

3. The filtrate is fractionated on an HPLC fitted with a 2-ml sample loop. The following HPLC protocol is used:

Column: PRP-1 (7 by 305 mm, Hamilton) with guard cartridge (4.6 by 10 mm, Alltech)

Flow rate: 2 ml/minute

Solvent A: 0.1 M TEAA (pH about 7.4) containing 5% acetonitrile

Solvent B: Acetonitrile

Gradient: 30 minutes for equilibration
             10 to 50% B over 25 minutes
             50 to 100% B over 5 minutes
             100% B for 5 minutes
             100 to 10% B over 5 minutes

Detection: UV at 260 nm

The trityl-containing fraction should elute last from the column as a major peak at about 18 minutes (see Fig. 4). A good synthesis usually results in about 40 A_{260} units of tritylated oligomer.

The HPLC profiles of the oligomers are almost identical over a wide range of sizes of oligomers (13 to 59 bases). The purified oligomer is

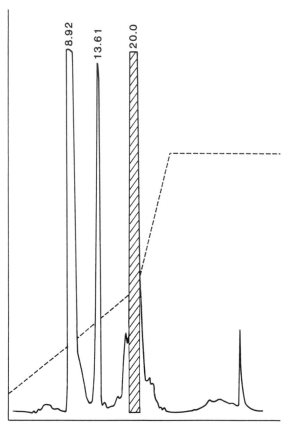

Figure 4   HPLC purification of S-tritylmercapto oligonucleotides. Vertical scale is absorbance at 260 nm. Retention times shown are in minutes. The dashed line shows the percentage of buffer B in the elution solvent. Details are given in the text.

aliquoted out for conjugation, and the remainder is dried and stored at −20°C. The oligomer is stable for at least a year under these conditions.

Removal of the Trityl Group from the Oligonucleotide

1. Dry an aliquot containing 10 A_{260} units of the HPLC-purified tritylthiol oligomer in a Speed Vac.
2. Dissolve the oligomer in 0.1 ml of 0.1 M TEAA buffer (pH 6.0).

3. Add 10 μl of 0.1 M silver nitrate stock solution (in water, protected from light) to the oligomer. The solution should turn cloudy with the appearance of a white precipitate, indicating the release of a trityl group. After briefly vortexing the mixture, incubate it at room temperature for 1 hour.

4. Add 10 μl of freshly prepared dithiothreitol (DTT) solution [0.15 M in TEAA buffer (pH 6.0)] to the mixture. A thick yellowish precipitate forms immediately. The reaction mixture is incubated for another hour at room temperature with occasional vortexing.

5. Dilute the mixture to 1 ml with TEAA buffer and transfer it to the top of a short, disposable G-25 Sephadex column (NAP-10 from Pharmacia) that has been pre-equilibrated with 30 ml of 0.1 M sodium phosphate buffer (pH 6.0). Allow the mixture to enter the bed of the column.

6. Apply 1.5 ml of phosphate buffer to the column and collect the eluate. (The thiolated DNA is contained in this 1.5-ml eluate.)

7. The detritylated 5'-thiolated oligomer is combined immediately with the maleimido-modified HRP prepared as follows.

Preparation of Maleimido-HRP

A heterobifunctional crosslinking reagent, i.e., mal-sac-HNSA (maleimido-6-aminocaproyl ester of 1-hydroxy-2-nitrobenzene-4-sulfonic acid sodium salt), is used to introduce maleimidyl moieties to HRP through amide linkages to its lysine residues. Mal-sac-HNSA can be prepared (Aldwin and Nitecki 1987) or purchased (Bachem Bioscience Inc., Philadelphia, Pennsylvania). HRP (Type VI, Sigma Chemical Co., St. Louis, Missouri) can be used without further purification.

The maleimido-HRP prepared according to the following protocol is sufficient for conjugation to four 5'-thiolated oligomers obtained from the previous procedure. Since better results are obtained when the modified HRP is used fresh, the following protocol should be carried out concurrently with the detritylation of protected thiolated oligomers.

1. HRP (40 mg) is dissolved in 1.5 ml of 0.1 M sodium phosphate buffer (pH 7.5) with gentle stirring. Keep the solution in the dark. The dissolution takes about 10 minutes.
2. Mal-sac-HNSA (3 mg) is added to the dark red-brown HRP solution. Stir for 30 minutes at room temperature.
3. Transfer the mixture to the top of a disposable G-25 Sephadex column (NAP-25, Pharmacia) that has been pre-equilibrated with 30 ml of 0.1 M sodium phosphate buffer (pH 6.0). Allow the dark solution to enter the bed of the column.
4. Apply about 3.5 ml of the phosphate buffer (pH 6.0) to the top of the column and collect the brown HRP-containing fraction (which elutes out of the column first) in about 2 ml of buffer.
5. Add a 0.5-ml aliquot of this maleimido-HRP solution to the 5'-thiolated oligomer obtained above in 1.5 ml of buffer. Mix this solution briefly and concentrate the volume to about 1 ml in a Speed Vac.
6. The reaction mixture is kept in a refrigerator (4°C) for 24 hours before purification.

Purification of the HRP-Oligomer Conjugate

Weak anion-exchange (DEAE) column chromatography is used to purify the conjugate because of (1) its high capacity: the conjugate prepared as described can be purified in one run; (2) mild running conditions: since a salt (sodium chloride) gradient is used to elute the conjugate from the column, no post-column manipulation is required for the use and storage of the conjugate; and (3) high resolution: the desired conjugate can be easily separated from the excess uncoupled HRP, which elutes at the void volume, or the unreacted oligomer, which elutes after the conjugate peak (see Fig. 5).

HPLC Purification Protocol

Column: Nucleogen DEAE 500–10 (6 by 125 mm, Macherey
            Nagel), with guard cartridge (4.6 by 10 mm, Alltech)
Flow rate: 1.5 ml/minute
Buffer A: 20 mM sodium phosphate (pH 6.0)
Buffer B: Buffer A plus 1 M sodium chloride

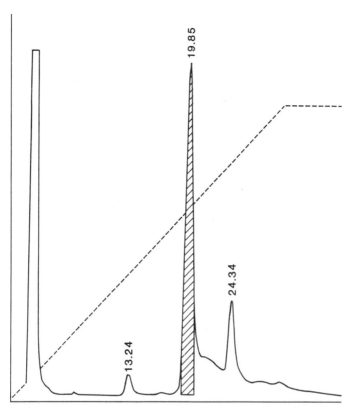

Figure 5 HPLC purification of HRP-labeled DNA probe. Vertical scale is absorbance at 260 nm. Retention times shown are in minutes. The dashed line shows the percentage of buffer B in the elution solvent. Details are given in the text.

Gradient: 15 minutes for equilibration with Buffer A
             0 to 100% B over 30 minutes
             100% B for 10 minutes
             100 to 0% B over 5 minutes
Detection: UV at 260 nm

Because of the anion-exchange nature of this purification method, the retention time of the conjugate is proportional to its negative charge. The uncoupled HRP is not retained by the column and elutes almost at the void volume of the column (at about 2.2 minutes). The conjugate elutes after the HRP according to the size of the oligomer (for instance, the observed retention times for several 15-mers ranged from 19 to 21 minutes). The unconjugated oligomers and their dimers (generated through disulfide formation) elute from the column after elution of the desired conjugate. The separation between the conjugate and the uncoupled oligomer on the column is usually 2 to 3 minutes. The dimers are eluted even later (Fig. 5).

The purified conjugates are stored at 4°C in the HPLC elution buffer without further manipulation. Under these conditions, the conjugates are stable for at least 12 months. The biological activity of the HRP is not adversely affected by either the conjugation to the oligomer or storage.

Stoichiometry and Quantitation of the Purified HRP–Oligomer Conjugate

The ratio of HRP to oligonucleotide can be determined from the ratio of UV absorbance at 402 and 260 nm. The major product formed in the conjugation reaction described above is a 1 : 1 adduct of enzyme to nucleic acid.

The concentration of the conjugate in the elution buffer is determined spectrophotometrically using the absorbance of the HRP at 402 nm. A molar absorption coefficient of 100,000 is used for this determination (Shannon *et al.* 1966).

Summary

Synthetic oligonucleotide probes and primers may be prepared to which are attached a variety of nonisotopic "reporter groups." Labeled primers (i.e., where the label is biotin or fluorescein) have been used successfully in otherwise conventional PCRs to yield PCR products that incorporate these labels, eliminating the need for radioactive materials. In the case of primers, the nature of the PCR precludes the attachment of labels to the 3' termini. Although labels attached to the bases would probably be tolerated, provided they do not interfere with hybridization of the primer to its template, we have used 5'-end-labeled probes and primers extensively, and it is their preparation that has been described here.

PCRs conducted with conventional (nonlabeled) primers yield products that can be probed using enzyme-labeled oligonucleotide probes. Using commercially available phosphoramidite reagents, one can produce oligomers containing functional groups (i.e., thiols or primary amines) at either terminus and can label them using the protocols described in this chapter. Once a primary amine (or thiol) has been introduced, one can select from a vast array of commercially available labeling reagents (most of which were originally designed to be used for protein modification). By using such reagents, biotin, fluorescent dyes, photoreactive groups, spin labels and other moieties may be attached to synthetic oligonucleotides. Indeed, primary amines that have been introduced can be converted to thiols by using such reagents (Bischoff *et al.* 1987). Control reactions using unmodified oligomers should always be conducted to ensure that side reactions are not occurring between the particular labeling reagent employed and the exocyclic amines on dA, dC, and dG.

Acknowledgments

The authors thank Lauri Goda and Dragan Spasic for their excellent and longstanding contributions to the work described here.

Literature Cited

Adarichev, V. A., G. M. Dymshits, S. M. Kalachikov, P. I. Pozdnyakov, and R. I. Salganik. 1987. Introduction of aliphatic amino groups into DNA and their labelling with fluorochromes in preparation of molecular hybridization probes. *Bioorganichekaya Khimiya* **13**:1066–1069.

Aldwin, L., and D. E. Nitecki. 1987. A water-soluble, monitorable peptide and protein crosslinking reagent. *Anal. Biochem.* **164**:494–501.

Ansorge, W., B. S. Sproat, J. Stegemann, and C. Schwager. 1986. A non-radioactive automated method for DNA sequence determination. *J. Biochem. Biophys. Methods* **13**:315–323.

Ansorge, W., B. Sproat, J. Stegemann, C. Schwager, and M. Zenke. 1987. Automated DNA sequencing: ultrasensitive detection of fluorescent bands during electrophoresis. *Nucleic Acids Res.* **15**:4593–4603.

Bischoff, R., J. M. Coull, and F. E. Regnier. 1987. Introduction of 5'-terminal functional groups into synthetic oligonucleotides for selective immobilization. *Anal. Biochem.* **164**:336–344.

Brosalina, E. B., and S. A. Grachev. 1986. The synthesis of 5'-biotin-labelled oligo- and polynucleotides and investigation of their complexes with avidin. *Bioorganischekaya Khimiya* **12**:248–256.

Chollet, A., and E. H. Kawashima. 1985. Biotin-labeled synthetic oligodeoxyribonucleotides: chemical synthesis and uses as hybridization probes. *Nucleic Acids Res.* **13**:1529–1541.

Chu, B. C. F., and L. E. Orgel. 1988. Ligation of oligonucleotides to nucleic acids or proteins via disulfide bonds. *Nucleic Acids Res.* **16**:3671–3691.

Connolly, B. A., and P. Rider. 1985. Chemical synthesis of oligonucleotides containing a free sulphydryl group and subsequent attachment of thiol specific probes. *Nucleic Acids Res.* **13**:4485–4502.

Connolly, B. A. 1987. The synthesis of oligonucleotides containing a primary amino group at the 5'-terminus. *Nucleic Acids Res.* **15**:3131–3139.

Coull, J. M., H. L. Weith, and R. Bischoff. 1986. A novel method for the introduction of an aliphatic primary amino group at the 5' terminus of synthetic oligonucleotides. *Tetrahedron Lett.* **27**:3991–3994.

Jablonski, E., E. W. Moomaw, R. H. Tullis, and J. L. Ruth. 1986. Preparation of oligodeoxynucleotide—alkaline phosphatase conjugates and their use as hybridization probes. *Nucleic Acids Res.* **14**:6115–6128.

Li, P., P. P. Medon, D. C. Skingle, J. A. Lanser, and R. H. Symons. 1987. Enzyme-linked synthetic oligonucleotide probes: non-radioactive detection of enterotoxigenic *Escherichia coli* in faecal specimens. *Nucleic Acids Res.* **15**:5275–5287.

Lin, T., and W. H. Prusoff. 1978. Synthesis and biological activity of several amino analogues of thymidine. *J. Medicinal Chem.* **21**:109–112.

Shannon, L. M., E. Kay, and Y. Y. Lew. 1966. Peroxidase isozymes from horseradish roots. *J. Biol. Chem.* **241**:2166–2172.

Smith, L. M., S. Fung, M. W. Hunkapiller, T. J. Hunkapiller, and L. E. Hood. 1985. The synthesis of oligonucleotides containing an aliphatic amino group at the 5' terminus: synthesis of fluorescent DNA primers for use in DNA sequence analysis. *Nucleic Acids Res.* **13**:2399–2412.

Smith, L. M., J. Z. Sanders, R. J. Kaiser, P. Hughes, C. Dodd, C. R. Connell, C. Heiner, S. B. H. Kent, and L. E. Hood. 1986. Fluorescence detection in automated DNA sequence analysis. *Nature* (London) **321**:674–679.

Smith, L. M., R. J. Kaiser, J. Z. Sanders, and L. E. Hood. 1987. The synthesis and use of fluorescent oligonucleotides in DNA sequence analysis. in *Methods Enzymol.* **155**:260–301.

Urdea, M. S., B. D. Warner, J. A. Running, M. Stempien, J. Clyne, and T. Horn. 1988. A comparison of non-radioisotopic hybridization assay methods using fluorescent, chemiluminescent and enzyme labeled synthetic oligodeoxynucleotide probes. *Nucleic Acids Res.* **16**:4937–4956.

Wachter, L., J.-A. Jablonski, and K. L. Ramachandran. 1986. A simple and efficient procedure for the synthesis of 5'-aminoalkyl oligodeoxynucleotides. *Nucleic Acids Res.* **14**:7985–7994.

14

INCORPORATION OF
BIOTINYLATED dUTP

Y-M. Dennis Lo, Wajahat Z. Mehal, and Kenneth A. Fleming

The production of vector-free inserts for use as probes is time-consuming and inefficient, involving restriction enzyme digestion, preparative gel electrophoresis, and elution by one of a variety of methods. PCR is an efficient way of producing vector-free inserts (Saiki *et al.* 1988), and, by including biotinylated-11-dUTP (bio-11-dUTP) during amplification, we have developed a one-step procedure for producing vector-free biotinylated probes (Lo *et al.* 1988). Nonradioactive label is preferable to radioactive label as large quantities of labeled probes can be produced in one experiment that are stable for considerable periods of time, e.g., 1 year. When used with appropriate internal primers, this method can be used to synthesize labeled reagents for probing PCR products to confirm the presence or absence of specific sequences following amplification of genomic DNA and provide a completely nonisotopic system for amplifying and detecting genomic PCR products for research or diagnostic purposes.

Protocol

Reagents

The components of a typical 100-μl reaction are as follows:

Target DNA	0.2 fmol (~1 ng of an 8-kb plasmid)
1 M KCl	5 μl
0.05 M MgCl$_2$	5 μl
1 M Tris (pH 8.3) at 37°C	1 μl
0.2% gelatin	5 μl
20 mM dATP	1 μl
20 mM dCTP	1 μl
20 mM dGTP	1 μl
15 mM TTP	1 μl
0.3 mM bio-11-dUTP*	17 μl
Primer 1 (100 μM)	1 μl
Primer 2 (100 μM)	1 μl
Taq polymerase (Cetus, 5 units/μl)	1 μl
Water	to make up to 100 μl
Paraffin oil (BDH)	100 μl

*0.3 mM bio-11-dUTP is obtained from Bethesda Research Laboratories, Inc. Bio-11-dUTP in powder form is available from Sigma Chemical Co. but requires prior dissolution in 100 mM Tris, 0.1 mM EDTA, pH 7.5.

Incorporation of Biotinylated dUTP

1. Incubate the mixture at 94°C for 10 minutes.
2. Perform 25 cycles of PCR. Our cycle profile consists of 2 minutes at 55°C (annealing), 3 minutes at 72°C (extension), and 2 minutes at 94°C (denaturation). At the end of cycle 25, increase the extension step by an additional 7 minutes.
3. Remove as much paraffin oil as is possible.
4. Add 1 μl of glycogen (20 mg/ml) (Boehringer), 10 μl of 2.5 M

sodium acetate (pH 5.2), and 220 μl of ethanol. Leave the mixture at −20°C overnight or −70°C for 1 hour.

5. Spin the mixture for 15 minutes at maximum speed on a Micro Centaur (MSE), remove the supernatant, and vacuum desiccate for 10 minutes.

6. Dissolve the precipitate in 100 μl of 10 mM Tris, 1 mM EDTA, pH 8.0.

7. Run 5 μl on an agarose gel. The biotinylated PCR product should migrate as a tight band with slightly decreased electrophoretic mobility when compared with the nonbiotinylated PCR product (Fig. 1).

8. The success of labeling can be checked by running 1 μl of the PCR product on a denaturing gel, Southern transferring to a nylon filter (Hybond-N, Amersham Corp.), and detecting the biotin label by using a streptavidin/alkaline phosphatase detection system (Chan *et al.* 1985) (Fig. 1). However, this may not be routinely necessary as the method described above is very reliable.

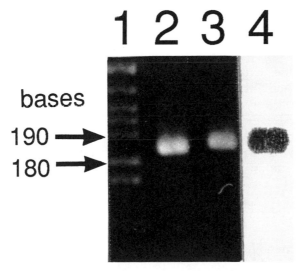

Figure 1 Production of an 185-bp biotinylated hepatitis B virus probe by using PCR. Tracks 1–3, 4% NuSieve Gel. (1) *Msp*I-digested pBR322 (marker). (2) Nonbiotinylated PCR product. (3) Biotinylated PCR product. (4) Southern transfer of biotinylated PCR product (from an alkaline denaturing gel).

9. As a guideline, for hybridization using a probe of 185 bp, 10 μl of the labeled PCR product is used per 50 cm^2 of nitrocellulose filter (2 ml of hybridization mix).

10. Store probe at either −20°C or −70°C in aliquots.

Discussion

To maintain the specificity of labeling, the starting DNA target should be kept to about 0.2 fmol (~1 ng of an 8-kb plasmid). The use of larger amounts of target DNA results in the production of significant quantities of labeled vector sequences. The cause of this is unclear, but it may be that during cycles of PCR, synthesis occurs by priming on the original starting template in addition to priming on the newly synthesized templates. Synthesis on the former produces DNA strands of various lengths outside the region delineated by the primers, resulting in nonspecific labeling. Another factor may be the nonspecific binding of primers at high target concentrations. Since the PCR plateau ranges from 2 to 5 pmol (Higuchi *et al.* 1988), the ratio of specific labeling to nonspecific labeling ranges from 400 to 1000 : 1, after 25 cycles of PCR. As can be seen in Fig. 2, the method is very specific, with the probe having negligible hybridization to the vector band. Reducing the target DNA concentration further may result in even higher ratios. However, we have not tested this.

We have estimated the sensitivity of probes generated in this fashion by using serial dilutions of a plasmid containing a full-length hepatitis B virus insert pHBV130 (Gough and Murray 1982). The PCR product is an 185-bp fragment generated from pHBV130 by using primers

GAGTGTGGATTCGCACTCCTC and
GATTGAGATCTTCTGCGACGC

As can be seen in Fig. 2, when used in Southern hybridization, 1 pg of target DNA could be detected by using a streptavidin/alkaline phosphatase detection system previously described (Chan *et al.* 1985). The method is very time efficient; microgram quantities of probe can be synthesized in less than 4 hours. The stability of biotinylated probes allows them to be stored for a long time (i.e., 1 year).

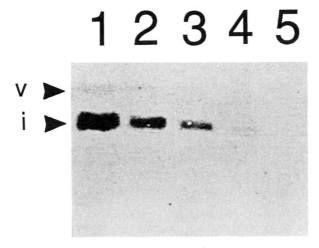

Figure 2 Sensitivity and specificity of PCR-generated biotinylated probe using serial dilutions of pHBV130 cleaved with *Xho*I to release insert. v = vector, i = insert. (1) 50 ng (1 ng). (2) 5 ng (100 pg). (3) 500 pg (10 pg). (4) 50 pg (1 pg). (5) 5 pg (0.1 pg). Figures in brackets are the amounts of target sequences.

The concentration ratio of TTP and bio-11-dUTP is chosen to be 3 : 1 (150 mM : 50 mM) because previous work (Chan *et al.* 1985) has shown that a percentage incorporation of bio-11-dUTP of 25 to 30% is optimal for hybridization.

We have attempted other procedures for labeling the PCR products. After a conventional PCR (without bio-11-dUTP), 10 μl of the reaction mixture was removed and added to a mixture consisting of 68 μl of water, 5 μl of 1 M KCl, 1 μl of 1 M Tris (pH 8.3) at 37°C, 5 μl of 0.2% gelatin, 5 μl of 0.05 M MgCl$_2$, and 15 μl of 0.3 mM bio-11-dUTP, with or without additional *Taq* polymerase (5 units). The mixture was heated at 94°C for 10 minutes, and the labeling reaction was allowed to continue overnight (~16 hours) at room temperature (24°C), 37°C, 55°C, or 72°C. We found that incubation at room temperature without further addition of *Taq* gave the best results. However, the protocol outlined is superior to this procedure both in efficiency and ease of operation. We have also attempted labeling at deoxyribonucleotide triphosphate concentrations considerably lower than that normally recommended for PCR: dATP, dCTP, and dGTP at 20 μM to 200 μM; TTP from 0.5 μM to 5 μM; and bio-11-dUTP from 20 μM to 40 μM. While these resulted in biotin-labeled amplified probes, the hybridization efficiency of these probes has not been investigated in detail. The method has general application for producing vector-free biotinylated probes if primers flanking the vector cloning site are used. Labels other than bio-11-dUTP may also be tried, although we have not attempted these.

A probe for detection of PCR products amplified from genomic DNA can be produced by using a pair of primers internal to that for the original amplification. This allows unequivocal, highly sensitive nonisotopic detection of specific bands among many "ghost" bands. It also greatly increases the sensitivity of the PCR.

Literature Cited

Chan, V. T.-W., K. A. Fleming, and J. O' D. McGee. 1985. Detection of subpicogram quantities of specific DNA sequences on blot hybridization with biotinylated probes. *Nucleic Acids Res.* **13**:8083–8091.

Gough, N. M., and K. Murray. 1982. Expression of the hepatitis B virus surface, core and E antigen genes by stable rat and mouse cell lines. *J. Mol. Biol.* **162**:43–67.

Higuchi, R., B. Krummel, and R. K. Saiki. 1988. A general method of *in vitro* preparation and specific mutagenesis of DNA fragments: study of protein and DNA interactions. *Nucleic Acids Res.* **16**:7351–7367.

Lo, Y-M. D., W. Z. Mehal, and K. A. Fleming. 1988. Rapid production of vector-free biotinylated probes using the polymerase chain reaction. *Nucleic Acids Res.* **16**:8719.

Saiki, R. K., D. H. Gelfand, S. Stoffel, S. J. Scharf, R. Higuchi, G. T. Horn, K. B. Mullis, and H. A. Erlich. 1988. Primer-directed enzymatic amplification of DNA with a thermostable DNA polymerase. *Science* **239**:487–491.

NONISOTOPIC DETECTION OF PCR PRODUCTS

Rhea Helmuth

The use of nonradioactive oligonucleotide probes to analyze PCR-amplified DNA for nucleotide sequence variation provides a simple and safe genetic test. Although radioactive ^{32}P-labeled probes represent an informative and general approach to genetic analysis, their structural instability and associated biohazard procedures have limited their routine clinical and forensic use. The convenience of highly stable, nonradioactively labeled probes for standardized testing is valuable to laboratories assaying large numbers of samples, as well as to labs not equipped to handle radioactivity. An additional benefit to the laboratory is sample turnaround time; the entire procedure from DNA sample to PCR product to diagnostic result can be performed in less than 5 hours.

Described here is a basic dot blot-based format, including the use of two types of nonradioactive labeled probes. This format has been used routinely for HLA-DQα typing as well as for many other dot blot-based PCR detection systems (Saiki *et al.* 1986; Saiki *et al.* 1988; Bugawan *et al.* 1988) of single- or double-stranded nucleic acids.

Protocols

Reagents

PCR-amplified product of sufficient quantity to be visualized on a 3% NuSieve plus 1% agarose gel by ethidium bromide stain. A 5- to 20-μl (50- to 100-ng) sample of product is needed per spot/probe.

Probes, labeled with HRP or biotin-psoralen. For HRP-labeled probes we typically use 10 ml of a 0.5 pmol/ml solution. For some probes a 2- to 3-fold higher concentration has given better results.

Denaturing solution: 0.4 N NaOH plus 25 mM EDTA.

TE buffer: 10 mM Tris–HCl, 1 mM EDTA (pH 8.0).

20× SSPE: 3.6 M NaCl, 200 mM NaH$_2$PO$_4$, 20 mM EDTA (pH 7.4).

Hybridization solution: SSPE (concentration determined by the T_m of the probe), 5× Denhardt's solution, 0.5% Triton X-100.

Wash solution: SSPE (concentration determined by the T_m of the probe), 0.1% Triton X-100.

Phosphate-buffered saline (PBS): 137 mM NaCl, 2.7 mM KCL, 1.5 mM KH$_2$PO$_4$, 8.0 mM Na$_2$HPO$_4$ (pH 7.4).

Buffer A: PBS, 100 mM NaCl, 5% Triton X-100.

Streptavidin horseradish peroxidase conjugate (Cetus) (Sheldon *et al.* 1986).

Buffer B: buffer A, 1 M urea, 1% dextran sulfate, stored at 4°C.

Buffer C: 100 mM sodium citrate (pH 5.0).

Tetramethylbenzidine (TMB): 3,3′, 5,5′ TMB (Fluka) 2 mg/ml in 100% ethanol. (This solution is light sensitive and should be stored in an opaque container at 4°C. The solution is stable for 2 months.)

Hydrogen peroxide, 3%.

Buffer D: 19 parts buffer C to 1 part 2 mg/ml TMB.

Buffer E: 2000 parts buffer D to 1 part 3% hydrogen peroxide. 0.18% Na$_2$SO$_4$.

Instruments/Supplies

Dot blot apparatus, such as "Bio-Dot" (Bio-Rad Laboratories).

Vacuum source and trap flask.

Round-bottom microtiter plates, 96-well.

Positively charged nylon membrane, such as Genatrans (Plasco).

Laundry pen or similar smear-proof marker.

Whatman 3MM paper.

Shortwave UV light transmitter, or vacuum oven.

Shaking water bath and/or dry-heat shaker.

Multichannel pipetting device and disposable tips.

Plastic heat-sealable bags (Dazey Seal-a-Meal) and heat sealer.

Dishes for washing membranes, such as crystallizing dishes (Pyrex).

An opaque dish for color development.

Aluminum foil for blocking out light.

Photography set-up.

Begin this procedure with PCR-amplified DNA product, checked for amplification on a 3% NuSieve plus 1% agarose gel. Positive control (one for each probe) and negative control reactions should also be included.

Preparation of Membranes

The PCR product will be spotted to a nylon membrane through a dot blot apparatus and then hybridized with a probe. Prepare one membrane per probe, spotting identical samples on each membrane.

1. Obtain positively charged nylon membrane (Genatrans), cut to the size of the dot blot apparatus (approximately the size of a standard microtiter plate, 9×12 cm).
2. Label the membrane (using a laundry marker or similar smear-proof pen) with the probe name and sample information.
3. Pre-wet the membrane by placing it in water for at least 1 minute.
4. Prepare PCR product for the dot blot:
   a. Denature 5 to 20 μl (50 to 100 ng) of PCR product per replicate spot in 100 μl of denaturing solution (0.4 N NaOH plus 25 mM EDTA). The amount of each PCR sample required for all the probes can be denatured in one mixture. This is done conveniently in a round-bottom microtiter plate.
   b. Complete denaturation is achieved within 5 minutes.

5. Assemble the pre-wet membrane into the dot blot apparatus. Add the 100 μl of the denatured PCR mix to the appropriate well. Turn on the vacuum and draw the solution through the membrane. Be wary of bubbles on the surface of the membrane that can cause unusual spotting patterns (rings, crescents, etc.).

6. "Wash" each sample by adding 100 μl of TE (or similar low-salt) buffer to the wells and pull through by vacuum.

7. Remove membrane and blot dry with 3MM paper.

8. Repeat blotting procedure until all replicate membranes are complete.

9. Fix the DNA onto the membranes either by baking in a vacuum oven at 80°C for 1 hour or by exposing the membrane to a UV light source for 60 to 120 mJ/cm^2 at 254 nm (i.e., placing the membranes on a standard shortwave UV transilluminator for 5 minutes) (Church and Gilbert 1984).

At this point the membranes are ready for hybridization. They can be probed immediately or stored dry in plastic for future hybridization.

Hybridization with Probes

Refer to Chapter 13 and to the references cited for more information on the preparation of nonradioactively labeled oligonucleotides.

Types of Nonradioactive Labels

Horseradish peroxidase (HRP): This label is prepared by covalently coupling synthetic oligomers containing a free sulfhydryl group at their 5' ends with thiol-specific, maleimido-derivitized HRP, as previously described (Bugawan et al. 1988; Connolly and Rider 1985; C. Chang, manuscript submitted).

Biotin-psoralen: This label is achieved by creating an oligonucleotide complex and labeling it with a psoralen-biotin labeling reagent, as previously described (Bugawan et al. 1988; Sheldon et al. 1986).

Hybridization and Washing Conditions

Each oligonucleotide probe has its own hybridization temperature and buffer conditions based on the T_m of the probe. The T_m is based on the length and G+C content of the oligonucleotide, as well as on the salt concentration of the hybridization buffer. As a general rule, the T_m is calculated by assigning each G-C pair a melting temperature of 4°C and each T-A pair a melting temperature of 2°C; these degree measurements are added to give the T_m of the oligonucleotide. The hybridization temperature can range from 5 to 20°C lower than the predicted T_m of the oligonucleotide in 1 M salt. The stringency of the system is increased by either lowering the salt concentration or increasing the hybridization temperature (Breslauer *et al.* 1986; Meinkoth and Wahl 1984). A wash solution generally has less salt, and the wash temperature is 5°C lower than that of the hybridization conditions.

Once the hybridization and wash conditions are established, proceed as follows.

1. Pre-wet spotted membranes in 2× SSPE.
2. Place the membranes in plastic heat-sealable bags and add hybridization solution (SSPE, 5× Denhardt's solution, 0.5% Triton X-100). For a full-size blot, approximately 8 ml of solution is needed. Seal the bag, taking care not to trap air bubbles.
3. Prehybridize the membranes in a water bath/dry-heat shaker for 5 minutes at the hybridization temperature of the probe.
4. Remove the membranes from the heat shaker and cut off one corner of each bag. Add the labeled probe at 1 pmol/ml of hybridization solution into the liquid in the bag. Reseal the bags and return them to the water bath/dry-heat shaker for 20 to 60 minutes. (Excessive hybridization time may result in irreversible, nonspecific binding of the probe. Again, specific conditions depend upon the nature of the particular probe.)
5. Remove the membranes from the bags. Rinse them in wash solution (SSPE plus 0.1% Triton X-100) one to two times at room temperature, with moderate shaking.
6. Wash the membranes in prewarmed wash solution at a temperature that will give good stringency (5°C less than the hybridization temperature) for 10 minutes, with moderate shaking.
7. If using biotinylated psoralen probes, incubate the membranes with streptavidin HRP conjugate (Sheldon *et al.* 1986) in buffer A (PBS, 100 mM NaCl, 5% Triton X-100) for 10 minutes at room temperature, with moderate shaking. (No conjugation step is necessary for probes labeled directly with HRP.)
8. Incubate the membranes in buffer B (buffer A, 1 M urea, 1%

dextran sulfate) for 5 minutes at room temperature, with moderate shaking.

Color Development

This section is standard for either type of labeled probe and does not vary with different hybridization/wash conditions. The chromogen (TMB) produces a positive signal of a deep blue/purple color (Sheldon *et al.* 1986). Other chromogens are available.

1. Wash the membranes (from buffer B) in buffer C [100 mM sodium citrate (pH 5.0)] for 5 minutes at room temperature, with moderate shaking.
2. During the above wash step, prepare buffer D by adding 1 part of the chromogen (TMB 2 mg/ml in EtOH) to 19 parts buffer C. The TMB must be kept under light exclusion to prevent photo-oxidation.
3. Incubate the membranes in buffer D for 10 minutes at room temperature, with moderate shaking. The dish must be covered to keep out strong light.
4. During the above wash step, prepare buffer E by mixing 2000 parts buffer D to 1 part 3% hydrogen peroxide. Keep out of strong light until use.
5. Develop the membranes in buffer E at room temperature with moderate shaking (covered). Development time varies depending on the amount of bound probe, but generally the time ranges from 1 to 15 minutes.
6. When the membranes are developed (good amount of color-to-background ratio), stop the color development by washing the membranes in buffer C. Change the buffer C solution two or three times within a 1-hour wash. (The membrane absorbs much of the TMB, and unless the chromogenic substrate is thoroughly removed, it may photo-oxidize over time to darken the background. The positive signal is sufficiently stable, so it is not necessary to limit wash time or volume.)
7. Photograph the membranes for permanent record. The membranes can be stored permanently in buffer C in heat-sealed bags under light exclusion with little fading of the signal.
8. If desired, the membranes can be decolorized in 0.18% Na_2SO_3, and the probe can be removed by washing for 1 hour in 65°C

water plus 0.5% SDS. After this, the membranes can be hybridized with another probe by using the same procedure as that previously described. This is a convenient way to save time in developing the appropriate hybridization/wash conditions of each probe.

Example of Use

A good illustration of the use of this protocol is HLA-DQα typing (see Fig. 1). Allele-specific oligonucleotide (ASO) probes are used to detect sequence variation in the DQα region of the HLA gene. In Fig. 1, four probes are used to detect the four major DQα haplotypes, and one probe is used as a positive control to detect all DQα types to ensure the accuracy of the results. The photographs of the membranes have been arranged to align the analogous spots of each

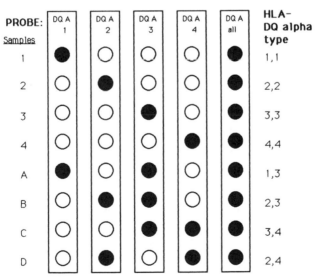

Figure 1   Illustration of HLA-DQα of PCR-amplified DNA using allele-specific oligo-nucleotide probes. Samples 1–4 are homozygous DNA-positive controls; samples A–D are heterozygous DNAs.

sample. In this format, the corresponding HLA-DQα type of each sample is easily read; for example, sample A is positive for probes DQA1, DQA3, and DQA"all" and therefore is typed as being heterozygous for DQα haplotypes 1 and 3.

ASO probes have also been routinely used in the analysis of β-globin genes, sickle cell anemia, and β-thalassemia, with applications in prenatal diagnosis and forensics (Saiki *et al.* 1986; Saiki *et al.* 1988; Bugawan *et al.* 1988).

Troubleshooting

The following is a list of potential problems that can be encountered in this system. One or more of the following suggestions may help to resolve the problem.

Faint Signal

1. Spot more PCR product (add more volume or amplify for a greater number of cycles).
2. Add a higher concentration of probe during hybridization.
3. Increase the hybridization time.
4. Decrease the stringency of hybridization/wash.
5. Prepare fresh TMB (in solution, TMB can be stored for 2 months at 4°C).

High Background Signal

1. Add a lower concentration of probe during hybridization.
2. Decrease the hybridization time.
3. Increase the stringency of hybridization/wash.
4. Decrease the time of color development.

High Background Color on Membrane

1. Decrease the hybridization time.
2. Wash the membranes longer in buffer C (excess dextran sulfate can increase membrane background).

3. Decrease the time of color development.
4. Be certain that membranes are kept under light exclusion during and after color development.

Negative Control Reaction with Positive Signal

1. PCR product may have been inadvertently spotted in the negative control reaction position. Repeat the entire detection procedure with the same PCR samples to verify the original results.
2. There may be PCR carry-over in the system. Repeat all amplifications with new preparation of reagents.

Conclusion

This procedure is simple, safe, and quick to perform, as well as straightforward to interpret. The use of allele- and sequence-specific oligonucleotide probes allows for detection of very minor genetic differences. These factors will allow this procedure to be very versatile and widely used throughout the research and clinical diagnostic communities.

Acknowledgments

I would like to acknowledge the following people for their help in preparing this chapter: T. Bugawan, R. Saiki, R. Madej, C. Gates, and P. Louie.

Literature Cited

Breslauer, K. J., R. Frank, H. Blocker, and L. A. Marky. 1986. Predicting DNA duplex stability from the base sequence. *Proc. Natl. Acad. Sci. USA* **83**:3746–3750.

Bugawan, T. L., R. K. Saiki, C. H. Levenson, R. M. Watson, and H. A. Erlich. 1988. The use of non-radioactive oligonucleotide probes to analyze enzymatically amplified DNA for prenatal diagnosis and forensic HLA typing. *Bio/technology* **6**:943–947.

Church, G. M., and W. Gilbert. 1984. Genomic sequencing. *Proc. Natl. Acad. Sci. USA* **81**:1991–1995.

Connolly, B. A. and P. Rider. 1985. Chemical synthesis of oligonucleotides containing a free sulphydryl group and subsequent attachment of thiol specific probes. *Nucleic Acids Res.* **13**:4485–4502.

Meinkoth, J., and G. Wahl. 1984. Hybridization of nucleic acids immobilized on solid supports. *Anal. Biochem.* **138**:267–284.

Saiki, R. K., T. L. Bugawan, G. T. Horn, K. B. Mullis, and H. A. Erlich. 1986. Analysis of enzymatically amplified β-globin and HLA-DQα DNA with allele-specific oligonucleotide probes. *Nature (London)* **324**:163–168.

Saiki, R. K., C.-A. Chang, C. H. Levenson, T. C. Warren, C. D. Boehm, H. H. Kazazian Jr., and H. A. Erlich. 1988. Diagnosis of sickle cell anemia and β-thalassemia with enzymatically amplified DNA and nonradioactive allele-specific oligonucleotide probes. *N. Engl. J. Med.* **319**:537–541.

Sheldon, E. L., D. E. Kellogg, R. Watson, C. H. Levenson, and H. A. Erlich. 1986. Use of nonisotopic M13 probes for genetic analysis: application to HLA class II loci. *Proc. Natl. Acad. Sci. USA* **83**:9085–9089.

16

THERMOSTABLE DNA POLYMERASES

David H. Gelfand and Thomas J. White

The availability of a thermostable DNA polymerase isolated from the thermophilic bacterial species *Thermus aquaticus* (*Taq*) has greatly simplified the polymerase chain reaction (PCR) method for DNA amplification and thus the application of PCR to molecular biology and other scientific fields (Saiki *et al.* 1988; White *et al.* 1989). *Taq* DNA polymerase can withstand repeated exposure to the high temperatures (94 to 95°C) required for strand separation; this property eliminated the need to add enzyme in each cycle (Mullis and Faloona 1987; Saiki *et al.* 1985) and simplified the instrumentation requirements for PCR (see Chapters 51 to 53 and Oste 1989).

Characterized Thermophilic Organisms

DNA polymerase activities from *Bacillus stearothermophilus* (Stenesh and Roe 1972; Kaboev *et al.* 1981), several *Thermus* species (Chien *et al.* 1976; Kaledin *et al.* 1980; Kaledin *et al.* 1981; Kaledin *et al.* 1982; Rüttimann *et al.* 1985), and several archaebacterial species (Rossi *et al.* 1986; Klimczak *et al.* 1986; Elie *et al.* 1989) have been reported and partially characterized. These organisms were grown in the laboratory at temperatures ranging from 60°C (*B. stearothermophilus*, Kaboev *et al.* 1981) to 87°C (*S. solfataricus*, Rossi *et al.* 1986). The most extensively purified archaebacterial DNA polymerase had a reported 15-minute half-life at 87°C (Elie *et al.* 1989). In addition, there are several extreme thermophilic eubacteria and archaebacteria that are capable of growth at very high temperatures (Bergquist *et al.* 1987; Kelly and Deming 1988). Some of these organisms may contain very thermostable DNA polymerases that exhibit amino acid sequence similarity to the *E. coli* DNA polymerase I gene family (Lawyer *et al.* 1989) or to the eukaryotic DNA polymerases (Bernad *et al.* 1987; Wang *et al.* 1989), or they may contain unrelated DNA polymerases. Although DNA polymerases from the former thermophilic microorganisms have been partially purified and characterized, almost all published work on the use of thermostable enzymes for PCR to date is limited to *Taq* DNA polymerase (Lawyer *et al.* 1989; Powell *et al.* 1987; Innis *et al.* 1988). Because DNA polymerases are very susceptible to proteolytic attack, care must be exercised during harvest of the cell material and during cell lysis and purification to minimize the chance of attributing different purification characteristics or enzymatic properties to enzyme fragments.

Thermus aquaticus

T. aquaticus strain YT1, a thermophile capable of growth at 70 to 75°C, was isolated from a hot spring in Yellowstone National Park (Brock and Freeze 1969). Previous workers had described DNA polymerase activities with an estimated molecular mass of 60 to 68 kDa and an inferred specific activity of 2000 to 8000 units/mg from this organism (Chien *et al.* 1976; Kaledin *et al.* 1980). Recently, Lawyer *et al.* (1989) cloned and sequenced the gene for a DNA polymerase activity from *T. aquaticus* with a specific activity of 200,000 units/mg and an inferred molecular weight of 93,910.

As observed for several DNA polymerase activities isolated from thermophilic microorganisms, 94-kDa *Taq* DNA polymerase has a relatively high temperature optimum (T_{opt}) for DNA synthesis. Depending on the nature of the DNA template, *Taq* DNA polymerase

Table 1

Half-life of *Taq* DNA Polymerase at
Elevated Temperatures

Temperature (°C)	$T_{1/2}$ (min)
92.5	130
95.0	40
97.5	5–6

has an apparent T_{opt} of 75 to 80°C with a specific activity approaching 150 nucleotides per second per enzyme molecule. Innis *et al.* (1988) reported high processivity and an extension rate of >60 nucleotides per second at 70°C and 24 nucleotides per second at 55°C with *Taq* DNA polymerase on M13 phage DNA when using a G + C-rich 30-mer primer. At lower temperatures, *Taq* DNA polymerase has extension activities of 1.5 nucleotides per second at 37°C and ~0.25 nucleotide per second at 22°C, and there is a marked attenuation in the apparent processivity. The temperature dependence of processivity could reflect an impaired ability of *Taq* DNA polymerase to extend through regions of template secondary structure or a change in the ratio of the forward rate constant to the dissociation constant. Very little DNA synthesis is seen at temperatures >90°C. DNA synthesis at such high temperatures *in vitro* must be limited by the stability of the primer template duplex.

Although *Taq* DNA polymerase has very limited activity above 90°C, the enzyme is relatively resistant to denaturation during exposure to high temperature (see Table 1 [R. Watson, unpublished]). Preliminary results indicate retention of 65% activity after a 50-cycle PCR when the upper limit temperature of the reaction mixture is 95°C for 20 seconds in each cycle.

Fidelity, Extension of Mismatched Primers, and Mutagenesis

Purified 94-kDa *Taq* DNA polymerase does not contain an inherent 3' to 5' exonuclease activity (Tindall and Kunkel 1988; S. Stoffel, in preparation). Single nucleotide incorporation/misincorporation, biochemical fidelity measurements have indicated that the ability of "nonproofreading" DNA polymerases (avian myeloblastosis virus reverse transcriptase and *D. melanogaster* DNA polymerase α) to misincorporate a deoxynucleotide triphosphate is determined critically by the concentration of that triphosphate (Mendelman *et al.* 1989). Similar data have been obtained with regard to extension of a mismatched primer/template, leading to a model of "K_m or V_{max} discrimination" that suggests how nonproofreading DNA polymerases may achieve high fidelity (Petruska *et al.* 1988). Although kinetic measurements of fidelity have yet to be made, *Taq* DNA polymerase appears to extend a mismatched primer/template significantly less efficiently than a correct primer/template [see Fig. 2 in Innis *et al.* (1988) and Table 2]. It is not known if *T. aquaticus* contains a separate 3' to 5' exonuclease activity that may be associated with the polymerase *in vivo*.

Kwok, Kellogg, Levenson, Spasic, Goda, and Sninsky (manuscript in preparation) have analyzed the effect of various primer/template mismatches on the efficiency of amplification of a region of the human immunodeficiency virus. They prepared a series of primer/template mismatches that included all possible pairwise mismatches between the 3' end of the primer and its template. This system allowed them to examine whether the effect of mismatches on PCR was symmetrical, e.g., whether a G·T (primer/template) mismatch would have the same effect as a T·G mismatch. Table 2 summarizes their results (at 0.8 mM total dNTP) for mismatches at the 3' terminus of the primer. These results are very useful in designing primers that provide for efficient amplification irrespective of mismatches, e.g., primers with 3' T's. Kwok *et al.* also provided useful information for designing primers that amplify only the desired target or a specific allele of the desired target (e.g., A·G, G·A, and C·C mis-

Table 2

Effects of Mismatches between Primer and Template on PCR Efficiency

		Primer 3'-base			
		T	C	G	A
	T	1.0	1.0	1.0	1.0
Template 3'-base	C	1.0	≤0.01	1.0	1.0
	G	1.0	1.0	1.0	≤0.01
	A	1.0	1.0	≤0.01	0.05

Relative PCR product yield after 30 cycles.

matches are strongly disfavored) and for the degree to which altering the triphosphate concentration affects the amplification of the desired target (see also Gibbs *et al.* 1989 and Ehlen and Dubeau 1989 for information on "competitive oligonucleotide priming" and "allele-specific" PCR). Kwok *et al.* also assessed the effect of mismatches 2, 3, or 4 bases internal to the 3' end of the primer with regard to reduced specificity of target amplification and the formation of artifacts.

Various groups have estimated the fidelity of *Taq* DNA polymerase. Saiki *et al.* (1988) observed a cumulative error frequency of about 0.25% (17/6692) after 30 cycles of PCR. The average mutation rate per cycle was calculated from the formula

$$\frac{2 \times \text{error frequency}}{\text{number of doublings}}$$

Thus, under the PCR conditions used to generate the 28 HLA DPβ clones (1.5 mM each dNTP, 10 mM MgCl$_2$, and 37°C annealing), the calculated mutation rate (assuming 30 doublings) was $\sim 1.7 \times 10^{-4}$ (1/6000 nucleotides polymerized per cycle). Tindall and Kunkel (1988) used several genetic fidelity assays to estimate *Taq* polymerase fidelity. They calculated a frameshift frequency of about 1/30,000 and a substitution frequency of about 1/8,000 as "single cycle" extensions on a gapped-duplex substrate. These studies employed 1 mM each dNTP and 10 mM MgCl$_2$.

More recent studies using different conditions have reported a significant increase in the fidelity of *Taq* DNA polymerase-mediated PCR. Goodenow *et al.* (1989) observed no mutations among 5400 nucleotides sequenced of 34 *env* and *gag* region clones from 30-cycle PCR-amplified HIV-1 plasmid sequences. Thus, the mutation rate was less than 1/81,000 misincorporations per nucleotide polymerized per cycle ($<1.2 \times 10^{-5}$). Finally, Fucharoen *et al.* (1989) amplified and cloned the entire β-globin genomic region. The entire sequence of the 30 cycle-amplified normal β-globin gene for five randomly picked clones was determined (14,990 nucleotides). No errors were detected. Thus, the cumulative error frequency is less than 6.6 $\times 10^{-5}$, and the average mutation rate is $\sim 5 \times 10^{-6}$ errors per nucleotide incorporated per cycle, assuming 25 cycles of doubling. In contrast to the earlier studies, both of these more recent reports used lower dNTP and Mg^{2+} ion concentrations (200 μM each dNTP and 1.5 mM MgCl$_2$) and 54 to 55°C for the annealing temperature. To the extent that *Taq* DNA polymerase exhibits "K_m discrimination" for both misinsertion and extension of a mispaired primer, minimizing anneal/extension time and dNTP and $MgCl_2$ concentrations is expected to enhance the fidelity of *Taq* DNA polymerase in the PCR.

For convenience, Saiki *et al.* (1988) assumed that the mutation rate does not change as a function of cycle number (product yield). This assumption requires validation. Amplification efficiency may be attenuated as a consequence of specific product yield (product strand renaturation imposing a requirement for strand-displacement synthesis that in turn leads to loss of processivity and to attenuation of extension rate [see Chapter 1]). Similarly, the fidelity of *Taq* DNA polymerase may decrease under strand-displacement synthesis conditions (high yield, late cycle). Finally, the calculation of the mutation rate requires knowing the exact number of product doublings, not just the number of cycles used in the PCR.

Recently, PCR has been used for random mutagenesis of a specific DNA region. Leung *et al.* (1989) have used conditions that reduce the fidelity of *Taq* DNA polymerase: high and pool-biased dNTP concentrations (1 mM each dTTP, dCTP, dGTP, 200 μM dATP), high $MgCl_2$ (6.1 mM) in the presence of $MnCl_2$ (0.5 mM), and 25 PCR cycles starting with 1 ng cloned plasmid target. These conditions resulted in a cumulative error frequency of 2% and a mutant yield (for targets over 300 bp) greater than 90%.

Template Requirements and Novel Properties

Several classes of *Taq* DNA polymerase-mediated events have required a reevaluation of the template requirements for a DNA-dependent DNA polymerase. "Primer-dimer" is a template-independent, duplex PCR product composed primarily of the primers employed in the PCR. In the major primer-dimer product, the extension added to one primer is the antiparallel complement of the other primer. Once initiated, primer-dimer products are amplified very efficiently and may become the predominant PCR product. Complementarity between the two 3' ends of a primer set has been shown to enhance primer-dimer formation (Watson 1989). In addition, low annealing temperatures and high enzyme and primer concentrations also increase the frequency of initiation of primer–dimer formation. Even in the absence of complementarity between the 3' ends of two primers, most primer sets yield primer–dimer products given a sufficient number of cycles (30 to 40), low annealing temperatures, and high enzyme and primer concentrations. *Taq* DNA polymerase has been observed to add a single nucleotide to a blunt-ended duplex DNA fragment in a nontemplate-directed manner (albeit at relatively low efficiency, see below). If a single-stranded primer is able to bind transiently to the polymerase primer binding site, and a second primer is able to bind transiently to the template site, then addition of a single nucleotide (template- or nontemplate-directed) might enhance the interaction/stability of the two primers and promote initiation of primer–dimer formation. The extraordinary ability of *Taq* DNA polymerase-mediated PCR to amplify rare targets will facilitate the detection of low-frequency events catalyzed by the polymerase.

Clark (1988) has described conditions under which *Taq* DNA polymerase is able to add a single nontemplate-directed nucleotide to a blunt-ended duplex DNA fragment at low efficiency. Although any of the four deoxyribonucleotide triphosphates could serve as a substrate, dATP was added more efficiently than any of the other three deoxyribonucleotide triphosphates were added. Further, the property of preferential dATP addition was observed for several other nonproofreading DNA polymerases. These results extend similar observations for the Klenow fragment of *E. coli* DNA polymerase I under conditions of 3' to 5' exonuclease inhibition or elimination (Clark *et al.* 1987; Beardsley *et al.* 1988). The ability of *Taq* DNA polymerase to add a single nontemplate-directed nucleotide to a blunt-ended duplex DNA fragment has significant implications for direct cloning of PCR products. Recently, Denney and Weissman (manuscript submitted) have shown that treating PCR products with Klenow fragment in the presence of dNTPs improves significantly the efficiency of blunt-end ligation. In contrast, similar treatment with T7 DNA polymerase, Sequenase 2.0 [genetically modified to eliminate the 3' to 5' exonuclease activity (Tabor and Richardson 1989)] failed to improve blunt-end ligation efficiency of PCR product. These results are consistent with a model in which PCR products may contain at least one 3' nontemplate-directed nucleotide extension. Finally, the DNA sequences of PCR products derived from the amplification of ancient (damaged) DNA samples are consistent with the hypothesis that *Taq* DNA polymerase may add dATP preferentially at an abasic site in a template strand (D. Irwin, personal communication; see also Mole *et al.* 1989).

Ionic Requirements, Solvents, and Inhibitors

Taq DNA polymerase activity is sensitive to the concentration of magnesium ion as well as to the nature and concentration of monovalent ions. With minimally activated salmon sperm DNA as template in a standard 10-minute assay (Lawyer *et al.* 1989), 2.0 mM MgCl$_2$ maximally stimulates *Taq* polymerase activity at 0.7 to 0.8 mM total dNTP. Higher concentrations of Mg^{2+} are inhibitory, with 40 to 50% inhibition at 10 mM MgCl$_2$. As deoxynucleotide triphosphates can bind Mg^{2+}, the exact magnesium concentration that is required to activate the enzyme maximally is dependent on the dNTP concentration. In addition, the synthesis rate of *Taq* polymerase decreases by as much as 20 to 30% as the total dNTP concentration is increased to 4 to 6 mM, despite optimization of magnesium concentration for each dNTP concentration. This phenomenon appears to represent substrate inhibition. Low, balanced concentrations of dNTPs have been observed to give satisfactory yields of PCR product, to result frequently in improved specificity, to facilitate labeling of PCR products with radioactive or biotinylated precursors, and to contribute to increased fidelity of *Taq* polymerase. In a 100-μl PCR with 40 μM each dNTP, there are sufficient nucleotide triphosphates to yield 2.6 μg of DNA when only half of the available dNTPs are incorporated into DNA. It is likely that very low dNTP concentrations may adversely affect the processivity of *Taq* DNA polymerase. Furthermore, the concentrations of free and enzyme-bound magnesium may affect the processivity of *Taq* polymerase, as has been inferred for calf thymus DNA polymerases α and δ (Sabatino *et al.* 1988).

Modest concentrations of KCl stimulate the synthesis rate of *Taq* polymerase by 50 to 60%, with an apparent optimum at 50 mM. Higher KCl concentrations begin to inhibit activity, and no significant activity is observed in a DNA sequencing reaction at $\geq$75 mM KCl (Innis *et al.* 1988) or in a 10-minute incorporation assay at >200 mM KCl.

The addition of either 50 mM ammonium chloride, ammonium acetate, or sodium chloride to a *Taq* DNA polymerase activity assay results in mild inhibition, no effect, or slight stimulation (25 to 30%), respectively.

Low concentrations of urea, DMSO, DMF, or formamide have no effect on the incorporation activity of *Taq* polymerase. (Table 3, Gel-

Table 3

Inhibitor Effects on *Taq* Pol I Activity

Inhibitor	Concentration	Activity (percent)[a]
Ethanol	≤3%	100
	10%	110
Urea	≤0.5 *M*	100
	1.0 *M*	118
	1.5 *M*	107
	2.0 *M*	82
DMSO	≤1%	100
	10%	53
	20%	11
DMF	≤5%	100
	10%	82
	20%	17
Formamide	≤10%	100
	15%	86
	20%	39
SDS	0.001%	105
	0.01%	10
	0.1%	0.1

[a]dNTP incorporated activity at 70°C with Salmon Sperm DNA/10 min.

Source: Gelfand, D. H. 1989. *Taq* DNA Polymerase. In *PCR technology: principles and applications for DNA amplification* (ed. H. A. Erlich), p. 17–22. Stockton Press, New York.

fand 1989). The presence of 10% DMSO (used previously in Klenow-mediated PCR, Scharf *et al.* 1986) inhibits DNA synthesis by 50% in a 70°C *Taq* polymerase activity assay. While several investigators have observed that inclusion of 10% DMSO facilitates certain PCR assays, it is not clear which parameters of PCR are affected. The presence of DMSO may affect the T_m of the primers, the thermal activity profile of *Taq* DNA polymerase, and/or the degree of product strand separation achieved at a particular "denaturation" or upper-limit temperature. Curiously, 10% ethanol fails to inhibit *Taq* activity, and 1.0 *M* urea stimulates *Taq* activity. These effects on incorporation activity may not reflect the degree to which these agents affect the PCR. For example urea at 0.5 *M* completely inhibits a PCR assay (C.-A. Chang, personal communication). Finally, the inhibitory effects of low concentrations of SDS can be completely reversed by high concentrations of certain nonionic detergents (e.g., Tween 20 and Nonidet P40 [NP40]). Thus 0.5% each Tween 20/NP40 instantaneously reverses the inhibitory effects of 0.01% SDS, and 0.1% each Tween 20/NP40 completely reverses the inhibitory effects of 0.01% SDS in the presence of DNA and Mg^{2+} (no dNTP) after 40 minutes at 37°C (S. Stoffel, in preparation).

Endogenous DNA

Preparations of *E. coli* DNA polymerase I and native and cloned *Taq* DNA polymerase obtained from various suppliers may contain small amounts of genomic DNA from the production microorganisms. In experiments that target bacterial genes for amplification, or that utilize "universal" primers based on sequences that may be invariant among homologous genes from eukaryotes, prokaryotes, and organelles, one occasionally observes an amplified DNA fragment in the "no DNA" controls. This observation may be due to product carryover (See Chapter 17) or amplification of the target fragment from residual *T. aquaticus* or *E. coli* DNA present in the DNA polymerase used in the PCR assays. This phenomenon is more apparent at 30 to 35 cycles and often can be circumvented by increasing the initial target concentration, by using native *Taq* DNA polymerase for PCR assays that target *E. coli* genes, or by designing primers that contain mismatches with the genes from *E. coli* or *T. aquaticus*.

Acknowledgments

We thank Chu-An Chang, David Irwin, Shirley Kwok, Susanne Stoffel, and Robert Watson for permission to cite their data prior to publication; Corey Levenson, Lauri Goda, and Dragan Spasic for the synthesis of many oligonucleotides; Dan Denney, David Goeddel, Maureen Goodenow, Shirley Kwok, and Lynn Mendelman for providing preprints prior to publication; Will Bloch and members of the Cetus PCR group for stimulating discussions, thoughtful advice, and suggestions. We also thank Jeff Price for his interest, patience, wisdom, and support of these efforts.

Literature Cited

Beardsley, G. P., T. Mikita, A. B. Kremer, and J. M. Clark. 1988. Oligonucleotides with site-specific structural anomalies as probes of mutagenesis mechanisms and DNA polymerase function. In *DNA replication and mutagenesis* (ed. R. E. Moses and W. C. Summers), p. 208–219. American Society for Microbiology, Washington, D.C.

Bergquist, P. L., D. R. Love, J. E. Croft, M. D. Streiff, R. M. Daniel, and W. H. Morgan. 1987. Genetics and potential biotechnological applications of thermophilic and extremely thermophilic microorganisms. *Biotech. & Genet. Eng. Rev.* **5**:199–244.

Bernad, A., A. Zaballos, M. Salas, and L. Blanco. 1987. Structural and functional relationships between prokaryotic and eukaryotic DNA polymerases. *EMBO J.* **6**:4219–4225.

Brock, T. D., and H. Freeze. 1969. *Thermus aquaticus* gen. n. and sp. n., a nonsporulating extreme thermophile. *J. Bacteriol.* **98**:289–297.

Chien, A., D. B. Edgar, and J. M. Trela. 1976. Deoxyribonucleic acid polymerase from the extreme thermophile *Thermus aquaticus. J. Bacteriol.* **127**:1550–1557.

Clark, J. M., C. M. Joyce, and G. P. Beardsley. 1987. Novel blunt-end addition reactions catalyzed by DNA polymerase I of *Escherichia coli. J. Mol. Biol.* **198**:123–127.

Clark, J. M. 1988. Novel non-templated nucleotide addition reactions catalyzed by procaryotic and eucaryotic DNA polymerases. *Nucleic Acids Res.* **16**:9677–9686.

Ehlen, T., and L. Dubeau. 1989. Detection of *ras* point mutations by polymerase chain reaction using mutation-specific inosine-containing oligonucleotide primers. *Biochem. Biophys. Res. Commun.* **160**:441–447.

Elie, C., A. M. DeRecondo, and P. Forterre. 1989. Thermostable DNA polymerase from the archaebacterium *Sulfolobus acidocaldarius*: purification, characterization and immunological properties. *Eur. J. Biochem.* **178**:619–626.

Fucharoen, S., G. Fucharoen, P. Fucharoen, and Y. Fukumaki. 1989. A novel ochre mutation in the β-thalassemia gene of a Thai. *J. Biol. Chem.* **264**:7780–7783.

Gelfand, D. H. 1989. *Taq* DNA polymerase. In *PCR technology: principles and applications for DNA amplification* (ed. H. A. Erlich), p. 17–22. Stockton Press, New York.

Gibbs, R., P.-N. Nguyen, and C. T. Caskey. 1989. Detection of single DNA base differences by competitive oligonucleotide priming. *Nucleic Acids Res.* **17**:2437–2448.

Goodenow, M., T. Huet, W. Saurin, S. Kwok, J. Sninsky, and S. Wain-Hobson. 1989. HIV-1 isolates are rapidly evolving quasispecies: evidence for viral mixtures and preferred nucleotide substitutions. *JAIDS*, **2**:344–352.

Innis, M. A., K. B. Myambo, D. H. Gelfand, and M. A. D. Brow. 1988. DNA sequencing with *Thermus aquaticus* DNA polymerase and direct sequencing of polymerase chain reaction-amplified DNA. *Proc. Natl. Acad. Sci. USA* **85**:9436–9440.

Kaboev, O. K., L. A. Luchkina, A. T. Akhmedov, and M. L. Bekker. 1981. Purification and properties of deoxyribonucleic acid polymerase from *Bacillus stearothermophilus. J. Bacteriol.* **145**:21–26.

Kaledin, A. S., A. G. Slyusarenko, and S. I. Gorodetskii. 1980. Isolation and properties of DNA polymerase from extremely thermophilic bacterium *Thermus aquaticus. Biokhimiya* **45**:644–651.

Kaledin, A. S., A. G. Slyusarenko, and S. I. Gorodetskii. 1981. Isolation and properties of DNA polymerase from the extremely thermophilic bacterium *Thermus flavus. Biokhimiya* **46**:1576–1584.

Kaledin, A. S., A. G. Slyusarenko, and S. I. Gorodetskii. 1982. Isolation and properties of DNA polymerase from the extremely thermophilic bacterium *Thermus ruber. Biokhimiya* **47**:1785–1791.

Kelly, R. M., and J. W. Deming. 1988. Extremely thermophilic archaebacteria: biological and engineering considerations. *Biotechnol. Prog.* **4**:47–62.

Klimczak, L. J., F. Grummt, and K. J. Burger. 1986. Purification and characterization of DNA polymerase from the archaebacterium *Methanobacterium thermoauto-trophicum*. *Biochemistry* **25**:4850–4855.

Lawyer, F. C., S. Stoffel, R. K. Saiki, K. Myambo, R. Drummond, and D. H. Gelfand. 1989. Isolation, characterization, and expression in *Escherichia coli* of the DNA polymerase gene from *Thermus aquaticus*. *J. Biol. Chem.* **264**:6427–6437.

Leung, D. W., E. Chen, and D. V. Goeddel. 1989. A method for random mutagenesis of a defined DNA segment using a modified polymerase chain reaction. *Technique* **1**:11–15.

Mendelman, L. V., M. S. Boosalis, J. Petruska, and M. F. Goodman. 1989. Nearest neighbor influences on DNA polymerase insertion fidelity. *J. Biol. Chem.* **264**: 14415–14423.

Mole, S. E., R. D. Iggo, and D. P. Lane. 1989. Using the polymerase chain reaction to modify expression plasmids for epitope mapping. *Nucleic Acids Res.* **17**:3319.

Mullis, K. B., and F. A. Faloona. 1987. Specific synthesis of DNA *in vitro* via a polymerase catalyzed chain reaction. *Methods Enzymol.* **155**:335–350.

Oste, C. 1989. PCR automation. In *PCR technology: principles and applications for DNA amplification* (ed. H. A. Erlich), p. 23–30. Stockton Press, New York.

Petruska, J., M. F. Goodman, M. S. Boosalis, L. C. Sowers, C. Cheong, and I. Tinoco, Jr. 1988. Comparison between DNA melting thermodynamics and DNA polymerase fidelity. *Proc. Natl. Acad. Sci. USA* **85**:6252–6256.

Powell, L. M., S. C. Wallis, R. J. Pease, Y. H. Edwards, T. J. Knott, and J. Scott. 1987. A novel form of tissue-specific RNA processing produces apolipoprotein-B48 in intestine. *Cell* **50**:831–840.

Rossi, M., R. Rella, M. Pensa, S. Bartolucci, M. DeRosa, A. Gambacorta, C. A. Raia, and N. D.-A. Orabona. 1986. Structure and properties of a thermophilic and thermostable DNA polymerase isolated from *Sulfolobus solfataricus*. *System. Appl. Microbiol.* **7**:337–341.

Rüttimann, O., M. Cotorás, J. Zaldívar, and R. Vicuña. 1985. DNA polymerases from the extremely thermophilic bacterium *Thermus thermophilus* HB-8. *Eur. J. Biochem.* **149**:41–46.

Sabatino, R. D., T. W. Myers, R. A. Bambara, O. Kwon-Shin, R. L. Marraccino, and P. H. Frickey. 1988. Calf thymus DNA polymerases α and δ are capable of highly processive DNA synthesis. *Biochemistry* **27**:2998–3004.

Saiki, R. K., S. Scharf, F. Faloona, K. B. Mullis, G. T. Horn, H. A. Erlich, and N. Arnheim. 1985. Enzymatic amplification of β-globin genomic sequences and restriction site analysis for diagnosis of sickle cell anemia. *Science* **230**:1350–1354.

Saiki, R. K., D. H. Gelfand, S. Stoffel, S. J. Scharf, R. Higuchi, G. T. Horn, K. B. Mullis, and H. A. Erlich. 1988. Primer-directed enzymatic amplification of DNA with a thermostable DNA polymerase. *Science* **239**:487–491.

Scharf, S. J., G. T. Horn, and H. A. Erlich. 1986. Direct cloning and sequence analysis of enzymatically amplified genomic sequences. *Science* **233**:1076–1078.

Stenesh, J., and B. A. Roe. 1972. DNA polymerase from mesophilic and thermophilic bacteria I. Purification and properties of DNA polymerase from *Bacillus licheniformis* and *Bacillus stearothermophilus*. *Biochim. Biophys. Acta* **272**:156–166.

Tabor, S., and C. C. Richardson. 1989. Selective inactivation of the exonuclease activity of bacteriophage T7 DNA polymerase by *in vitro* mutagenesis. *J. Biol. Chem.* **264**:6447–6458.

Tindall, K. R., and T. A. Kunkel. 1988. Fidelity of DNA synthesis by the *Thermus aquaticus* DNA polymerase. *Biochemistry* **27**:6008–6013.

Wang, T. S-F., S. W. Wong, and D. Korn. 1989. Human DNA polymerase α: predicted functional domains and relationships with viral DNA polymerases. *FASEB J.* **3**:14–21.

Watson, R. 1989. The formation of primer artifacts in polymerase chain reactions. *Amplifications* **2**:5–6.

White, T. J., N. Arnheim, and H. A. Erlich. 1989. The polymerase chain reaction. *Trends in Genetics* **5**:185–189.

17

PROCEDURES TO MINIMIZE PCR-PRODUCT CARRY-OVER

Shirley Kwok

The ability of PCR to produce large numbers of copies of a sequence from minute quantities of DNA necessitates that extreme care be taken to avoid false positives. Although false positives can result from sample-to-sample contamination, a more serious source of false positives is the carry-over of DNA from a previous amplification of the same target. Because of the large numbers of copies of amplified sequences, carry-over of even minute quantities of a PCR sample can lead to serious contamination problems. The following is a list of procedures that will help to minimize the carry-over of amplified DNA.

Physical Separation of Pre- and Post-PCR Amplifications

To prevent carry-over of amplified DNA sequences, reactions should be set up in a separate room or containment unit such as a biosafety cabinet. A separate set of supplies and pipetting devices should be

PCR Protocols: A Guide to Methods and Applications

dedicated for the specific use of setting up reactions. Care must be taken to insure that amplified DNA is not brought into this area. Reagents and supplies should be taken directly from storage cabinets and must never be taken from an area where PCR analyses are being performed. Similarly, devices such as pipettors should never be taken into the containment area after use on amplified material.

Aliquot Reagents

All reagents used in the PCR must be prepared, aliquotted, and stored in an area that is free of PCR-amplified product. Similarly, oligonucleotides used for amplification should be synthesized and purified in a PCR-product-free environment. Reagents should be aliquotted to minimize the number of repeated samplings. It is advisable to record the lot(s) of reagents used so that if carry-over occurs, it can be more easily traced.

Positive Displacement Pipettes

Contamination of pipetting devices can result in cross-contamination of samples. For example, the barrels of pipetting devices are often contaminated with radioisotopes as a result of aerosolization. To eliminate cross-contamination of samples by pipetting devices, positive-displacement pipettes are recommended. Pipettors such as those manufactured by Rainin (Microman) have both disposable tips and plungers. The units are completely self-contained.

Meticulous Laboratory Techniques

Although carry-over of amplified sequences contributes to the majority of the false positives, cross-contamination between samples can also be a factor. Consequently, precautions must be taken not only in setting up the amplification reactions but in all aspects of sample handling, from sample collection to sample extraction. The following are additional precautions that should be taken:

1. Change gloves frequently.
2. Quick spin tubes before opening them.
3. Uncap and close tubes carefully to prevent aerosols.
4. Minimize sample handling.
5. Add nonsample components (mineral oil, dNTPs, primers, buffer, and enzyme) to the amplification reactions before the addition of sample DNA. Cap each tube after the addition of DNA before proceeding to the next sample.

Judicious Selection of Controls

First, for use as a positive control, select a sample that amplifies weakly but consistently. The use of strong positives will result in the unnecessary generation of a large amount of amplified sequences. If plasmid DNA containing the target sequence is used as a positive control, it should be substantially diluted. Depending on the detection system used, as few as 100 copies of target will suffice as a positive control. Second, use well-characterized negative controls. The extreme sensitivity of PCR may enable the detection of nucleic acid sequence from a sample that is negative by all other criteria. Third, include multiple reagent controls with each amplification. Because the presence of a small number of molecules of PCR product in the reagents may lead to sporadic positive results, it is important to perform multiple reagent controls. The reagent controls should contain all the necessary components for PCR but without the addition of template DNA. This system has proved to be extremely sensitive in detecting the presence of contaminants, as the absence of exogenous DNA enables the efficient amplification of just a few molecules of contaminating sequence.

Although amplified products are most problematic, other potential sources of contamination/carry-over need to be considered, especially when additional manipulations of the amplified DNA are performed. The cloning of amplified product is a case in point. Often, the amount of target generated from an amplification is insufficient for direct cloning and requires re-amplification of the target. To minimize re-amplification of nonspecific products, the band of interest is first separated on a gel, excised, eluted in a buffer, and used to re-seed a subsequent amplification. Each of these additional steps can potentially result in cross-contamination and thereby jeopardize the authenticity of the product. Precautions must be taken to minimize contamination. For example, gel apparatus and combs can be soaked in 1 N HCl to depurinate any residual DNA. A new razor blade or similar device should be used to excise each band of interest. Because the surfaces of UV transilluminators are potentially contaminated, a sheet of plastic wrap should be used to physically separate the gel from the surface of the UV transilluminator. Positive controls that have been amplified with the same primer pair as that of the sample of interest should not be electrophoresed on the same preparative gel.

The list below highlights other potential sources of contamination:

1. Plasmid or phage DNA containing target sequence
2. Purified restriction fragment of target sequence
3. Dot blot apparatus
4. Microtome blades
5. Centrifuges
6. Speed Vacs/vacuum bottles
7. Dry ice/ethanol baths

Other sources of contamination as well as precautionary procedures (the preparation of samples for PCR must be handled with similar care) will most certainly be identified, but the suggestions here will serve as a guide in implementing procedures that will help minimize if not eradicate carry-over.

18

SAMPLE PREPARATION FROM BLOOD, CELLS, AND OTHER FLUIDS

Ernest S. Kawasaki

There are a large number of different protocols for the isolation of nucleic acids. Most of the methods are geared toward the isolation of highly purified samples, and DNA or RNA obtained by many of these procedures is suitable for use in PCR amplifications. Several good protocols have been described in detail (Davis *et al*. 1986; Ausubel *et al*. 1987; Berger and Kimmel 1987). In this chapter I describe a few rapid methods that do not involve extensive purification of the nucleic acids yet yield preparations that can be used for amplification of DNA and/or RNA sequences.

Protocols

Use autoclaved or sterilized tubes, pipettes, solutions, etc., whenever possible. Wear gloves to prevent contamination from nucleases and extraneous (your own) nucleic acids. All the procedures must be done with techniques that avoid possible carry-over from other

PCR Protocols: A Guide to Methods and Applications

samples or laboratory equipment and supplies. Remember to include proper precautions and established guidelines when pathogenic organisms are involved.

Reagents for DNA and RNA Isolation

Proteinase K: Dissolved at 20 mg/ml in 10 mM Tris-Cl (pH 7.5). Obtained from Boehringer-Mannheim or Bethesda Research Labs.

Fungi-Bact: Obtained from Irvine-Scientific.

TE: 10 mM Tris-Cl, 1 mM EDTA (pH 7.5 or 8.0).

PBS: Phosphate buffered saline.

PCR buffer without gelatin or bovine serum albumin: 50 mM KCl, 10–20 mM Tris-Cl, 2.5 mM MgCl$_2$ (pH 8.3).

Detergents: Laureth 12 (Mazer Chemicals, Gurnee, Illinois), NP-40, and Tween 20, highest purity obtained from any reputable source.

K buffer: PCR buffer without gelatin or bovine serum albumin, containing 1% Laureth 12 or 0.5% Tween 20, and 100 μg/ml of *fresh* Proteinase K.

IHB: Isotonic high-pH buffer with 140 mM NaCl, 10 mM Tris-Cl, 1.5 mM MgCl$_2$ (pH 8.0).

DEP: Diethylpyrocarbonate obtained from Sigma.

Ethanol: Reagent grade at 95 to 100%.

Preparation of DNA from Clinical Swabs

The following protocol was designed to obtain samples from genital areas but can be used in any situation where cells are easily collected by soft abrasion. This method is also excellent for use with tissue culture cells. Cervical, vulvar, or penile samples are collected with a cotton swab (pre-wet with saline) or cytobrush. The specimen is placed in a 10- to 15-ml conical tube containing 2 ml PBS plus a 2× concentration of Fungi-Bact. If the sample is to be processed within 24 hours, it can be kept at room temperature. Refrigeration at 4°C is recommended for longer storage times.

Pellet the cells by centrifuging for 5 minutes at 2000 to 3000 rpm in a tabletop centrifuge and carefully remove the supernatant with an aspirator. If any blood is present, resuspend the cell pellet in 1 ml of TE and transfer to a 1.5-ml microfuge tube. Pellet the intact cells (or nuclei, depending on the cell type involved) by centrifuging 10 seconds. Remove and discard the supernatant, which contains the lysed RBC components. Repeat the rinses, if necessary, until the pellet is clean.

Resuspend the pellet in 50 to 300 μl K buffer, depending on the size of the initial cell pellet (~4 volumes). The resuspension volume should make the concentration about 100 to 1000 cell equivalents per μl. Transfer to a 1.5-ml microfuge tube and incubate at 55°C for 1 hour. Heat at 95°C for 10 minutes to inactivate the protease, and store the samples at −20°C. For PCR, thaw and vortex the sample and use 5 to 10 μl in a 100-μl reaction.

Preparation of DNA from White Blood Cells or Whole Blood

Fractionated Cells: Mononuclear cells from 1 to 2 ml of blood are isolated by density gradient centrifugation (such as Ficoll-Hypaque) and washed twice by centrifugation in 10 ml of PBS. Count the cells in a hemacytometer and resuspend an aliquot in 100 μl of K buffer at about 5000 cells per μl. Incubate 45 minutes at 56°C to digest the cells, then 10 minutes at 95°C to inactivate the protease. Use 10 μl in a 100-μl PCR.

Whole Blood: Mix 50 μl of whole blood with 0.5 ml of TE in a 1.5-ml microfuge tube, and then spin for 10 seconds at 13,000 × g. Resuspend the pellet in 0.5 ml of TE by vortexing and pellet as before. Repeat this procedure twice more and then resuspend final pellet in 100 μl of K buffer and incubate as above. Use 10 μl in a PCR. "Normal" blood contains on the order of 5000 white cells per μl, so 50 μl of whole blood will contain about 250,000 or more nucleated cells.

Preparation of RNA from Blood Cells

Fractionate 1 to 2 ml of blood by any standard method such as with Ficoll-Hypaque. Place mononuclear cells in a 2-ml screw-cap microfuge tube, dilute with PBS, and pellet cells at 500 × g for 5 minutes. Wash cells one time with PBS, and then resuspend in 200 to 400 μl of ice-cold IHB containing 0.5% NP-40 and 0.01% *fresh* DEP. A 10% solution of DEP is made in ethanol just before use and diluted a thousandfold in the IHB lysis buffer. The cells are vortexed and the nuclei are pelleted for 10 seconds in a microfuge. The nuclei may be saved for DNA preparation if desired but should be washed to remove the DEP. Transfer the postnuclear supernatant to another microfuge tube and place in a humidified 37°C water bath for 20 minutes, and then a humidified 90°C water bath for 10 minutes. Keep the microfuge caps loose to allow degraded DEP gases to escape. Pellet any precipitate that forms and transfer the supernatant to another tube. Use 5 to 10 μl in reverse transcriptase/PCRs as described in Chapter 3. If your positive controls do not work by this method, try reheating the sample at 90°C for 5 minutes. Sometimes there is residual DEP left that will inhibit any enzymatic reaction. This method also works with tissue culture cells. Use five to ten times fewer tissue culture cells than white blood cells because the tissue culture cells have much more RNA per cell cytoplasm than do unstimulated white blood cells.

Preparation of DNA or RNA from Viral Pellets

Pellet 5 ml of cells from tissue culture supernatants or blood at 500 × g for 5 minutes. Remove supernatant and centrifuge at 10,000 × g for 10 minutes to remove large particulates. Carefully remove this supernatant and pellet viruses by centrifugation at 50,000 rpm for 45 minutes in an SW 50.1 rotor. Use PBS for filling or balancing the tubes if necessary. Remove and discard the supernatant, and dissolve the viral pellet in 100 to 500 μl of K buffer or 100 to 500 μl of TE containing 1% NP-40 and 100 μg/ml of Proteinase K. Transfer the dissolved pellet to a microfuge tube and incubate at 55°C for 30 to 60 minutes and then heat kill protease at 95°C for 10 minutes. Chill samples and pellet any debris.

For DNA viruses use 5 to 10 μl in a 100-μl PCR; for RNA viruses use a similar amount for cDNA synthesis in RNA/PCR amplifications.

Preservation of Samples When You Are "In the Field"

If you happen to be a field biologist and molecular biologist at the same time, you may want to have an easy and good way to preserve biological samples for subsequent DNA or RNA analysis. Since backpacking a Revco is out of the question and ice tends to melt in one day, another method is highly desirable. One of the simplest and perhaps best methods is also one of the oldest, namely, placing your sample in 50% ethanol. Although DNA or RNA can be extracted from formalin-fixed tissues (Goelz et al. 1985; Rupp and Lacker 1988), the least degraded and the highest yield of DNA can be isolated from ethanol-fixed samples (Smith et al. 1987; Bramwell and Burns 1988). DNA or RNA suitable for hybridizations and enzymatic reactions can also be isolated from methanol-fixed cells (Fey et al. 1987; Kawasaki, unpublished results). The following method is adapted from that found in Smith et al. (1987) and Bramwell and Burns (1988). Rinse tissue sample once in saline and cut into small pieces (<1 cm on a side). Resuspend in an equal volume of saline (exact volume not crucial), and add slowly an equal volume of ethanol with gentle shaking. The volume of ethanol added is equal to the volume of the saline plus tissue sample so that the final concentration is about 50%. For the detailed method of isolating DNA from the ethanol-fixed samples, see Smith et al. (1987). The fixed cells are suitable for DNA extraction for many days when left at room temperature (Bramwell and Burns 1988) and more than 6 years at 4°C (Smith et al. 1987). Thus, this protocol is ideal for sample preservation by the field molecular biologist. If one is studying animal blood samples, heparinized or citrated blood can be used for DNA analysis after 1 or 2 days at room temperature, maybe longer. Fixing the blood as a cell suspension as previously described may work, but we have not seen its use in the literature. A good way to preserve blood samples for DNA analysis is to simply make air-dried blood smears on slides (Fey et al. 1987). Ethanol or methanol fixing the cells on the slides also works but may not be necessary. DNA from 13-year-old air-dried bone marrow slides has been used successfully in Southern blots (Fey et al. 1987). Whole blood has fewer nucleated cells than bone marrow does, so you may want to make several blood smears per "volunteer." If the blood contains about 5000 nucleated cells per μl, a 100-μl blood smear will contain about 500,000 cells, or ~3 μg of nuclear DNA for mammalian species.

Discussion and Summary

We have presented methods that come under the "quick and easy" label for the isolation of DNA and RNA suitable for use in PCRs. We have purposely avoided describing protocols that involve CsCl centrifugations and/or phenol extractions, since these methods are already widely described in the literature and in laboratory manuals. Our simplified methods involve the use of nonionic detergents and Proteinase K to solubilize DNA from cells in PCR buffer or TE. Laureth 12 and Tween 20 at 0.5% are compatible with *Taq* polymerase. Reverse transcriptase works well in the presence of up to 1% NP-40, but we have not tested it against Laureth 12 or Tween 20. NP-40 should be used with caution with *Taq* polymerase, because it may be inhibitory even at a level of 0.1%. Unfortunately, the commonly used ionic detergent sodium dodecyl sulfate (SDS) cannot be used with either polymerase at any useful level, although SDS is an excellent detergent for use with Proteinase K. Proteinase K is a very good protease for digesting nuclei or whole cells to release DNA or RNA into a form readily accessible to the polymerases. It has the advantage of being relatively heat stable in the mid-temperature range (50 to 60°C), yet can be thermally inactivated easily at 95°C.

Lysis of cells in the presence of DEP prevents the degradation of DNA and RNA by nucleases, but one must be very careful in its use. Higher levels of DEP can easily destroy the template activity of both the DNA and RNA. It is possible to use Proteinase K to digest cytoplasmic supernatants from some tissue culture lines because these cells have low ribonuclease activity. However, white blood cells and certain tissues such as pancreas contain too much RNase for the protease to inactivate before the nucleases destroy most of the RNA templates.

As described, DNA can be isolated from tissues or cells that have been stored for long periods in 50% ethanol or as dried smears on slides. This fact is especially helpful for those who do extensive collecting in the field. This will also prove useful to those designing long-term medical studies of a molecular biological nature, since one can depend on the reliability of obtaining good DNA from biopsies that are several years old.

Finally, DNA isolated by the detergent/protease methods works about as well as purified samples do, so the sensitivity of detection of PCR-amplified products should be sufficient for most purposes. For RNA the digestion protocol works very well on isolated viral particles, with the detection limit being about 100 to 1000 viruses. The DEP protocol is more problematical and probably should be used mainly in cases where the target sequence is not too low in abundance and the highest sensitivity is not required. In general, one should always have a purified DNA or RNA sample to compare to the impure nucleic acids, especially when sequences are being amplified for the first time.

Acknowledgments

We wish to thank M. M. Manos and R. Higuchi for providing methods for DNA isolations.

Literature Cited

Ausubel, F. M., R. Brent, R. F. Kingston, D. D. Moore, J. G. Seidman, J. A. Smith, and K. Struhl (ed.). 1987. *Current protocols in molecular biology*. Greene Publishing Associates and Wiley-Interscience, New York.

Berger, S. L., and A. R. Kimmel. 1987. Guide to molecular cloning techniques. *Methods Enzymol.* **152**:215–304.

Bramwell, N. H., and B. F. Burns. 1988. The effects of fixative type fixation time on the quantity and quality of extractable DNA for hybridization studies on lymphoid tissue. *Exp. Hematol.* **16**:730–732.

Davis, L. G., M. D. Dibner, and J. F. Battey. 1986. *Basic methods in molecular biology*. Elsevier Science Publishing Co., Inc., New York.

Fey, M. F., S. P. Pilkington, C. Summers, and J. S. Wainscoat. 1987. Molecular diagnosis of haematological disorders using DNA from stored bone marrow slides. *Brit. J. Haematol.* **67**:489–492.

Goelz, S. E., S. R. Hamilton, and B. Vogelstein. 1985. Purification of DNA from formaldehyde fixed and paraffin embedded human tissue. *Biochem. Biophys. Res. Commun.* **130**:118–126.

Rupp, G. M., and J. Locker. 1988. Purification and analysis of RNA from paraffin-embedded tissues. *Biotechniques* **6**:56–60.

Smith, L. J., R. C. Braylan, J. E. Nutkis, K. B. Edmundson, J. R. Downing, and E. K. Wakeland. 1987. Extraction of cellular DNA from human cells and tissues fixed in ethanol. *Anal. Biochem.* **160**:135–138.

SAMPLE PREPARATION FROM PARAFFIN-EMBEDDED TISSUES

Deann K. Wright and M. Michele Manos

The ability to study preserved tissues at the molecular level makes possible retrospective studies on large numbers of patients and may permit the tracking, over long periods of time, of genetic changes or infectious agents that are associated with diseases.

The most common method for preserving human tissue is fixation in formalin followed by paraffin embedding. Several research groups have performed Southern blot analyses on DNA extracted from such specimens (Goelz et al. 1985; Dubeau et al. 1986). While this approach can provide invaluable information, it is extremely laborious and therefore not suitable for the examination of large numbers of samples. DNA from old or improperly fixed samples is often degraded and cannot be analyzed by Southern blotting.

The use of PCR to examine the DNA in fixed, paraffin-embedded tissues provides a relatively simple and extremely sensitive method for examining large numbers of samples. (Impraim et al. 1987; Shibata et al. 1988a) Additionally, PCR analysis of archival tissues can be accomplished with 5- to 10-μm sections, whereas Southern blot analysis requires larger amounts of tissue. PCR does not require high-molecular-weight DNA and therefore allows the analysis of deteriorated specimens that may be inappropriate for Southern blot analyses. This includes tissue specimens ranging from several years old to over 40 years old. (Shibata *et al.* 1988b; Impraim *et al.* 1987).

Shibata *et al.* (1988a) have reported a simple method for the preparation of paraffin-embedded tissue sections for PCR. The method allowed amplification of 100-bp products from sections that were deparaffinized and boiled before PCR. We have developed improved methods that will have a broader spectrum of uses (Manos *et al.* 1989 and this chapter). The procedures provide increased product yield and have been used to amplify products of over 800 bp from intact samples.

Protocols

Reagents

Tissue sections (prepared as follows)
Octane or xylene
100% ethanol
HPLC-grade acetone (optional)
Proteinase K (20 mg/ml stock solution)
Digestion buffer [50 mM Tris (pH 8.5), 1 mM EDTA, plus 1% Laureth 12 (Mazer Chemicals, Gurnee, IL) or 0.5% Tween 20]

Tissue Section Preparation

Prepare sections (5 to 10 μm wide) from blocks of fixed (preferably with buffered formalin), embedded tissue. If possible, trim excess paraffin from the block before slicing. Cut the sections and remove them from the microtome dry (if the sections are hydrated, the paraffin is extracted less efficiently). Handle the sections with clean tweezers or toothpicks and place (one per tube) into 1.5-ml microfuge tubes. To avoid cross-contamination of samples, the microtome blade, tweezers, and anything else in the cutting area that the samples may have touched should be carefully cleaned with xylene between each block. Label the tubes on the caps with permanent ink.

Deparaffinizing Sections

Each section is extracted twice with octane to remove the paraffin. Alternatively, mixed xylenes can be used for this extraction. This organic extraction is followed by two washes with 100% ethanol to remove the solvent. The ethanol is removed by drying the samples under vacuum or by rinsing the sample with acetone.

1. Add approximately 1 ml of octane to each tube. Close tubes and mix at room temperature for about 30 minutes.
2. Pellet the tissue and any residual paraffin by centrifugation (3 to 5 minutes in microfuge at full speed).
3. Remove the octane from each sample with a clean (preferably plugged) Pasteur pipette. (Very old or fragile tissues often fragment when the paraffin is removed, so care must be taken to avoid losing tissue in this step.)
4. Repeat steps 1, 2, and 3.
5. Add approximately 0.5 ml of 100% ethanol to each tube. Close and mix by inverting.
6. Pellet as in step 2.
7. Remove the ethanol as done with the octane in step 3.
8. Repeat steps 5, 6, and 7. Remove as much ethanol as possible with the pipette.
9. Dry the samples under vacuum until the ethanol has evaporated completely. Before drying the samples, cover the tubes by stretching Parafilm across the top of the tube and poking several holes in it. This precaution helps to avoid cross-contamination of samples and contamination from the vacuum bottle.

OR

Add 2 to 3 drops of acetone to each tube. Keeping the tubes open, place them carefully in a heating block or water bath (37 to 50°C) to promote the evaporation of the acetone.

Proteinase Digestion

1. Add 100 μl of digestion buffer (above) containing 200 μg/ml of Proteinase K to the extracted, dried samples. Samples containing large amounts of tissue should be resuspended in 200 μl of digestion buffer.

2. Incubate for 3 hours at 55°C (alternatively, 37°C overnight).
3. Spin the tubes briefly to remove any liquid from the cap.
4. Incubate at 95°C for 8 to 10 minutes to inactivate the protease. Avoid heating longer than 10 minutes.
5. To use for PCR, pellet any residual paraffin or tissue by centrifuging for about 30 seconds. Use an aliquot of the supernatant for the amplification (typically 1 to 10 μl).
6. Store prepared samples at −20°C.

PCR

The amount of supernatant that is optimal for an amplification reaction is determined by numerous factors and may be specific to each sample. It is useful to test several concentrations of each sample in each amplification (e.g., 1 versus 10 μl per 100-μl PCR). Samples extracted from sections that contained large amounts of tissue may necessitate the use of a smaller fraction of the supernatant (e.g., 0.1 μl). The amount of supernatant that is "tolerated" in a PCR may be affected by residual fixation chemicals, excessive tissue debris, etc.

Amplification from prepared paraffin-embedded tissue is less efficient than amplification from other types of templates such as purified DNA or extracts from fresh clinical material. To compensate for this reduced efficiency, we modify the thermocycling parameters by increasing the number of cycles and lengthening the time at each temperature within the cycle.

A comparison of thermocycling conditions for amplification of the human papillomavirus L1 product (see Chapter 42) from purified DNA (or fresh clinical material) versus amplification from a paraffin sample follows.

	Purified DNA	Paraffin sample
95°C	30 seconds	1 minute
55°C	30 seconds	1 minute
72°C	1 minute	2 minutes
	30 cycles	40 cycles

3 to 5 minutes	72°C	(final extension)

Discussion

Preparation of paraffin-embedded tissue sections for PCR involves a great deal of manual manipulation (e.g., Eppendorf tubes are opened and closed and transferred from centrifuge to rack, relatively large volumes of solvents are added to and removed from the samples, tubes may be covered and uncovered with Parafilm). This provides ample opportunity for cross-contamination between samples or contamination from extraneous PCR products or plasmids within the work area. In the extraction/digestion procedure, common vehicles for such contamination are the experimenter's gloves (or fingers). Caps and rims of microfuge tubes are most susceptible to this. We recommend frequent glove changes. When opening and closing tubes, do not touch the rims or insides of caps and be careful to keep tubes adequately spaced in racks. If samples are dried under vacuum, cover tubes and caps with Parafilm to minimize cross-contamination. After it has been removed from the tubes, the Parafilm should be considered contaminated waste.

Amplification of extracted, digested samples that have been stored is not always as successful as is amplification of samples that have been amplified immediately after digestion. Possibly storage or freeze/thawing somehow promotes breaking or nicking of the DNA.

We have noticed a length limit on amplification products from extracted, digested tissue from formalin-fixed, embedded sections. PCR products in the range of 450 to 650 bp can be amplified from most sections of this type (unless the blocks are very old). Larger products can be amplified from some, but not all, of these samples. The integrity of the DNA (or the extent of modification or crosslinking) is likely related to the fixative used and the fixation time (Dubeau et al. 1986). Particular fixatives (other than buffered formalin) may modify DNA and thus reduce the amplification "competency" of the section. Alternatively, particular fixatives may contain chemicals that inhibit PCR. We are currently investigating the effects of a variety of fixatives. The inhibitory effects of certain fixatives can be overcome by further purification (phenol-chloroform extraction and ethanol precipitation after the protease digestion) of the tissue extracts. Since such further purification involves much manipulation and therefore more potential for contamination, it is not recommended routinely.

Because of the increased number of amplification cycles and the extended annealing and polymerization times used in PCR of paraffin samples, nonspecific products are often produced in addition to the specific product(s). Such nonspecific products are often visible by ethidium bromide staining, and they are most often smaller than the specific product. However, on a Southern blot, these products typically do not hybridize with a probe specific for the desired product. The presence of nonspecific products complicates purification of the desired product and may affect procedures such asymmetric PCR or direct sequencing.

Acknowledgments

We thank D. Shibata and N. Arnheim (University of Southern California), M. Cornelissen (University of Amsterdam), H. Fox (University of California at San Francisco), and L. Villa (Ludwig Institute, São Paulo, Brazil) for valuable discussions. We are grateful to C. Greer and A. Lewis for their work on optimizing PCR from paraffin-embedded tissues. Many thanks to D. Camp (Lawrence Livermore National Laboratory) for recommending alkanes as solvents for paraffin.

Literature Cited

Dubeau, L., L. A. Chandler, J. R. Gralow, P. W. Nichols, and P. A. Jones. 1986. Southern blot analysis of DNA extracted from formalin-fixed pathology specimens. *Cancer Res.* **46**:2964–2969.

Goelz, S. E., S. R. Hamilton, and B. Vogelstein. 1985. Purification of DNA from formaldehyde fixed and paraffin embedded human tissue. *Biochem. Biophys. Res. Commun.* **130**:118–126.

Impraim, C. C., R. K. Saiki, H. A. Erlich, and R. L. Teplitz. 1987. Analysis of DNA extracted from formalin-fixed, paraffin-embedded tissues by enzymatic amplification and hybridization with sequence-specific oligonucleotides. *Biochem. Biophys. Res. Commun.* **142**:710–716.

Manos, M. M., Y. Ting, D. K. Wright, A. J. Lewis, T. R. Broker, and S. M. Wolinsky. 1989. The use of polymerase chain reaction amplification for the detection of genital human papillomaviruses. *Cancer Cells (Molecular Diagnostics of Human Cancer)* **7**:209–214, Cold Spring Harbor Laboratory.

Shibata, D. K., N. Arnheim, and W. J. Martin. 1988a. Detection of human papilloma virus in paraffin-embedded tissue using the polymerase chain reaction. *J. Exp. Med.* **167**:225–230.

Shibata, D., W. J Martin, and N. Arnheim. 1988b. Analysis of forty-year-old paraffin-embedded thin-tissue sections: a bridge between molecular biology and classical histology. *Cancer Res.* **48**:4564–4566.

20

AMPLIFYING ANCIENT DNA

Svante Pääbo

Early work on ancient DNA (Higuchi *et al.* 1984; Pääbo 1985) involved the molecular cloning of DNA extracted from museum specimens and archaeological finds. However, several problems made the cloning of old DNA difficult, in particular, the fact that a large proportion of the extracted DNA is heavily modified (Pääbo 1989). This reduces the cloning efficiencies dramatically and introduces the serious risk that the sequences determined may contain cloning artifacts introduced during repair and replication of the old DNA in bacteria (see Pääbo and Wilson 1988a). The advent of PCR has opened up the possibility of isolating DNA sequences from a few copies of intact DNA present in extracts where the majority of the molecules are damaged and degraded to an extent that precludes analysis by other molecular techniques. However, the amplification of ancient DNA poses some problems to the investigator that are due to mainly two circumstances. First, the great sensitivity of PCR makes it liable to pick up even very minor contaminations consisting of contemporary, undamaged DNA. Second, the modifications present in the old template DNA as well as other components in extracts of old tissues may inhibit the DNA polymerase and influence the results obtained.

Our present methods for dealing with these problems are summarized in this chapter.

All reagents and laboratory supplies used for the extractions need to be devoid of contaminating DNA that may stem from the reagents as they are delivered from the manufacturer, from laboratory personnel, or from previous amplifications of the sequences under study. Therefore, disposable plastic laboratory supplies should be used whenever possible. To check for contamination in the reagents, it is advisable to do an initial control extraction in which no tissue is added to the extraction buffer and the whole extraction procedure is performed as outlined. A few amplifications from that "mock extract" should then be performed to make sure that no specific products are obtained. If contamination occurs and is not confined to one particular set of primers, different sources of reagents as well as water should be tried. Note that autoclaving does not degrade DNA sufficiently to make impossible amplifications of around 100 bp. To minimize the risk of contamination, our protocols for extraction and amplification have been set up with a minimum number of ingredients.

Protocols

Collection of Samples

The amount of tissue samples that can be obtained from old specimens is generally dictated by considerations such as scarcity of material and museological appearance. Generally, it is sufficient to remove samples of 0.1 to 0.5 g of dried soft tissues. Ideally, at least two samples should be taken from each specimen, and these should be extracted as well as amplified on different occasions to demonstrate the reproducibility of the results. Invariably, the tissues that seem the best preserved macroscopically will yield the best results. During all handling of the specimens, plastic gloves should be worn. Clean surgical blades are suitable for the removal of samples, which should be put into sterile plastic tubes (e.g., 50-ml Falcon tubes) that can be used for the subsequent DNA extraction. Every effort should be made to minimize the handling of samples. In the case of museum specimens that have been extensively handled prior to sampling, it is recommended that the least superficial parts of the sample be used for analysis, since the most superficial parts of the sample can be expected to be contaminated by human cell debris (e.g. Thomas *et al.* 1989).

Extraction of DNA

Extraction buffer containing 10 m*M* Tris–HCl (pH 8.0), 2 m*M* EDTA, 10 mg/ml of DTT, 0.5 mg/ml of Proteinase K, 0.1% SDS is added directly to the tubes containing the samples. Extraction buffer is also added to one or more empty tubes, and these control extracts are processed in parallel with the tissue extracts throughout the procedure. For samples of 0.1 to 0.2 g, a volume of 4 ml is generally sufficient; for larger samples correspondingly larger volumes can be used. The sample is incubated overnight at 37°C with slow agitation. After this, most of the tissue is in solution. For samples such as skin, which contain a large amount of collagen, a preincubation for 2 hours at 37°C in Tris–HCl, EDTA, and 0.5 mg/ml of collagenase before the addition of the other reagents may be useful.

An equal volume of phenol saturated with 1 *M* Tris–HCl (pH 8.0) is added to the extract. After slow agitation for 5 to 10 minutes, the aqueous phase is separated by centrifugation from the phenol and solid-tissue remains. The aqueous phase is extracted once more by phenol and once by chloroform-isoamyl alcohol (24 : 1 v/v). A sample (1 to 2 ml) of the aqueous phase is then concentrated and purified from remaining chloroform, salts, and other small molecules by Centricon 30 microconcentrators (Amicon Inc., Danvers, Massachusetts) by using distilled water in two consecutive cycles of dilutions and concentrations. The use of the Centricon devices results in greater recovery of DNA than that possible with ethanol precipitation and has the additional advantage of allowing the components of the extracts to remain in solution. The remainder of the aqueous phase can be stored at −70°C without concentration for future experiments.

The concentrated extract is almost invariably brown in color, most likely because of the Maillard products of reducing sugars. These contaminants give a blue fluorescence in UV light and may be misinterpreted as nucleic acids if the extract is analyzed by agarose gel electrophoresis and visualized by ethidium bromide staining, which gives a pinkish-red fluorescence when intercalated into nucleic acids. It is therefore advisable to take a photograph of the gel before as well as after staining, to differentiate between the spontaneously fluorescing contaminants and nucleic acids fluorescing after the intercalation of ethidium bromide.

Analyses of the extracted DNA may be offset by the wide array of modifications present in ancient DNA (Pääbo 1989). For example, the extracts frequently contain components that exhibit peak absorption at approximately 215 nm, which makes impossible the determination of concentration of DNA by absorption at 260 nm. Further, estimates of the size distribution of the extracted DNA may be affected by the presence of DNA-DNA cross-links. Also, many enzymes can show aberrant or unusual activities owing to modifications of the DNA. For example, ancient DNA is frequently sensitive to pancreatic RNase because of a high frequency of baseless sites and shows only limited sensitivity to DNase. Thus, many of the techniques that the molecular biologist uses to characterize extracted nucleic acids cannot be used for ancient DNA.

Amplification of DNA Sequences

The following reaction buffer has proved suitable for amplification of old DNA sequences: 67 mM Tris–HCl (pH 8.8), 2 mM MgCl$_2$, 250 μM each of dATP, dCTP, TTP, and dGTP, 2 μg/ml of bovine serum albumin, and 0.5 units of Taq DNA polymerase (Perkin-Elmer Cetus, Norwalk, Connecticut). The albumin is useful in overcoming an inhibitory activity of unknown origin that is present in many old extracts. Initially, one or a few microliters of the extract can be added to amplifications of 25 μl. Forty cycles of PCR are then performed. This high number of cycles will allow for the amplification up to detectable levels of very few or even single copies of intact template molecules. It is often informative to try amplifications that differ in length from <100 bp to around 500 bp. In parallel, the following control amplifications should be performed for each primer pair used in the experiment: (1) a control extract amplification, using as template a control extract prepared in parallel with the tissue extract and (2) a no-template control, where no extract is added to the amplification.

Aliquots of the amplifications can be analyzed by gel electrophoresis using low-melting-temperature agarose (see Fig. 1). The control extract amplifications should be devoid of any specific prod-

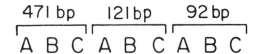

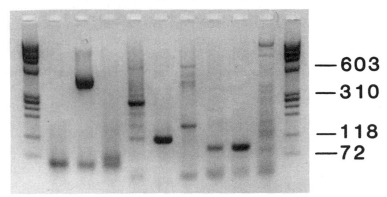

Figure 1 Amplification of mitochondrial DNA from a 7000-year-old human brain from Florida as well as from contemporary human DNA analyzed by gel electrophoresis using low-melting-temperature agarose. Gels were stained with ethidium bromide and bands were visualized with UV light. Lanes A show amplifications from the ancient brain, lanes B show amplifications from 1.2 ng of contemporary human DNA, and lanes C show amplifications from a control extract prepared in parallel with the brain extract. It can be seen how in the case of the short amplification of 92 bp, one major product appears in amplifications from both the contemporary and the ancient DNA. When a 121-bp amplification is performed, the specific product is only a minor component in the amplification from the brain, and when a 471-bp amplification is attempted, only the contemporary DNA yields a specific product. For the 471-bp amplification we used the following primers:

(M14725) 5'CGAAGCTTGATATGAAAAACCATCGTTG-3'-14724 and
(cytb2) 5'-AAACTGCAGCCCCTCAGAATGATATTTGTCCTCA-3'-15149

For the 121-bp amplification we used

(primer A) 5'-ATGCTAAGTTAGCTTTACAG-3'-8297 and
(primer B) 5'-ACAGTTTCATGCCCATCGTC-3'-8215

For the 92-bp amplification we used

(primer A and primer C) 5'-ATTCCCCTAAAAATCTTTGA-3'-8244

Numbers at 3' ends refer to Anderson *et al.* (1981). The low-molecular-weight products are dimers of primers. The sizes of molecular weight markers are indicated in numbers of base pairs. See Pääbo et al. (1988) for experimental details.

uct, thus demonstrating that the extracts as well as the amplification reagents are free from contaminating DNA. If products should appear in the control extract amplifications, the no-template controls will allow you to decide if a particular primer, the amplification reagents, or the extracts are contaminated.

If no specific products are obtained in the amplifications from the tissue extract, this may be due to the presence of contaminants or damaged DNA that inhibit the DNA polymerase. In this case, no dimers of primers will be seen in the amplifications of the old DNA but may be visible in the controls. Such heterodimers of the two primers used for PCR are almost always seen in reactions where no or little *bona fide* template DNA is present. Therefore, if no dimers are seen, it may indicate that the enzymatic activity of the *Taq* DNA polymerase has been inhibited. The amount of extract added should in this case be progressively diluted with the hope of being able to dilute away the contaminant while still having enough DNA templates left. Also, the amount of enzyme added can be increased. If no specific products are obtained, but dimers of primers are present in the amplification from the ancient DNA, it is possible that not enough template molecules are present and that more extract should be added to the amplification.

Amplification products of sizes different from those expected are often seen in the old extracts. These products do not contain the relevant sequences and presumably stem from minute amounts of DNA (e.g., of fungal origin) introduced into the amplification by the sample or the chemicals used for the extraction. They often decrease in incidence if collagenase is not included in the extraction procedure, and they disappear if small amounts of intact template DNA are added to the amplification.

Bands of interest can be cut out of the low-melting-temperature gel, melted, diluted, and used as templates for a second amplification, which, by the unbalanced-priming method (Gyllensten and Erlich 1988), yields single-stranded templates suitable for sequencing. The amplification primer that has been limiting in the unbalanced-priming reaction can be used for the subsequent sequencing by the dideoxynucleotide chain termination method (Sanger *et al.* 1977). For mitochondrial sequences, where homoplasmy is the rule, the sequences obtained should be unequivocal. Double sequence at any position is almost invariably due to contamination. Furthermore, several independent amplifications from the same extract should give identical sequences, as should amplifications from extracts of different samples removed from the same individual.

Discussion

When amplifications of various length are tried, there is generally an inverse correlation between the efficiency of the amplification and the size of the fragment that is amplified (see Fig. 1). This is due to the degraded and damaged state of the old DNA. In ancient samples, the maximum length of mitochondrial DNA segments that has been amplified is between 100 and 200 bp (Pääbo et al. 1988b; Pääbo 1989) whereas better-preserved specimens such as study skins from zoological collections may yield fragments of up to 500 bp (W. K. Thomas and S. Pääbo, unpublished observation). In our experience, any amplifications of old DNA that are longer than this probably represent contaminating DNA.

We have shown that in the absence of any template molecules that are long enough to serve as a template for the primers used, the two primers may be sequentially extended during the first cycles of PCR by using shorter DNA molecules as templates. After a number of such extensions, the 3' ends of the extended primers will overlap, and an exponential amplification will ensue (Pääbo et al. 1989). This process of "jumping PCR" allows for the amplification of DNA segments that are actually longer than the longest intact template present in the extract. However, the enzyme frequently inserts incorrect bases when it "jumps" from one template to the other (D. M. Irwin and S. Pääbo, unpublished observation). This is not seen in direct sequencing of the amplification product but will distort sequences obtained from the cloned amplification product. In general, for this and other reasons (see Pääbo and Wilson 1988), it is highly preferable to sequence the amplification products directly rather than to clone them and subsequently sequence multiple clones. However, sequencing of cloned amplification products may complement the direct sequencing by providing an indirect estimate of the amount of damage present in the template DNA. Finally, it should be noted that the "jumping PCR" causes in vitro recombination that may give rise to mosaic sequences when nuclear genes are studied in heterozygous individuals (see also Chapter 28 for discussion of polymerase halt-mediated linkage of primer [PHLOP]).

Acknowledgments

Many colleagues and friends have contributed to the development of molecular archaeology. I thank Dr. Allan Wilson for providing invaluable advice, generous support, and excellent working facilities. I also want to specifically mention Dr. Kelley Thomas, whose constructive criticism is instrumental to much of my work. The European Molecular Biology Organization provides financial support through a long-term fellowship (ALTF 76–1986).

Literature Cited

Anderson, S., A. T. Bankier, B. G. Barrell, M. H. L. de Bruijn, A. R. Coulson, J. Drouin, I. C. Eperon, D. P. Nierlich, B. A. Roe, F. Sanger, P. H. Schreier, A. J. H. Smith, R. Staden, and I. G. Young. 1981. Sequence and organization of the human mitochondrial genome. *Nature (London)* **290**:457–465.

Gyllensten, U. B., and H. A. Erlich. 1988. Generation of single-stranded DNA by the polymerase chain reaction and its application to direct sequencing of the *HLA-DQA* locus. *Proc. Natl. Acad. Sci. USA* **85**:7652–7656.

Higuchi, R., B. Bowman, M. Freiberger, O. A. Ryder, and A. C. Wilson. 1984. DNA sequences from the quagga, an extinct member of the horse family. *Nature (London)* **312**:282–284.

Pääbo, S. Molecular cloning of Ancient Egyptian mummy DNA. 1985. *Nature (London)* **314**:644–645.

Pääbo, S., and A. C. Wilson. 1988a. Polymerase chain reaction reveals cloning artefacts. *Nature (London)* **334**:387–388.

Pääbo, S., J. A. Gifford, and A. C. Wilson. 1988b. Mitochondrial DNA sequences from a 7000-year old brain. *Nucleic Acids Res.* **16**:9775–9787.

Pääbo, S. 1989. Ancient DNA: extraction, characterization, molecular cloning, and enzymatic amplification. *Proc. Natl. Acad. Sci. USA* **86**:1939–1943.

Pääbo, S., R. G. Higuchi, and A. C. Wilson. 1989. Ancient DNA and the polymerase chain reaction. The emerging field of molecular archaeology. *J. Biol. Chem.* **264**:9709–9712.

Sanger, F., S. Nicklen, and A. R. Coulson. 1977. DNA sequencing with chain-terminating inhibitors. *Proc. Natl. Acad. Sci. USA* **74**:5463–5467.

Thomas, R. H., W. Schaffner, A. C. Wilson, and S. Pääbo. 1989. DNA phylogeny of the extinct marsupial wolf. *Nature (London)* **340**:465–467.

Part Two

RESEARCH APPLICATIONS

21

IN VITRO TRANSCRIPTION OF PCR TEMPLATES

Michael J. Holland and Michael A. Innis

The experiments presented here illustrate the application of PCR for screening mutants and accessing the *in vitro* template activity of a collection of yeast 35S ribosomal ribonucleic acid (rRNA) spacer promoter mutants generated by *in vitro* mutagenesis in M13.

There are 100 to 200 tandemly repeated ribosomal cistrons per haploid yeast genome. Transcription of these genes by RNA polymerase I yields a 35S rRNA precursor that is processed to form 18S, 5.8S, and 25S rRNAs. Each ribosomal cistron is separated from the next by a 2.4-kb segment of spacer rDNA. Sequences within the spacer region support RNA polymerase I-dependent selective initiation of transcription *in vitro* (Swanson and Holland 1983). Deletion mapping analysis showed that a 22-bp region of spacer rDNA located 2.2 kb upstream from the 35S rRNA gene promoter is required for and is sufficient for selective initiation of transcription by RNA polymerase I *in vitro* (Swanson *et al.* 1985). The 22-bp spacer promoter is an essential component of a 160-bp region of spacer rDNA that acts as an enhancer of 35S rRNA synthesis *in vivo* (Mestel *et al.* 1989). To begin to define the mechanism of enhancement of 35S rRNA synthesis, we investigated the relationship between the 22-bp spacer promoter activity and the enhancer activity. The experimental approach involved saturation mutagenesis of the 22-bp spacer promoter element followed by examination of the effects of these point mutations on promoter activity *in vitro* and enhancer activity *in vivo*. We describe here saturation mutagenesis of the 22-bp spacer promoter and the use of PCR to amplify the mutant segments of rDNA for *in vitro* transcription assays, subcloning into vectors for analyzing enhancer function *in vivo*, and DNA sequence analysis.

Protocols

Saturation Mutagenesis of the Spacer Promoter

A 193-bp fragment of yeast spacer rDNA extending from an *Eco*RI site to a *Sal*I site was cloned into the polylinker region of the replicative form of bacteriophage M13mp10amber. The location of the 22-bp spacer promoter within this fragment of spacer rDNA is indicated in Fig. 1A. Site-directed mutagenesis of the spacer promoter essentially followed the highly efficient gap repair method (Kramer *et al.* 1984) that selects against amber mutations in the parental vector.

Three mutagenic oligonucleotides complementary to the spacer promoter were designed to generate single transversion mutations within the 22-bp element (McNeil and Smith 1985). The nucleotide sequences of the mutagenic oligonucleotides are shown in Fig. 1B. Primer I contained C-to-A transversions, primer II contained T-to-G transversions, and primer III contained A-to-C transversions. In separate reactions, each of the primers was annealed with the gapped circular phage DNA, and the gaps were repaired with T4 DNA polymerase and T4 DNA ligase. The repaired circular DNAs were used to transform *Escherichia coli* strain HB2154 (*sup°,mutL*). The *mutL* mutation prevents mismatch repair, and the lack of a suppressor tRNA in this strain provides a strong genetic selection against phage derived from the parental strand carrying amber mutations.

Phage stocks were prepared on HB2151 (*sup°*) and plated to yield approximately 200 plaques per 12.5-cm petri dish. Plaques were lifted onto nitrocellulose disks and prepared for hybridization according to standard procedures. A 22-bp oligonucleotide complementary to wild-type spacer promoter sequences in the recombinant phage

A   AAAGATGGGTTAAAAGAGAAGG

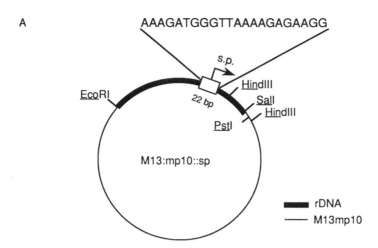

B   Primer I   GTGAAAGCXXTTXTXTTTTAAXXXATXTTTGCAACG
X=C>A, 30:1

Primer II   GTGAAAGCCCXXCXCXXXXAACCCAXCXXXGCAACG
X=T>G, 40:1

Primer III   GTGAAAGCCCTTCTCTTTTXXCCCXTCTTTGCAACG
X=A>C, 12:1

Wild-type   CCTTCTCTTTTAACCCATCTTT

Figure 1   (A) Restriction endonuclease cleavage map of M13mp10::sp. A 193-bp *Eco*RI/*Sal*I fragment of yeast rDNA was ligated into the polylinker of bacteriophage M13mp10. The location of the spacer promoter (s.p.) within the rDNA sequences is indicated by the *arrow*. The nucleotide sequence of the 22-bp spacer promoter is shown above the restriction map. (B) The nucleotide sequences of three 36-nucleotide mutagenic primers as well as a 22-nucleotide probe that is complementary to the wild-type spacer promoter sequence. Primer I contains C-to-A transversions, primer II contains T-to-G transversions, and primer III contains A-to-C transversions. The molar ratios of the wild-type versus mutant nucleotides used to synthesize each primer are indicated.

DNA (Fig. 1B) was labeled with polynucleotide kinase and [γ³²P]ATP and annealed with the nitrocellulose filter lifts at 57°C in a buffer containing: 6× SSC, 5× Denhardt's solution, and 0.1% SDS. These hybridization conditions permit annealing of the probe to the wild-type 22-bp sequence; however, phage DNA containing even a single-base-pair mismatch would not be expected to hybridize under these conditions. The observed frequencies of nonhybridizing plaques for each of the primers were as follows: primer I, 1.25%; primer II, 13%; and primer III, 0.6%. Plaques that did not hybridize with the probe were picked, and phage stocks were prepared from each plaque.

Amplification of Ribosomal DNA Sequences from M13 Phage Using the Polymerase Chain Reaction

The failure of a plaque to hybridize to the wild-type 22-bp rDNA probe could be the result of one or more base-pair mismatches within the spacer promoter sequence in the M13 phage genome or to the presence of a plaque derived from a "miniphage" that carries a deletion of the rDNA sequences. To distinguish between these possibilities and to prepare sufficient quantities of the rDNA fragments for transcription assays and subcloning, PCRs were performed on nonhybridizing M13 phage by using primers that are complementary to M13 sequences located immediately upstream and downstream from the polylinker region. PCRs were performed in 100 μl containing 20 mM Tris (pH 8.3), 50 mM KCl, 2.5 mM MgCl₂, 0.05% NP-40, 0.05% Tween 20, 100 mg/ml of gelatin, 20 pmol each of the M13 primers

RG05 (5'-AGGGTTTTCCCAGTCACGAC-3') and
RG02 (5'-GTGTGGAATTGTGAGCGGAT-3'),

2.5 units of *Taq* DNA polymerase, and 1 ml of intact M13 bacteriophage (10⁹ PFU). The thermal profile involved 20 cycles of denaturation at 95°C for 30 seconds, primer annealing at 50°C for 1 minute, and extension at 72°C for 1 minute.

The PCR products were visualized by ethidium bromide staining following electrophoresis on 1.5% TBE agarose gels. The presence of rDNA sequences in the phage genome was indicated by a DNA fragment of the size expected for phage containing the 193-bp rDNA insert. The observed frequencies of M13 phage that served as tem-

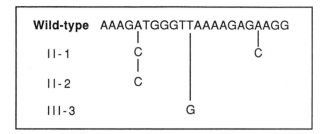

Figure 2 Nucleotide sequences of representative mutants within the 22-bp spacer promoter. The base substitutions in mutants II-1, II-2, and III-1 are indicated.

plate for the correct-size PCR product among the phage that did not hybridize with the wild-type 22-base oligonucleotide probe were primer I, 15/40; primer II, 35/60; and primer III, 9/20. To confirm that these latter phage carried transversion mutations within the spacer promoter sequence of their respective genomes, the nucleotide sequences of the rDNA regions of two representative phage generated with primer II and one representative generated with primer III were determined. In each case, the nucleotide sequences were identical except within the 22-bp spacer promoter. As illustrated in Fig. 2, the phage generated with primer II contained A-to-C transversion mutations, while the phage generated with primer III contained a T-to-G transversion mutation. Two of the phage contain single-base-pair mutations while one contains two substitution mutations. These results confirm that the mutagenesis strategy yielded the expected mutations within the 22-bp spacer promoter.

In Vitro Transcription Assays Using Templates Generated by the Polymerase Chain Reaction

The effects of the mutations within the spacer promoter regions of the three phage described in Fig. 2 on template activity in an *in vitro* transcription assay were tested by using the PCR products as templates. Transcription assays were performed using an RNA polymerase I-dependent yeast whole-cell extract (Swanson and Holland 1983). PCR products were used as templates without purification. The amounts of each of the templates used in the transcription assays were normalized (Fig. 3A). Transcription reactions were per-

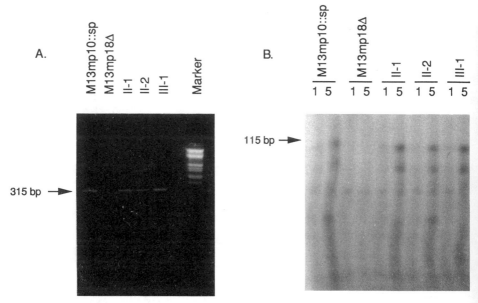

Figure 3 (A) Agarose gel electrophoresis of PCR products synthesized from the following bacteriophage templates: M13mp10::sp, M13mp18 (a bacteriophage carrying a complete deletion of the RGO-2 and RGO-5 primer sites as well as the polylinker in M13mp18), and the mutant bacteriophage II-1, II-2, and III-1, respectively. The *arrow* indicates the position of the full-length (315 bp) PCR product. (B) Polyacrylamide gel electrophoresis of transcripts synthesized *in vitro* from 10 (1) and 50 (5) ng of PCR template synthesized from M13mp10::sp, M13mp18, II-1, II-2, and III-1, respectively. The *arrow* indicates the position of the full-length (115-nucleotide) run-off transcript.

formed in 75 μl containing 50 mM Tris–HCl (pH 7.9), 5 mM MgCl$_2$, 6 mM NaF, 0.5 mM dithiothreitol, 50 mM KCl, 0.6 mM ATP, GTP, and CTP, 0.01 mM UTP, 5 μCi [α-^{32}P]UTP (specific activity 410 Ci/mM), 0.2 to 5 μl of PCR product (without purification), and 50 μg of yeast whole-cell extract (S100). Transcripts were analyzed on 7 M urea-polyacrylamide step gels (30 × 13.5 × 0.15 cm) consisting of a lower 10% (3.5 cm) and an upper 3.0% (26.5 cm) polyacrylamide layer (Swanson *et al.* 1985).

As illustrated in Fig. 3B, PCR templates generated from the wild-type parental phage, M13mp10::sp, and each of the three mutant phage directed the synthesis of a 115-nucleotide transcript. The size of this transcript was consistent with initiation at the spacer promoter and truncation at the end of the PCR template. The template activities of the PCR products synthesized from all three mutant phage were very similar to those of the wild-type template, demonstrating that none of the mutations had a measurable effect on spacer promoter activity *in vitro*. Similar results were obtained after sub-cloning of the mutant rDNA sequences into plasmid DNA templates, confirming that PCR products can be used to access template activity *in vitro* without prior purification. For the analysis described here, the ability to assay the PCR products directly permitted the rapid screening of a large number of prospective mutant templates. A potential drawback of this method is that a number of premature termination transcripts were observed when PCR products were used as templates (Fig. 3B). We have observed premature termination in similar assays with plasmid DNA templates when the rate of the transcription reaction is slowed by inhibitors or limiting substrates. Since PCR products were used as templates without purification, it is likely that the PCR mixture contains inhibitors of the transcription reaction, perhaps deoxynucleotides, that cause increased premature termination of transcription. In spite of the presence of premature termination products, however, assays of the PCR products did give a reliable indication of template activity.

Applications

The ability to assay PCR products synthesized from M13 phage carrying promoter mutations directly, using *in vitro* transcription assays, represents a rapid method for identification of interesting mutants in a large collection. Similar methods could be used to screen for mutations that interfere with the binding of a protein to a regulatory sequence, by using gel shift assays or footprinting assays. Utilization of the PCR products for direct analysis avoids subcloning, DNA isolation, and template or probe constructions that make screening experiments analogous to those described in this report impractical.

Literature Cited

Kramer, W., V. Drutsa, H.-W. Jansen, B. Kramer, M. Pflugfelder, and H.-J. Fritz. 1984. The gapped duplex DNA approach to oligonucleotide-directed mutation construction. *Nucleic Acids Res.* **12**:9441–9456.

McNeil, J. B., and M. Smith. 1985. *Saccharomyces cerevisiae CYC1* mRNA 5'-end positioning: analysis by in vitro mutagenesis, using synthetic duplexes with random mismatch base pairs. *Mol. Cell. Biol.* **5**:3545–3551.

Mestel, R., M. Yip, J. P. Holland, E. Wang, J. Kang, and M. J. Holland. 1989. Sequences within the spacer region of yeast rRNA cistrons that stimulate 35S rRNA synthesis in vivo mediate RNA polymerase I-dependent promoter and terminator activities. *Mol. Cell. Biol.* **9**:1243–1254.

Swanson, M. E., and M. J. Holland. 1983. RNA polymerase I-dependent selective transcription of yeast ribosomal DNA. *J. Biol. Chem.* **258**:3242–3250.

Swanson, M. E., M. Yip, and M. J. Holland. 1985. Characterization of an RNA polymerase I-dependent promoter within the spacer region of yeast ribosomal cistrons. *J. Biol. Chem.* **260**:9905–9915.

22

RECOMBINANT PCR

Russell Higuchi

Mutagenesis of PCR fragments via mismatched primer or "5'-add-on" sequences is restricted to the ends of the fragments by the primer lengths that can be achieved by chemical DNA synthesis (see Chapter 11). A means of combining PCR products to generalize primer-mediated mutagenesis has been described (Higuchi *et al.* 1988). Here I describe a method by which such changes can be moved to any part of a PCR fragment.

Shown in Fig. 1 are two PCR products that overlap in sequence; both contain the same mutation introduced as part of the PCR primers. These overlapping, primary products can be denatured and allowed to reanneal together, producing two possible heteroduplex products. The heteroduplexes that have recessed 3' ends can be extended by *Taq* DNA polymerase to produce a fragment that is the sum of the two overlapping products. A subsequent reamplification of this fragment with only the right- and left-most primers ("outside" primers) results in the enrichment of the full-length, secondary product. In this way fragments containing the mutations far away from the fragment ends can be made by using PCR.

As with add-on mutagenesis (see Chapter 11), the mutation can be a base substitution, a small insertion, or a deletion (Vallette *et al.*

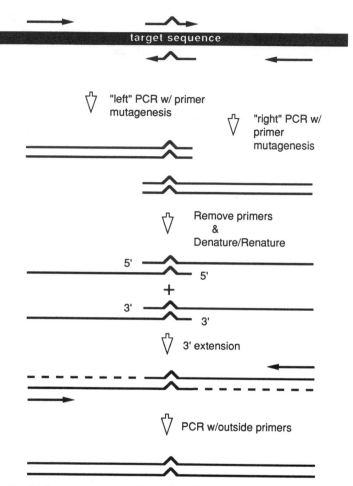

Figure 1   Combining two separate PCR products with overlapping sequence into one longer product. The two overlapping primers ("inside" primers) are shown containing a mismatched base to the target sequence.

1989). To produce a deletion, the primer sequence contains the deletion near the 5' end such that anything 5' of the deletion point is effectively add-on (see Fig. 2). Sheffield *et al.* (1989) have reported add-on sequences of forty-five bases. The mutation needs only to be mirrored in each of the two primary PCR fragments. Note also that two previously unrelated DNA sequences can be joined this way at almost any position by using the 5'-add-on sequences as "adapters" (see, for example, Fig. 3), in which a hypothetical promoter sequence and gene are joined as precisely as desired by having the sequence of

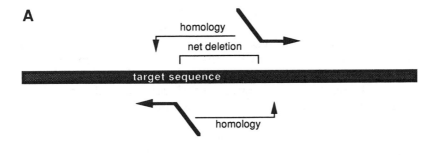

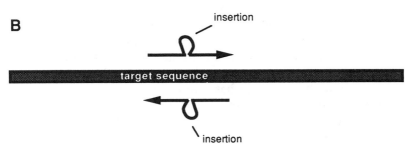

Figure 2 Inside primers for the creation of deletions (A) or small insertions (B).

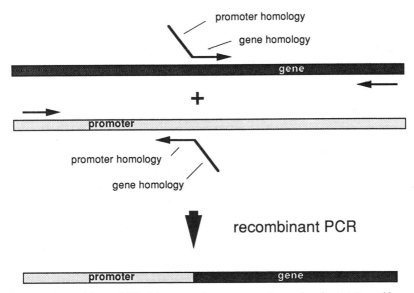

Figure 3 Recombinant PCR. Primers and sequences shown for the joining of hypothetical gene and promoter sequences.

the junction region shared by the "inside" primers. The process has been called "recombinant PCR" (thanks to A. C. Wilson). For more complex DNA constructs, serial applications of this principle (Mullis *et al.* 1986) provide a convenient alternative to cycles of conventional fragment purification, ligation, cloning, and screening and growing clones.

To make a base substitution, inside primers are mismatched to the target sequence at the substituted base. The effect of this mismatch on PCR will be less the more 5' the mismatch is. However, the more 5' the mismatch is in one primer, the more 3' it will be in the other inside primer (see Fig. 1). Therefore the primers I have used to create substitutions in this way have had the mismatch at their middle and have been about 20 nucleotides long. Other aspects of PCR primer design discussed in Chapter 1 apply.

To do add-on mutagenesis to create deletions or small insertions (see Fig. 2), or to combine sequences (Fig. 3), I would recommend inside primers with at least 15 bases of target sequence homology (20 if amplifying from complex genomic DNA rather than, say, plasmid DNA) that are 3' to add-on sequences of 15 to 20 bases. A longer insertion may be possible with a longer add-on sequence. Very large insertions may be done by combining three PCR fragments, two flanking and one comprising the insertion sequence, using serial applications of recombinant PCR.

Protocols

Primary PCRs

The two primary PCRs are performed separately, using standard PCR conditions as appropriate to the amplification desired. For optimal replication fidelity (see following), dNTP concentration should be between 50 to 200 μM each, and annealing temperatures should be as high as possible while still providing an efficient PCR. If a low initial annealing temperature is needed, it can probably be followed by a higher annealing temperature in subsequent cycles, as mismatches under the primers are resolved (see Chapter 11). As discussed as follows, preventing unwanted mutagenesis through misincorporations by *Taq* DNA polymerase argues for starting with as much template

DNA as possible. However, unless one gel purifies the primary PCR products, the relative amount of mutagenized and unmutagenized secondary PCR product one obtains is proportional to the relative molar amounts of primary PCR product and original template that are carried over, since both can be amplified with the outside primers. An acceptable level of unmutagenized product depends on the subsequent application.

Removal of Excess Primers

Remove the inside primers after the first round of PCR and before recombinant PCR to favor the production of the full-length product. This in theory can be accomplished merely by diluting by a large factor the primary PCRs into the secondary reaction such that the added outside primers are in large excess. Again, however, to limit unwanted mutagenesis, it may be preferable to start the secondary PCR with as much of the primary PCR product as possible. Thus a physical separation of PCR product and primers is called for.

The simplest way to remove PCR primers of less than 40 nucleotides is by selective filtration of the product on a Centricon 100 (Amicon; Higuchi et al. 1988). A protocol I have used is as follows:

1. Add 10 to 50 μl of each primary PCR to 2 ml of TE (0.01 M Tris–HCl (pH 8.0), 0.1 mM EDTA) that is in the upper reservoir of the Centricon 100. Cover with Parafilm and invert to mix.
2. Spin the device in a fixed-angle rotor at 1000 $\times$ g for 25 minutes. The oligonucleotides pass through the filter, while the PCR product is retained. Greater centrifugal force than 1000 $\times$ g may result in excessive passage of product DNA through the filter.
3. Add another 2 ml of TE to the upper reservoir and repeat the centrifugation.
4. Recover the retentate containing the PCR product (about 40 μl) per the manufacturer's instructions.

One alternative is purification of the products by gel electrophoresis through acrylamide or agarose, which, as discussed, has the advantage of separating out the original unmutagenized templates. Another is ammonium acetate/isopropanol precipitation as discussed in Chapter 10.

Secondary PCR

As a positive control, it is a good idea to verify, if possible, that the secondary amplification works well on unmutagenized template. After mixing together the primary PCR products, the creation and 3' extension of the heteroduplex templates will take place under PCR conditions. Annealing at lower temperatures for the first few cycles may be preferable for the formation of the recombinant heteroduplex molecules, followed by a higher annealing temperature in subsequent cycles. After the Centricon 100 treatment described above, mix 10 to 40 μl of the eluate into a 0.1-ml final volume PCR containing the outside primers and perform 8 to 12 PCR cycles under the conditions that work for the positive control. Depending on the subsequent application, gel purification of the secondary PCR product may be required.

Misincorporation by *Taq* DNA Polymerase

In a mutation reversion assay, *Taq* DNA polymerase has been estimated to incorporate an incorrect nucleotide once every 9000 nucleotides polymerized and produce frameshift errors once every 41,000 nucleotides (Tindall and Kunkel 1988). As discussed in Chapter 1 and Innis *et al.* (1988), however, misincorporated bases can promote chain termination in PCR, so that not all such errors necessarily propagate in subsequent PCR cycles. The frequency of errors in PCR product was earlier reported as one per 400 bases after 20 doublings (Saiki *et al.* 1988); under more optimum conditions one error per 4000 to 5000 bases was reported (Innis *et al.* 1988). If the PCR product is used in the aggregate, the fact that at any one position a tiny minority of molecules has an incorrect base is probably of little consequence. If the PCR product is to be put through a cloning step, however, it is important to keep in mind that the one molecule selected for amplification by cloning may have an error. The frequency of such mutations in PCR products should accumulate linearly with the number of doublings of DNA copy number (Saiki *et al.* 1988). To limit the number of such errors, the fewer doublings required, the better; thus the recommendation to start with as much template DNA as is reasonable. Note, of course, that the difference between 10 and 20 doublings is a theoretical twofold less error frequency, but a thousandfold difference in degree of amplification.

Literature Cited

Higuchi, R., B. Krummel, and R. K. Saiki. 1988. A general method of *in vitro* preparation and specific mutagenesis of DNA fragments: study of protein and DNA interactions. *Nucleic Acids Res.* **16**: 7351–7367.

Innis, M. A., K. B. Myambo, D. H. Gelfand, and M. A. D. Brow. 1988. DNA Sequencing with *Thermus aquaticus* DNA polymerase and direct sequencing of polymerase chain reaction-amplified DNA. *Proc. Natl. Acad. Sci. USA* **85**: 9436–9440.

Mullis, K., F. Faloona, S. Scharf, R. Saiki, G. Horn, and H. Erlich. 1986. Specific enzymatic amplification of DNA in vitro: The polymerase chain reaction. *Cold Spring Harbor Symp.* **51**: 263–273.

Saiki, R. K., D. H. Gelfand, S. Stoffel, S. J. Scharf, R. Higuchi, G. T. Horn, K. B. Mullis, and H. A. Erlich. 1988. Primer-directed enzymatic amplification of DNA with a thermostable DNA polymerase. *Science* **239**: 487–491.

Sheffield, V. C., D. R. Cox, L. S. Lerman, and R. M. Myers. 1989. Attachment of a 40-base-pair G+C rich sequence (GC clamp) to genomic DNA fragments by the polymerase chain reaction results in improved detection of single-base changes. *Proc. Natl. Acad. Sci. USA* **86**: 232–236.

Tindall, K. R., and T. A. Kunkel. 1988. Fidelity of DNA synthesis by the *Thermus aquaticus* DNA polymerase. *Biochemistry* **27**: 6008–6013.

Vallette, F., E. Mege, A. Reiss, and M. Adesnik. 1989. Construction of mutant and chimeric genes using the polymerase chain reaction. *Nucleic Acids Res.* **17**: 723–733.

23

DNase I FOOTPRINTING

Barbara Krummel

PCR can be used to prepare microgram quantities of double-stranded DNA fragments for use in experiments such as gel shift assays, DNase I footprinting, and *in vitro* transcription (Higuchi *et al.* 1988; Krummel and Chamberlin, in press). The use of PCR obviates the need for convenient restriction sites, large amounts of plasmid DNA as starting material, and gel isolation of products. The PCR product can be radioactively labeled at only one or both of the 5' ends. A simple procedure for the preparation of a uniquely end-labeled DNA fragment for use in DNase I footprinting is described (Figure 1).

Protocols

DNA To Be Amplified

The DNA starting material can be supercoiled or linearized plasmid DNA prepared by a CsCl banding procedure or from a small-scale

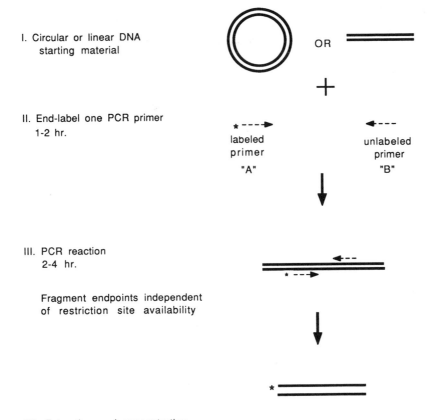

I. Circular or linear DNA
   starting material

OR

II. End-label one PCR primer
   1-2 hr.

labeled
primer
"A"

unlabeled
primer
"B"

III. PCR reaction
   2-4 hr.

Fragment endpoints independent
of restriction site availability

IV. Extraction and concentration

   1 hr.

Figure 1 Steps required to obtain a uniquely 5′-end-labeled DNA fragment. The times indicated are approximate and depend on the exact procedure used.

lysis procedure (Maniatis *et al.* 1982); DNA fragments are also suitable.

End-Labeled Primer

Choosing the primer: Typically primers 20 to 25 nucleotides in size with at least a 50% G+C content and no obvious self-complementarity are used (see Chapter 1).

   Size of fragment to be amplified: If one fragment will be used to obtain a footprint from each strand, the region of interest should ideally be close to the center of the fragment and not more than approximately 100 bp from either end. The distance between the footprint region and the fragment end is limited by the resolution available on the denaturing gel system.

Incorporation of ^{32}P at the 5' end: The amount of primer "A" necessary for the PCR is determined (see following) and then ^{32}P-labeled at the 5' end using [γ-^{32}P]ATP (NEN; 6000 Ci/mmol) and T4 polynucleotide kinase (Maniatis *et al.* 1982, p. 22). The completed reaction is extracted once with phenol. The unincorporated [γ-^{32}P]ATP can be removed by centrifugal gel filtration using Sephadex G-10 (Pharmacia) or ethanol precipitation with the addition of 10 mM MgCl$_2$ (final concentration) to the standard conditions of 0.3 M sodium acetate and 2.5 volumes of 100% ethanol.

PCR

Reactions (final volume 200 μl) are performed in a 0.5-ml microfuge tube.

250 ng of template DNA (e.g., 4-kb plasmid)

20 μl of 10× PCR buffer (1× concentrations: 50 mM KCl, 10 mM Tris–HCl (pH 8.3), 2.5 mM MgCl$_2$, 0.01% gelatin w/v)

50 pmol end-labeled primer "A" (from labeling protocol)

50 pmol primer "B"

3 μl of 10 mM dNTPs (a mixture of all four dNTPs at 10 mM each)

5 units of *Taq* DNA polymerase (Perkin-Elmer Cetus)

Cycling reactions are performed in a DNA thermal cycler.

Temperature	Time
94°C (denaturation)	60 seconds
37°C (annealing)	40 seconds
72°C (extension)	60 seconds

The total number of cycles is 20 to 25, and a last cycle with a 10-minute extension time is included. This increased extension is intended to allow complete reannealing and/or extension so that little single-stranded DNA remains. A higher annealing temperature (55°C instead of 37°C) may disrupt potential secondary structure in the primers and can result in more specific amplification (Saiki *et al.* 1988). Larger fragments may require longer extension times. The typical yield of a 300-bp fragment from a 200-μl reaction is 2 to 3 μg with a specific activity of 5 × 10^6 cpm/pmol.

Extraction and Precipitation of PCR

Before extraction, 1% of the total reaction volume is electrophoresed on an agarose minigel to evaluate the reaction. If the plasmid template and unused primer will not interfere significantly with binding experiments, the reaction is extracted once with phenol and once with chloroform, then ethanol-precipitated. A Centricon 100 microconcentrator (Amicon) can be used to remove unused primer, while removal of both primer and plasmid starting material is accomplished by nondenaturing PAGE. To quantitate the DNA, an aliquot is electrophoresed on an agarose minigel with size standards of known concentration. If the unused primer has not been removed, the cpm/ng of the fragment product cannot be determined by simply counting an aliquot. The band of interest can be excised from the agarose gel and dried onto filter paper; the total cpm can be determined by scintillation counting.

Discussion

PCR provides a rapid and simple method for the preparation of double-stranded DNA fragments. The procedure outlined here is for the preparation of a uniquely ^{32}P-end-labeled fragment for use in DNase I footprinting. The same PCR protocol is followed for preparing unlabeled DNA fragment or DNA labeled at both ends. Clearly this procedure could also be used to incorporate nonradioactive molecules such as biotin, specifically at the 5' end (see Chapter 13). No problems have been encountered because of misincorporation by *Taq* DNA polymerase that would lead to a population of fragments with a mixed DNA sequence. However, the product of the PCRs described here is not always a homogeneous preparation of full-length double-stranded DNA molecules. When the ^{32}P-end-labeled product is electrophoresed on a denaturing polyacrylamide gel, the result may be a "ladder" of fragments ranging in size from full length to that of the original primer. Every position appears to be represented, but the full-length material predominates (90 to 95%). We believe that these shorter fragments are due to the presence of incompletely replicated molecules. Nondenaturing PAGE does not provide separation of the two populations. Reducing the total number of cycles or increasing the annealing temperature to 55°C (closer to the optimum temperature of the enzyme) results in no significant change in the product.

This phenomenon has not interfered with the use of PCR-generated DNA in DNase I footprinting experiments. However, PCR has not reproducibly provided a suitable substrate for the hydroxyl radical cleavage reaction, which requires a DNA template with a very low background of cleaved or nicked molecules (Tullius *et al.* 1987).

Literature Cited

Higuchi, R., B. Krummel, and R. K. Saiki. 1988. A general method of *in vitro* preparation and specific mutagenesis of DNA fragments: study of protein and DNA interactions. *Nucleic Acids Res.* **16**:7351–7367.

Krummel, B., and R. Chamberlin. 1989. RNA chain initiation by *Escherichia coli* RNA polymerase. Structural transitions of the enzyme in early ternary complexes. *Biochemistry.* In press.

Maniatis, T., E. Fritsch, and J. Sambrook. 1982. In *Molecular cloning: a laboratory manual.* Cold Spring Harbor Laboratory, Cold Spring Harbor, New York.

Saiki, R. K., D. H. Gelfand, S. Stoffel, S. J. Scharf, R. Higuchi, G. T. Horn, K. B. Mullis, and H. A. Erlich. 1988. Primer-directed enzymatic amplification of DNA with a thermostable DNA polymerase. *Science* **239**:487–491.

Tullius, T. D., B. A. Dombroski, M. E. A. Churchill, and L. Kam. 1987. Hydroxyl radical footprinting: a high-resolution method for mapping protein-DNA contacts. *Methods Enzymol.* **155**:537–558.

24

SEQUENCING WITH *Taq* DNA POLYMERASE

Mary Ann D. Brow

In addition to being useful in PCR, *Taq* DNA polymerase has proven to be highly advantageous for the dideoxynucleotide chain-termination method of DNA sequencing of both conventional and single-stranded PCR templates (Sanger *et al.* 1977; Innis *et al.* 1988). The basic sequencing protocol described here involves (1) annealing an oligonucleotide primer to a single-stranded template; (2) labeling the primer in a short, low-temperature polymerization reaction in the presence of one α-labeled dNTP and three unlabeled dNTPs, all at low concentration; and (3) extending the labeled primer in four separate base-specific, high-temperature reactions, each in the presence of higher concentrations of all dNTPs and one chain-terminating ddNTP. If 5'-end-labeled primers are used, step (2) is eliminated. The helix-destabilizing base analog 7-deaza-2'-deoxyguanosine-5'-triphosphate (c^7dGTP) can be incorporated to prevent gel compressions (Barr *et al.* 1986). The products of these reactions are then separated by high-resolution polyacrylamide-urea gel electrophoresis and visualized by autoradiography (Fig. 1) or by nonisotopic detection methods (Smith *et al.* 1986; Prober *et al.* 1987; Ansorge *et al.* 1987).

In many experiments using PCR as a tool of inquiry, it is desirable to determine the nucleotide sequence of the product, preferably without adding the complication of a cloning step. In other cases, a

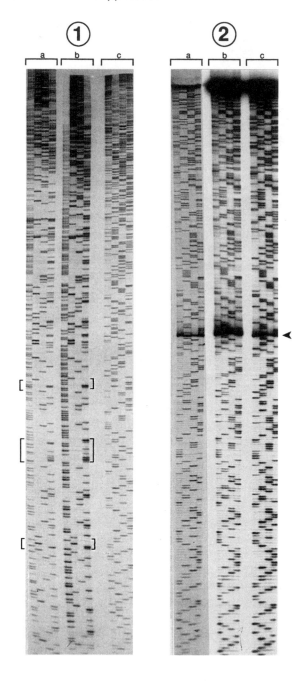

DNA of interest is cloned but is in a vector that is not amenable to DNA sequencing. A modified method of PCR, termed "asymmetric" PCR, is used to prepare single-stranded template specifically for the purpose of sequence analysis [see Chapter 10 and Gyllensten and Erlich (1988)]. Two methods for applying the basic sequencing protocol to the analysis of these single-stranded PCR products are described here. The first involves a selective alcohol precipitation of the PCR products to remove residual primers, nucleotides, and salts. The purified DNA can then serve as the template in a sequencing reaction incorporating α-labeled nucleotides as tags. The second is a direct PCR-to-sequencing procedure, performed without intervening purification, using 5'-end-labeled primers for visualization (Fig. 1).

Basic Sequencing Protocol

Reagents

Taq DNA polymerase	5 units/μl (Perkin-Elmer Cetus)
5× Reaction buffer	250 mM Tris–HCl (pH 8.8)
	35 mM MgCl$_2$

← **Figure 1** Autoradiograph of polyacrylamide-urea gels comparing sequencing with *Taq* DNA polymerase on M13-based single-stranded and asymmetric PCR-derived templates. (1a) and (1b) show the products of sequencing a cloned G + C-rich template with dGTP or c^{7}dGTP reaction mixes, respectively. The brackets indicate some of the areas of compression resolved by the c^{7}dGTP substitution. (1c) shows a cloned template of average base composition sequenced with dGTP mixes. All of the reactions were labeled by incorporation of {α[^{35}S]thio}dATP. (2) compares sequencing of asymmetric PCR-derived templates using ^{35}S-incorporation (a) or ^{32}P-labeled primers (b and c) for visualization. The PCR products shown in (2a) and (2c) were purified by isopropanol precipitation prior to sequencing; the template in (2b) was sequenced directly from the PCR without intervening purification. The arrow indicates an example of a gel artifact commonly seen in PCR sequencing products. Bands of this type vary in intensity from gel to gel, and the relative position of migration varies with the gel temperature during the run. This gel was run at 45°C, and the band comigrated with products of approximately 180 bases in length. At 50–55°C the band is more diffuse and has an apparent mobility of >300 bases (data not shown). Asymmetric PCR was performed as described (see Chapter 10), with the nucleotides reduced to 20 μM each dNTP. The target DNA was 5 μl of a liquid phage stock of 10^8 PFU/ml, and the amplification was performed for 35 cycles. Purification of the PCR product, DNA sequencing, and direct sequencing were performed as described. The products were resolved on buffer gradient sequencing gels (Biggin *et al.* 1983). Reaction sets were loaded G, A, T, C.

Taq dilution buffer	25 m*M* Tris–HCl (pH 8.8)
	0.1 m*M* EDTA
	0.15% Tween 20
	0.15% Nonidet P-40
Labeling mix	1.5 μ*M* dGTP*
	1.5 μ*M* dCTP
	1.5 μ*M* dTTP
{α-[35S]thio}dATP, >1000 Ci/mmol	

Termination mixes   (All concentrations are in μ*M*)

Mix:	G	A	T	C
dGTP*	20	20	20	20
dATP	20	20	20	20
dTTP	20	20	20	20
dCTP	20	20	20	20
ddGTP	60	—	—	—
ddATP	—	800	—	—
ddTTP	—	—	800	—
ddCTP	—	—	—	400
Specific dd:d ratios	3 : 1	40 : 1	40 : 1	20 : 1

*NOTE: 7-deaza-2'-dGTP can be used by substituting 1:1 in place of dGTP in *all* of the reaction mixes.

Stop solution	95% deionized formamide
	20 m*M* EDTA (pH 8.0)
	0.05% bromophenol blue
	0.05% xylene cyanol FF

4 *M* ammonium acetate

2-propanol

70% ethanol

TE buffer	10 m*M* Tris–HCl (pH 8.0)
	0.1 m*M* EDTA

Glass-distilled water

Annealing Template and Primer

1. Combine in a microcentrifuge tube:
   7 μl of template DNA (0.5 pmol)*
   1 μl of primer (0.5 pmol)
   2 μl of 5× reaction buffer

2. Heat at 70 °C for 3 minutes, then at 42°C for 10 minutes.

> *NOTE: Impurities in agar and agarose can copurify with phage DNA isolated from plate lysates and can inhibit subsequent sequencing with *Taq* DNA polymerase. While this effect can be overcome by increasing the enzyme two- to fourfold, liquid phage culture is recommended for this protocol.

Labeling Reaction

1. To the annealed primer-template add:
   0.5 μl of {α-[^{35}S]thio}dATP
   2 μl of labeling mix
   2 μl of *Taq* DNA polymerase (diluted to 1 unit/μl)
   3 μl of dH$_2$0
2. Vortex briefly to mix, centrifuge to collect drops, and incubate at 42°C for 5 minutes; then place the reactions at room temperature.

If a labeled primer is used, the dATP, the labeling mix, and the incubation are omitted. The volume is made up with distilled water, and the reaction proceeds directly to the termination reactions.

Termination Reactions

These reactions are performed in microcentrifuge tubes or in microtiter plates (Falcon #3911).

1. For each labeling reaction, label four tubes or wells: G, A, T, and C.
2. Dispense 4 μl of the appropriate termination mix into each tube or well. This should be done before the labeling reactions are started.
3. Transfer 4-μl aliquots of the completed labeling reaction into the termination mixes. Incubate at 70°C for 5 minutes. During heating, the reaction vessels should be in direct contact with water, oil, metal, glass beads, or another proven thermal conductor. Incubation in open air gives insufficient heat transfer.
4. Cool the reactions to room temperature. Add 4 μl of stop solution to each tube.
5. Immediately before electrophoresis, heat the samples to 70°C

for 5 minutes; load 1 to 2 μl of the sample in each lane. Unused samples may be stored at $-20°C$ for up to a week.

Sequencing of PCR Products

The single-stranded products of an asymmetric PCR can be sequenced by using the protocol previously described, which incorporates labeled nucleotides during sequencing primer extension, or by using prelabeled primers in the reaction. In either case, the primer that was limited in the asymmetric PCR or a third primer may be used for sequencing. If the PCR has any background products, which may contain the amplification primer sequences, an internal primer will provide an added degree of specificity in the sequencing step.

Purification of Asymmetric PCR for Sequencing

To use a reaction that incorporates an α-labeled dNTP, the residual primers and dNTPs from the PCR must be removed. This is done with a microconcentrator [e.g., Centricon 100, Amicon (Gyllensten and Erlich 1988)] or by selective alcohol precipitation, described here.

1. Combine 100 μl (total) of PCR mixture and 100 μl of 4 M ammonium acetate (NH_4OAc).
2. Add 200 μl of 2-propanol.
3. Mix well; incubate at room temperature for 10 minutes. Spin in a microcentrifuge at room temperature for 10 minutes. Carefully remove the supernatant.
4. Add 500 μl of 70% ethanol.
5. Spin in a microcentrifuge for 1 minute. Carefully remove the supernatant. Dry the pellet under vacuum.
6. Dissolve the dry DNA in 10 μl of TE buffer.
7. Sequence as described, using one-fifth to one-half of the recovered DNA in each annealing reaction.

Direct Sequencing of Asymmetric PCR without Purification

By using a slight variation of the basic protocol for sequencing with *Taq* DNA polymerase, it is possible to couple the asymmetric PCR and the sequencing reaction without intervening purification. If lower concentrations of dNTPs (i.e., 20 μM each dNTP) are used in the PCR, the residual nucleotides will not interfere with the sequencing reaction. However, the presence of these dNTPs precludes the use of an α-label-incorporation reaction, so coupled sequencing must be done with labeled primers to visualize the results.

1. Perform the asymmetric PCR as described, using 20 μM each dNTP. Label microcentrifuge tubes or a microtiter plate for the termination reactions, as previously described.

2. Distribute 5 μl of the appropriate termination mix into each tube or well.

3. Combine 10 to 17 μl of PCR mixture, 2 μl of 5× reaction buffer, 1 μl of labeled primer (0.5 to 1 pmol), 2 μl of *Taq* DNA polymerase (diluted to 1 unit/μl), and dH$_2$O to 22 μl.

4. Vortex briefly to mix, and centrifuge to collect drops.

5. Distribute the reaction mix into the termination mixes in 5-μl aliquots. Incubate the reactions at 70°C for 5 minutes. Cool the reactions to room temperature, and add 4 μl of stop solution. Store at -20°C until ready to use.

6. Immediately before electrophoresis, denature as previously described.

Acknowledgments

I thank Peter C. McCabe for sharing data prior to publication, Michael A. Innis for help and advice, both for critical reading of the manuscript, and Eric Ladner for photography.

Literature Cited

Ansorge, W., B. Sproat, J. Stegemann, C. Schwager, and M. Zenke. 1987. Automated DNA sequencing: ultrasensitive detection of fluorescent bands during electrophoresis. *Nucleic Acids Res.* **15**:4593–4602.

Barr, P. J., R. M. Thayer, P. Laybourn, R. C. Najarian, F. Seela, and D. R. Tolan. 1986. 7-Deaza-2'-deoxyguanosine-5'-triphosphate: enhanced resolution in M13 dideoxy sequencing. *BioTechniques* **4**:428–432.

Biggin, M. D., T. J. Gibson, and G. F. Hong. 1983. Buffer gradient gels and ^{35}S label as an aid to rapid DNA sequence determination. *Proc. Natl. Acad. Sci. USA* **80**:3963–3965.

Gyllensten, U. B., and H. A. Erlich. 1988. Generation of single-stranded DNA by the polymerase chain reaction and its application to direct sequencing of the HLA-DQA locus. *Proc. Natl. Acad. Sci. USA* **85**:7652–7656.

Innis, M. A., K. B. Myambo, D. H. Gelfand, and M. A. D. Brow. 1988. DNA sequencing with *Thermus aquaticus* DNA polymerase and direct sequencing of polymerase chain reaction-amplified DNA. *Proc. Natl. Acad. Sci. USA* **85**:9436–9440.

Prober, J. M., G. L. Trainor, R. J. Dam, F. W. Hobbs, C. W. Robertson, R. J. Zagursky, A. J. Cocuzza, M. A. Jensen, and K. Baumeister. 1987. A system for rapid DNA sequencing with fluorescent chain-terminating dideoxynucleotides. *Science* **238**:336–341.

Sanger, F., S. Nicklen, and A. R. Coulson. 1977. DNA sequencing with chain-terminating inhibitors. *Proc. Natl. Acad. Sci. USA* **74**:5463–5467.

Smith, L. M., J. Z. Sanders, R. J. Kaiser, P. Hughes, C. Dodd, C. R. Connell, C. Heiner, S. B. H. Kent, and L. E. Hood. 1986. Fluorescence detection in automated DNA sequence analysis. *Nature (London)* **321**:674–679.

25

DIRECT SEQUENCING WITH THE AID OF PHAGE PROMOTERS

Steve S. Sommer, Gobinda Sarkar, Dwight D. Koeberl, Cynthia D. K. Bottema, Jean-Marie Buerstedde, David B. Schowalter, and Joslyn D. Cassady

RNA amplification with transcript sequencing (RAWTS) (Sarkar and Sommer 1988; Sarkar and Sommer 1989) and genomic amplification with transcript sequencing (GAWTS) (Stoflet *et al.* 1988) are methods of direct sequencing that utilize a phage promoter sequence 5' to at least one of the PCR primers. RAWTS is a four-step procedure that includes: (1) cDNA synthesis with oligo(dT), random primers, or an mRNA-specific oligonucleotide primer; (2) a PCR in which one or both oligonucleotides contain a phage promoter attached to a sequence complementary to the region to be amplified; (3) transcription with a phage polymerase; and (4) dideoxy sequencing with reverse transcriptase (Fig. 1). The procedure for GAWTS is identical except that genomic DNA is the input to step 2. RAWTS and GAWTS have a number of advantages that include: (1) *the transcription step produces an additional level of amplification that obviates the need for purification subsequent to PCR*, (2) *the amplification afforded by transcription can compensate for a suboptimal PCR*, and (3) the generation of a single-stranded template provides a more reproducible sequence than that obtained from a double-stranded template. Disadvantages of the technique are the limited number of different sequencing enzymes available and the added expense of attaching

197

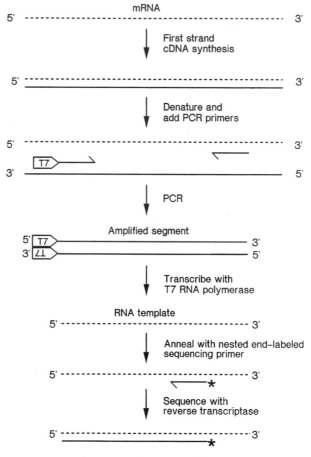

Figure 1 Schematic of RAWTS.

phage promoters to the PCR primers. As with all direct sequencing methods, RAWTS and GAWTS are insensitive to the error rate of *Taq* polymerase because the sequence generated is the average for a population of molecules.

In our laboratory, GAWTS has been applied to measure the rate of polymorphism in regions of functional significance in the factor IX gene (Koeberl *et al.* 1989). It has also been used to delineate the causative mutations in 35 hemophiliacs by sequencing eight regions of the factor IX gene, which total 2.46 kb per individual. By performing a nested series of PCR amplifications, RAWTS has been used to determine that a low level of expression of "tissue-specific" mRNAs occurs in many and possibly all tissues (Sarkar and Sommer 1989). By performing low-stringency PCR using human factor IX primers, one can directly obtain the factor IX cDNA sequence of other species (Sarkar and Sommer 1989). Finally, a translational initiation signal can be added 3' to the phage promoter sequence, thereby allowing the amplified and transcribed product to be translated *in vitro* (Sarkar and Sommer 1989).

The great versatility of PCR-based methodology has lead to a proliferation of oligonucleotide primers of diverse design. To aid in communication, we have proposed a nomenclature that is presented after the protocol for RAWTS and GAWTS (Stoflet *et al.* 1988; Sarkar and Sommer, 1989). A brief outline of some other phage promoter-based methods is presented in the Discussion.

Protocols for RAWTS

(For GAWTS, genomic DNA is added during the PCR step at a final concentration of 10 ng/μl. Currently, PCR for GAWTS is routinely performed in a total volume of 25 μl.)

First-Strand cDNA Synthesis

A 20-μl sample of 50 μg/ml heat-denatured total RNA or mRNA, 50 mM Tris–HCl (pH 8.3), 8 mM MgCl$_2$, 30 mM KCl, 1 mM DTT, 2 mM each dATP, dCTP, dGTP, dTTP, 50 μg/ml of oligo(dT) 12 to 18, 1000 units/ml of RNasin, and 1000 units/ml of AMV reverse transcriptase were incubated at 42°C for 1 hour, followed by 65°C for 10 minutes. Subsequently, 30 μl of H$_2$O was added, generating a final volume of 50 μl.

PCR

A 1-μl sample of the above was added to 40 μl of 50 mM KCl, 10 mM Tris–HCl (pH 8.3), 1.0 to 2.5 mM MgCl$_2$ (empirically determined for each set of primers), 0.01% (w/v) gelatin, 200 μM each dNTP, and typically 0.03 to 0.1 μM each primer (modified Perkin-Elmer Cetus protocol). After 10 minutes at 94°C, 0.5 unit of *Taq* polymerase was added and 30 to 40 cycles of PCR were performed (denaturation: 1 minute at 94°C; annealing: 2 minutes at 50°C; elongation: 3 minutes at 72°C) with the Perkin-Elmer Cetus automated Thermal Cycler. One primer included a T7 promoter as previously described (Stoflet *et al.* 1988). After the last cycle of PCR, a final 10-minute elongation was performed.

Transcription

A 3-μl sample of the amplified material was added to 17 μl of the RNA transcription mixture. The final mixture contains 40 mM Tris–HCl (pH 7.5), 6 mM MgCl$_2$, 2 mM spermidine, 10 mM sodium chloride, 0.5 mM of the four ribonucleoside triphosphates, RNasin (1.0 unit/μl), 10 mM DTT, 10 units of T7 RNA polymerase, and diethylpyrocarbonate-treated H$_2$O. Samples were incubated for 1 hour at 37°C, and the reaction was stopped by the addition of 12.5 mM EDTA (final concentration) or by freezing the sample.

Sequencing Protocol I [modification of Geliebter (1987)]

For most applications, Sequencing Protocol I is preferable (see Discussion). A 2-μl sample of the transcription reaction and 1 μl of ^{32}P-end-labeled (see following) reverse transcriptase primer were added to 10 μl of annealing buffer [250 mM KCl, 10 mM Tris–HCl (pH 8.3)]. The samples were heated at 80°C for 3 minutes and then annealed for 45 minutes at 45°C (approximately 5°C below the denaturation temperature of the oligonucleotide). Microfuge tubes were labeled with A, C, G, and T. The following was added: 3.3 μl of reverse transcriptase buffer [24 mM Tris–HCl (pH 8.3), 16 mM MgCl$_2$, 8 mM DTT, 0.8 mM dATP, 0.4 mM dCTP, 0.8 mM dGTP, and 1.2 mM dTTP] containing 1 unit of AMV reverse transcriptase, 1 μl of a dideoxyribonucleoside triphosphate (1 mM ddATP, 0.25 mM ddCTP, 1 mM ddGTP, or 1 mM ddTTP), and, finally, 2 μl of the primer RNA template solution. The sample was incubated at 55°C for 45 minutes and the reaction was stopped by adding 2.5 μl of 85% formamide with 25 mM EDTA, 0.1% bromophenol blue, and xylene cyanol FF. Samples were incubated at 94°C for 3 minutes, placed on ice, and 1.5 μl was loaded onto a 100-cm sequencing gel separated by electrophoresis for approximately 14,000 volt-hours. Subsequently, autoradiography was performed.

End labeling of the reverse transcriptase primer was performed by incubating a 0.1-μg sample of oligonucleotide in a 13-μl volume containing 50 mM Tris–HCl (pH 7.4), 10 mM MgCl$_2$, 5 mM DTT, 0.1 mM spermidine, 100 μCi [γ-^{32}P]ATP (5000 Ci/mmol), and 10 units of T4 polynucleotide kinase for 30 minutes at 37°C. The reaction was heated to 65°C for 5 minutes, and 7 μl of water was added for a final concentration of 5 ng of oligonucleotide per microliter. Labeled oligonucleotide (1 μl) was added per sequencing reaction without removal of the unincorporated mononucleotide.

Sequencing Protocol II [modification of Graham *et al.* (1986)]

A 2-μl sample of the transcription reaction was mixed with 5 μl of ddH$_2$O, 15 pmol oligonucleotide primer, and 0.8 μl of 10× reaction buffer [500 mM NaCl, 60 mM MgCl$_2$, 50 mM dithiothreitol, 340 mM Tris–HCl (pH 8.3)] and incubated for 15 minutes at 37°C. Five μCi [α-^{32}P]dATP (>410 Ci/mmol) was dried in a tube. Five μl of dN mix (0.5 mM each of dGTP, dCTP, dTTP), 0.8 μl of 10× reaction buffer, and the hybridized template-primer solution were added to the [^{32}P]dATP tube. Four tubes for each sequencing reaction were set up with the following solutions:

Tube	[^{32}P]dATP Mix (μl)	ddNTP Solution	AMV Reverse Transcriptase (5 Units/μl 1× Reaction Buffer) (μl)
A	3	1 μl ddATP (0.01 mM)	1
T	3	1 μl ddTTP (0.5 mM)	1
G	3	1 μl ddGTP (0.125 mM)	1
C	3	1 μl ddCTP (0.25 mM)	1

The tubes were incubated at 42°C in a water bath for 15 minutes. AMV reverse transcriptase (1.5 units) in 2 μl of chase solution (1.25 mM each of dATP, dTTP, dGTP, dCTP in 1× reaction buffer) was added to each tube and incubated for an additional 15 minutes at 42°C. The reactions were terminated by adding 3.2 μl of stop buffer (see Sequencing Protocol I) to each tube. Samples were incubated at 94°C for 3 minutes just prior to loading the sequencing gel. Samples were then placed on ice, and 1.7 μl was loaded per lane.

Nomenclature

The following nomenclature provides the relevant information on the oligonucleotide and readily allows the determination of the size of the amplified fragment and the origin and direction of the sequence generated. It is of the form: G(O)-(I-L)R(C)-SD, where G = gene abbreviation, O = organism, I = identifier(s) for the noncomplementary 5' bases, L = length of the noncomplementary bases, R = region of the gene, C = location of the 5' complementary base, S = total size of the oligonucleotide, and D = 5' to 3' direction of the oligonucleotide. The region of the gene (R) is abbreviated by 5', the region upstream of the gene; by E followed by exon number; by I followed by intron number; or by 3', the region downstream of the gene. The direction of the oligonucleotide is either U (upstream) or D (downstream). If a transcript has been defined, D is the sense direction and U is the antisense direction. Otherwise, the directions can be arbitrarily defined. Thus, BP(Hs)-(T7–29)E5(1453)-44U is an oligonucleotide specific for the blue pigment gene of *Homo sapiens* that has a T7 promoter (includes a six-base clamping sequence at the 5' end) of 29 base pairs. It is complementary to a sequence in exon 5 that begins at base 1453, using the numbering system of Nathans *et al.* (1986). The oligonucleotide is a 44-mer, which heads upstream relative to blue pigment mRNA. If there is no chance of confusion with other sequences, the name can be abbreviated in routine use by omitting the understood designations such as "BP," "(Hs)," and/or "(T7–29)." As another example, F9(Hs)-5'(-120)-15D is an oligonucleotide specific for human factor IX. It is complementary to a 15-base sequence 5' to the human factor IX gene that begins at base

−120. The oligonucleotide is a 15-mer, and the sequence heads downstream relative to *in vivo* transcription.

Discussion

Almost all our experience has been with sequencing protocol I. A 6- to 16-hour exposure is generally sufficient to obtain a strong signal on the autoradiogram. The intensity of the signal is unaffected by the threefold reduction of reverse transcriptase from our previous protocol (Sarkar and Sommer 1988), and the quality of the sequencing reaction is improved.

Sequencing protocol II provides an even stronger signal because of the incorporation of radiolabeled dATP, *but self-priming by RNA can be a problem*. Optimization of PCR to eliminate spurious amplification products will decrease the chance of such a problem. A 1-hour room temperature exposure is sometimes sufficient with Protocol II. By increasing the amount of label sixfold, a 10-minute exposure can suffice.

For sequencing X-linked genes in males, or for sequencing a cloned allele of a previously sequenced gene, a sequence change will produce both a new band and the absence of the expected band. Despite occasional shadow bands (especially at a subset of the "T bands"), it has been possible to obtain an unequivocal sequence on 2.46 kb of the factor IX gene as indicated by the delineation of a putative mutation in 35 of 36 families with hemophilia B.

For sequencing autosomal genes in which a sequence change on one chromosome will likely be paired with a normal chromosome, the best approach is to incorporate different phage promoters (e.g., T7 and Sp6, or T7 and T3) on each PCR primer and to sequence both strands. Alternately, by sequencing only one strand, it is possible to unequivocally rule out heterozygosity for more than 99% of the positions. However, the elimination of even faint shadow bands is critical. Reducing the concentration of reverse transcriptase may help, but another as yet undefined parameter(s) is important because, for different individuals performing GAWTS on the same region, the incidence of significant shadow bands in a region can vary from essentially 0 to 3% of the bands. Performing a chase with terminal deoxynucleotidyl transferase may also help to eliminate shadow bands (DeBorde *et al.* 1986).

Most of our experience has been with a 29-base T7 phage promoter sequence that contains the canonical 23-base promoter sequence

(TAATACGACTCACTATAGGGAGA)

3' to a six-base sequence (GGTACC), which positions the promoter sequence away from the end of the PCR product. Our current data suggest that this six-base sequence is not necessary for either the T7 or Sp6 promoters. The canonical SP6 and T3 promoter sequences are

AATTAGGTGACACTATAGAATAG   and
AATTAACCCTCACTAAAGGGAAG

respectively.

To generate PCR primers that allowed *in vitro* translation of the phage transcript, CCACCATG was added directly 3' to the phage promoter sequence (Sarkar and Sommer 1989). Before *in vitro* translation, it is important to eliminate certain inhibitors present in the PCR. This can be done by purifying the appropriate PCR product with GENECLEAN (Bio101) from an agarose gel before transcription. Subsequent to transcription, the sample is digested with DNase I (0.02 units/μl), extracted with phenol/chloroform, chromatographed through a G50 Sephadex column, ethanol-precipitated in the presence of ammonium acetate, rinsed with 70% ethanol, suspended in 10 mM Tris-HCl (pH 8.0)/1 mM EDTA, and translated in a reticulocyte lysate as recommended by the manufacturer (Promega).

GAWTS and RAWTS require the generation of PCR primers that are specific for the sequences on *both* sides of a region. Promoter ligation and transcript sequencing (PLATS) is a direct method for rapidly obtaining novel sequences that utilizes generic primers and only requires knowledge of the sequence on *one* side of a region (Schowalter *et al.*, in press). PLATS involves restriction digestion of DNA, ligation with a phage promoter, and then GAWTS using promoter sequences as the PCR primers. PLATS is a rapid and economical method of sequencing lambda clones or, potentially, unknown regions of genomic DNA adjacent to regions of known sequence. It uses a limited set of oligonucleotides, and it is potentially amenable to automation because it does not require *in vivo* manipulations.

The inclusion of phage promoters onto PCR primers allows labeled RNA probes to be generated from any PCR-amplified segment (Schowalter and Sommer 1989). A direct comparison between "PCR labeled" DNA probes and "transcript labeled" RNA probes indicates that single-stranded RNA does not generate a stronger signal on Southern blots than does double-stranded DNA of the same specific activity.

Acknowledgments

This research was aided by March of Dimes grant 5–647. Fellowship support was received from March of Dimes Birth Defect Foundation 18–3 (DDK), NIH CA09441–06Q (CDKB), and the Deutsche Forschungsgemeinschaft (J-MB).

Literature Cited

DeBorde, D. C., C. W. Naeve, M. L. Herlocher, and H. F Maassab. 1986. Resolution of a common RNA sequencing ambiguity by terminal deoxynucleotidyl transferase. *Anal. Biochem.* **157**:275–282.

Geliebter, J. 1987. Dideoxynucleotide sequencing of RNA and uncloned cDNA. *Focus* **9**:5–8.

Graham, A., J. Steven, D. McKechnie, and W. J. Harris. 1986. Direct DNA sequencing using avian myeloblastosis virus and Moloney murine leukemia virus reverse transcriptase. *Focus* **8**:4–5.

Koeberl, D. D., C. D. K. Bottema, J.-M. Buerstedde, and S. S. Sommer. 1989. Functionally important regions of the factor IX gene have a low rate of polymorphism and CpG is a dramatic hotspot of germline mutation. *Am. J. Hum. Genet.* **45**:448–457.

Nathans, J., D. Thomas, and D. S. Hogness. 1986. Molecular genetics of human color vision: the genes encoding blue, green, and red pigments. *Science* **232**:193–202.

Sarkar, G., and S. S. Sommer. 1988. RNA amplification with transcript sequencing (RAWTS). *Nucleic Acids Res.* **16**:5197.

Sarkar, G., and S. S. Sommer. 1989. Access to a messenger RNA sequence or its protein product is not limited by tissue or species specificity. *Science* **244**:331–334.

Schowalter, D. B., and S. S. Sommer. 1989. The generation of radiolabeled DNA and RNA probes with polymerase chain reaction. *Anal. Biochem.* **177**:90–94.

Schowalter, D. B., D. O. Toft, and S. S. Sommer. A method of sequencing without subcloning and its application to the identification of a novel ORF with a sequence suggestive of a transcription regulator in the water mold, *Achlya ambisexualis. Genomics*, in press.

Stoflet, E. S., D. D. Koeberl, G. Sarkar, and S. S. Sommer. 1988. Genomic amplification with transcript sequencing. *Science* **239**:491–494.

26

IDENTIFYING DNA POLYMORPHISMS BY DENATURING GRADIENT GEL ELECTROPHORESIS

Val C. Sheffield, David R. Cox, and Richard M. Myers

Denaturing gradient gel electrophoresis (DGGE) is a technique that allows the separation of DNA molecules differing by single base changes (Fischer and Lerman 1983; Myers *et al.* 1985a; Myers *et al.* 1985b; Myers *et al.* 1985c; Myers and Maniatis 1986). The separation is based on the fact that DNA molecules differing by a single base change have slightly different melting properties, which cause them to migrate differently in a polyacrylamide gel containing a linear gradient of DNA denaturants (Fischer and Lerman 1980; Fischer and Lerman 1983). DGGE can be used to detect about 50% of all possible single base changes in DNA fragments up to about 1000 bp in length. Furthermore, DGGE can be used to detect polymorphisms not identifiable by other techniques, such as RFLP analysis. These features make DGGE useful for detecting DNA polymorphisms in genetic linkage studies. The polymerase chain reaction (PCR) (Saiki *et al.* 1985; Saiki *et al.* 1986; Mullis *et al.* 1986; Mullis and Faloona 1987) has made it possible to amplify genomic DNA samples suitable for direct analysis by DGGE without the need for a labeled probe (Sheffield *et al.* 1989). In this chapter we present two simplified approaches for using DGGE to identify polymorphisms in PCR-amplified genomic DNA samples. The approaches described

PCR Protocols: A Guide to Methods and Applications

here are designed for cases that require detection of some but not all of the genetic variation in genomic DNA.

Strategies for Identifying Polymorphisms by DGGE

There are two types of denaturing gradient gels: (1) parallel gels, which contain a linearly increasing gradient of DNA denaturants from top to bottom in the gel and (2) perpendicular gels, which contain a linear gradient of denaturants from left to right across the gel. Both gel types can be used independently to identify DNA polymorphisms in PCR-amplified genomic DNA. Parallel denaturing gradient gels have the advantage of being useful for analysis of multiple samples on a single gel. Thus parallel DGGE is particularly suited for searching for polymorphisms in multiple members of a large pedigree simultaneously. Perpendicular DGGE can be used to efficiently identify polymorphisms in DNA from a single individual. The perpendicular gel can thus be used to search for polymorphisms in a key individual in a pedigree (i.e., an affected parent with multiple offspring). Once a polymorphism is found in a probe by perpendicular DGGE, multiple individuals can be examined by parallel DGGE. Strategies for using each of these types of gels to identify polymorphisms will be discussed here. These approaches have several advantages over previously described DGGE protocols, including (1) the ability to screen relatively large regions of DNA for base changes without a labeled probe, (2) no requirement for preliminary DNA melting determinations, (3) standardized gel denaturant concentrations and electrophoresis conditions, and (4) increased sensitivity, allowing the use of small amounts of genomic DNA. In addition, the use of DGGE to identify polymorphisms in PCR-amplified genomic DNA has a major advantage over other techniques that require a labeled probe. Such techniques are often limited to the use of probes that are free of repeat sequences (especially low- to medium-copy repeat sequences), since repeat sequences hybridize to multiple sites within the genome. The techniques described here examine DNA that has been amplified between specific oligonucleotide primers, so repeat sequences lying within the amplified region of DNA do not interfere with the analysis.

Parallel Denaturing Gradient Gel Approach

The strategy for using parallel DGGE to identify DNA polymorphisms is summarized in Fig. 1. A 1- to 5-kb region of genomic DNA is amplified by PCR and the resulting DNA product is digested with frequently cutting restriction endonucleases. Aliquots of the digested DNA are then run on two different denaturing gradient gels with overlapping gradient ranges (i.e., 10 to 50% and 40 to 80%). This approach allows the analysis of DNA from up to twenty individuals on a single gel. Polymorphisms are recognized by the appearance of bands in some DNA samples not seen in other samples, or by the shift in position of a band in one or more samples compared to other samples. An example of a polymorphism identified by this approach in a probe that recognizes a locus on human chromosome 21 is shown in Fig. 2.

We estimate that by digesting PCR-amplified DNA with two different restriction enzymes and by analyzing the products of the two digestions on the two different parallel denaturing gradient gels, between 50 and 80% of possible base changes are detected. Therefore, at least 1.5 kb is screened for base changes in a 3.0-kb fragment examined by parallel DGGE. Furthermore, the parallel DGGE approach described here is simplified compared to previous approaches in two significant ways: (1) it does not require a labeled probe because the PCR-amplified DNA can be visualized directly with ethidium bromide staining and (2) it eliminates the need for preliminary experiments to determine optimal gel conditions for each DNA fragment by using standardized gel conditions. One drawback of this approach is that it is not always apparent which DNA bands on the gel are allelic. Alleles may be determined by analyzing multiple members of the same family and demonstrating Mendelian inheritance of the specific DNA bands.

Perpendicular Gel Approach

Perpendicular denaturing gradient gels contain a linear gradient of DNA denaturants across the gel, perpendicular to the direction of electrophoresis. When a DNA sample is loaded across the entire width of a perpendicular denaturing gradient gel and electrophoresed, a melting curve for each DNA fragment is obtained. On the left side of the gel where the concentration of denaturant is low, the DNA fragment migrates at a rate dependent only on its molecular weight.

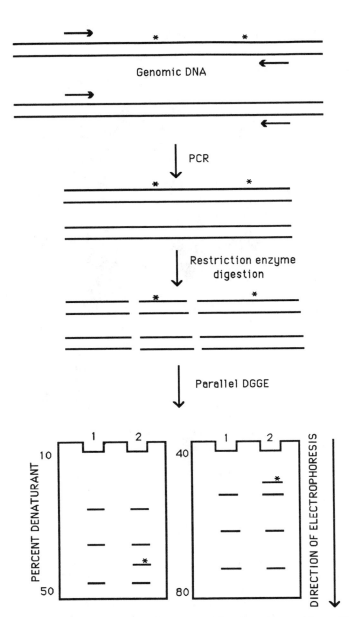

Figure 1 Strategy for using parallel DGGE to identify polymorphisms. The upper drawing represents genomic DNA with two polymorphisms within a 3-kb region (*). Arrows indicate the location of primers for PCR. Genomic DNA samples from different individuals are PCR-amplified in separate reactions, and the products of PCR are digested with a restriction enzyme. The resulting restriction fragments are run on two denaturing gradient gels, each having a different range of denaturant concentrations. An extra band is present in sample 2 in each of the gels, indicating the presence of polymorphisms. Note that the different conditions for each gel allow the identification of a different polymorphism.

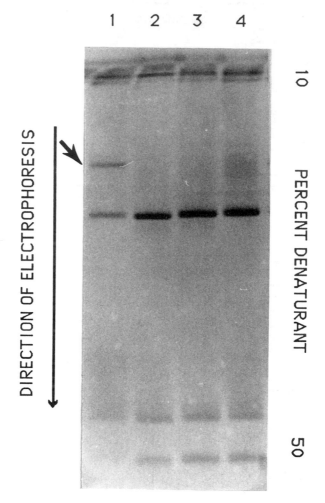

Figure 2   Identification of a DNA polymorphism by the parallel DGGE approach. DNA from several individuals was PCR-amplified with primers from a probe located on chromosome 21. The resulting 1.3-kb band was digested with the restriction enzyme *Rsa*I and electrophoresed on a 10–50% denaturant gel. An extra band (arrow) is seen in lane 1, indicating a polymorphism in this DNA sample. The picture is a negative image of an ethidium bromide-stained gel.

On the right side where the denaturant concentration is high, the fragments remain near the top of the gel because of the formation of a partially single-stranded molecule, which has reduced mobility in the gel compared to a completely double-stranded DNA molecule. One or more steep transitions in the DNA curve are seen between

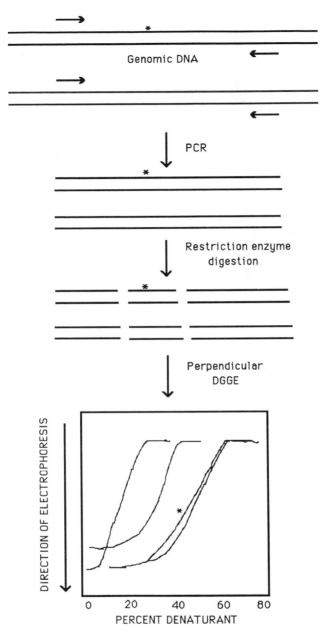

Figure 3 Strategy for using perpendicular DGGE to identify polymorphisms. The upper drawing illustrates genomic DNA from an individual heterozygous for a single base substitution (*). Arrows represent primers used for PCR. The products of PCR are restriction digested with restriction enzymes and electrophoresed on a perpendicular denaturing gradient gel. The lower drawing illustrates the gel patterns observed for each of three restriction fragments. One gel pattern (*) shows a splitting of the line of transition, indicating a single base difference between the two alleles.

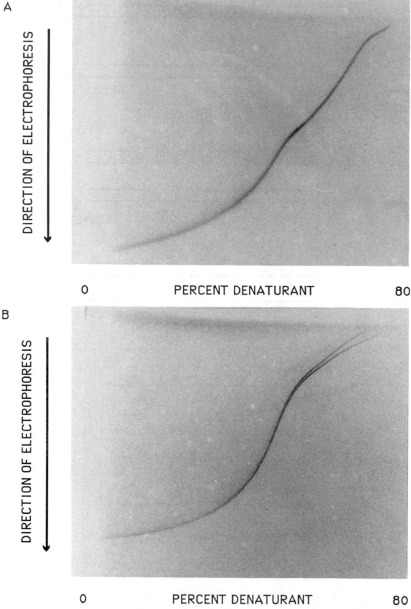

Figure 4 Identification of a DNA polymorphism by the perpendicular DGGE approach. (A) Perpendicular DGGE of a 700-bp fragment PCR-amplified with primers from a DNA sequence on chromosome 21. The individual from whom this DNA was amplified was not polymorphic in this region because only a single line of transition is observed. (B) The same sequence PCR-amplified from a different individual and electrophoresed on a perpendicular denaturing gradient gel. Two diverging lines of transition are observed at the top of the gel, indicating that the individual from whom this sample was amplified carries a base change within one of the melting domains of this fragment (i.e., the individual is heterozygous for a single base polymorphism). The picture is a negative image of an ethidium bromide-stained gel.

these two regions, resulting in a sigmoid-shaped pattern. When two DNA fragments differing by only a single base change are run together on a perpendicular gel, the transition zone for one of the fragments is shifted slightly to the right or left compared to that of the other fragment because of the slight difference in the melting properties of the two fragments. Usually, the difference is noticeable only if the two fragments are run together on the same gel. Generally, perpendicular denaturing gradient gels have been used to determine the melting-domain structure of DNA fragments to optimize conditions for analyzing them by parallel DGGE (Myers *et al.* 1987; Myers *et al.* 1988). However, in a new approach described here, perpendicular denaturing gradient gels can also be used to identify rare polymorphisms in DNA fragments (Fig. 3). The approach is to PCR-amplify specific DNA fragments from 1 to 5 kb in length, digest the DNA with frequently cutting restriction enzymes, and run the DNA fragments on a perpendicular denaturing gradient gel. If an individual is heterozygous for a single base change within the region of the amplified DNA fragment, two fragments differing by a single base pair are amplified from that individual's DNA. When the amplified sample from the polymorphic individual is electrophoresed on a 0 to 80% perpendicular denaturing gradient gel, a split in one of the curves is seen (Fig. 4). Samples amplified from a nonpolymorphic individual appear as single DNA curves that do not split into two. Similar to parallel DGGE, perpendicular DGGE can be used to identify more than 50% of all base changes within a DNA fragment. However, when attempting to identify polymorphisms in a single individual, perpendicular DGGE is often more efficient, because it utilizes single standardized gel and electrophoresis conditions, and the entire DNA sample is exposed to a very broad range of denaturant concentrations.

Protocols

Preparation of PCR-Amplified DNA Samples for DGGE

PCR has made it possible to analyze amplified DNA samples prepared from small amounts of genomic or cloned DNA templates on denaturing gradient gels by using only ethidium bromide staining. To use this approach, it is important that the amplified DNA is relatively free of spurious DNA fragments that sometimes result from the PCR procedure. Amplification of a pure single DNA fragment from genomic DNA has been accomplished with nearly all primer pairs we have tried by using established PCR protocols (Saiki *et al.* 1985; Saiki *et al.* 1986; Mullis *et al.* 1986; Mullis and Faloona 1987) and by optimizing the hybridization temperature. It is often necessary to empirically determine the optimum hybridization temperature for a given pair of primers. We have generally found that the use of relatively high temperatures limits the amplification of spurious DNA fragments (also, see Chapter 1).

In choosing primers for PCR amplification of genomic DNA, we select opposing primers of 20 to 25 nucleotides in length with similar G + C contents (ideally at least 50% G + C). It is important that primers not contain inverted repeat sequences or sequences repeated in the genome. In general, we design experiments such that about 3 kb of DNA is amplified. It is likely that larger regions of DNA can be amplified by PCR, and based on experiments with cloned DNA (Myers and Maniatis 1986), the DGGE approaches that we describe here are effective with amplified DNA up to about 5 kb when digested with frequently cutting restriction enzymes.

When a pair of primers is used for PCR amplification for the first time, it is important to confirm that the correct genomic DNA sequence is amplified. This can be done by demonstrating that the amplified DNA fragment is the predicted size, has the predicted restriction digestion pattern, and hybridizes to the correct cloned DNA sequence.

PCR amplification of genomic DNA can be performed with any of the published protocols (Saiki *et al.* 1985; Saiki *et al.* 1986; Mullis *et al.* 1986; Mullis and Faloona 1987). We generally include the following in a 100-μl PCR sample: 10 μl of 10× PCR buffer [670 mM Tris–HCl (pH 8.8), 67 mM $MgCl_2$, 160 mM ammonium sulfate, 100 mM 2-mercaptoethanol], 10 μl of dimethyl sulfoxide, 10 μl of dNTP mix (12.5 mM each dNTP), template DNA (50 to 500 ng of genomic DNA), 50 pmol of each oligonucleotide primer, 1 to 1.5 units of *Thermus aquaticus* DNA polymerase, and water to 100 μl. Approximately 100 μl of mineral oil is layered over each sample, and samples are incubated at 93°C, 50 to 65°C, and 70°C for 1 minute, 1 minute, and 1 to 4 minutes, respectively. During the polymerization incubation at 70°C, we allow 1 minute for each length of 500 bp between primers. The incubations are repeated for 30 to 40 cycles. When PCR-amplifying genomic DNA, it should be remembered that the DNA is diploid, and if the individual is heterozygous, two DNA fragments differing by a single base change will be amplified. In these cases, the melting and reannealing steps in the PCR procedure produce four species of amplified fragments: Two homoduplexes and two heteroduplexes (Sheffield *et al.* 1989). Heteroduplexes become a significant portion of the amplified DNA product after many cycles of PCR when the concentration of amplified DNA becomes high. At this point, the high concentration of full-length amplified DNA competes with the oligonucleotide primers during the reannealling step of PCR, allowing heteroduplexes to form. Heteroduplex formation is advantageous, since heteroduplexes (containing single base mismatches) usually separate from the wild-type homoduplex band to a greater extent than from the homoduplex "mutant" band on denaturing gradient gels. In some PCR samples the amplified DNA does not reach a high enough concentration to significantly compete with the oligonucleotide primers during the hybridization step; thus heteroduplexes make up only a minor portion of the total PCR-amplified product. If heteroduplexes are desired in greater amounts, the sample can be denatured and allowed to reanneal after the *Taq* polymerase activity has been destroyed by phenol extraction.

PCR-amplified samples are digested with one or more frequently cutting restriction enzymes to produce fragments ranging from 100 to 500 bp in length. We recommend the use of *Sau*3A and *Hae*III, although other enzymes may be equally useful. All samples are extracted with phenol, precipitated with ethanol, and resuspended in nondenaturing loading buffer (Maniatis *et al.* 1982) before analysis by DGGE.

Gel Preparation

The DGGE system including materials, reagents, gel preparation and electrophoresis procedures have been described in detail in Myers *et al.* (1987, 1988).

Parallel Denaturing Gradient Electrophoresis

The procedure to identify DNA polymorphisms by using parallel DGGE is as follows:

1. Perform PCR amplification of genomic DNA as previously described. Electrophorese a small aliquot (5%) of the amplified sample on an agarose gel to test for purity and success of the amplification.

2. Digest the amplified DNA with one or more frequently cutting restriction enzymes. Ideally, the resulting DNA fragments should range from about 100 to 500 bp in length.

3. Extract the DNA with phenol and precipitate with ethanol. Resuspend each sample in 25 to 50 μl of nondenaturing loading buffer (Maniatis *et al.* 1982).

4. Prepare two 8% polyacrylamide parallel denaturing gradient gels containing overlapping ranges of denaturants (Myers *et al.* 1987; Myers *et al.* 1988). We recommend that one gel contain a 10 to 50% gradient of denaturants and the other gel contain a 40 to 80% gradient of denaturants.

5. Load 5 to 10 μl of each sample into wells of each of the two parallel denaturing gradient gels and electrophorese for 6 to 10 hours at 60°C and 150 V. It is sometimes more fruitful to electrophorese the samples for two different time periods (i.e., 6 and 10 hours); this can be readily accomplished by loading each sample twice during the electrophoretic run. Results obtained with a particular digest of a fragment indicate whether longer or shorter electrophoresis times are best, and subsequent experiments can usually be optimized for a single electrophoresis time.

6. After electrophoresis, stain the gel with ethidium bromide by standard protocols and photograph under UV transillumination. Polymorphisms appear as either additional bands or a shift in position of an existing band (Figs. 1 and 2).

Perpendicular Denaturing Gradient Electrophoresis

The procedure for identifying DNA polymorphisms by using perpendicular DGGE is as follows:

1. PCR-amplify genomic DNA from a single individual as described previously for parallel DGGE.

2. Digest the amplified DNA with restriction enzymes. Fragments ranging in size from 100 to 800 bp are well suited for analysis on perpendicular denaturing gradient gels.

3. Extract the DNA with phenol, precipitate with ethanol, and resuspend in 150 μl of nondenaturing loading buffer.

4. Prepare an 8% polyacrylamide perpendicular denaturing gradient gel containing a 0 to 80% range of denaturants as described in Myers *et al.* (1987, 1988).

5. Load the entire DNA sample into the single large well at the top of the gel. Electrophorese the sample for 6 hours at 60°C and 150 V. After electrophoresis, stain the gel with ethidium bromide and photograph under UV transillumination. A polymorphism is readily identified as a splitting of a single transition line into two lines within the gel (Figs. 3 and 4).

Conclusions

The approaches that have been described in this chapter can be used to identify DNA polymorphisms for genetic linkage studies. Although we have identified polymorphisms by using both perpendicular and parallel DGGE, our data are currently limited, and we cannot predict which approach will ultimately prove the more useful for identifying polymorphisms. A major advantage of the two approaches described here is that they can be used to identify polymorphisms in regions of DNA containing repeat sequences. It should be emphasized that neither of these strategies makes it possible to identify every existing single base change. For those cases where it is desirable to detect all possible base changes, we recommend another approach involving the use of PCR to attach a 40-bp, G + C-rich sequence to amplified genomic DNA before analysis by DGGE. This approach has been previously described in Sheffield *et al.* (1989).

Acknowledgments

We thank Leonard Lerman, Tom Maniatis, Ezra Abrams, and Kary Mullis for their suggestions and advice. This work was supported by the Wills Foundation, the Searles Scholars Program, and a grant from the NIH to R.M.M. and D.R.C. V.C.S. was supported by NIH Postdoctoral Training Grant GM07085.

Literature Cited

Fischer, S. G., and L. S. Lerman. 1980. Separation of random fragments of DNA according to properties of their sequences. *Proc. Natl. Acad. Sci. USA* **77**: 4420–4424.

Fischer, S. G., and L. S. Lerman. 1983. DNA fragments differing by single base-pair substitutions are separated in denaturing gradient gels: correspondence with melting theory. *Proc. Natl. Acad. Sci. USA* **80**: 1579–1583.

Maniatis, T., E. F. Fritsch, and J. Sambrook. 1982. In *Molecular cloning: a laboratory manual*, p. 160. Cold Spring Harbor Laboratory, Cold Spring Harbor, New York.

Mullis, K. B., F. A. Faloona, S. J. Scharf, R. K. Saiki, G. T. Horn, and H. A. Erlich. 1986. Specific enzymatic amplification of DNA in vitro: the polymerase chain reaction. *Cold Spring Harbor Symp. Quant. Biol.* **51**: 263–273.

Mullis, K. B., and F. A. Faloona. 1987. Specific synthesis of DNA in vitro via a polymerase catalyzed chain reaction. *Methods Enzymol.* **155**: 335–350.

Myers, R. M., N. Lumelsky, L. S. Lerman, and T. Maniatis. 1985a. Detection of single base substitutions in total genomic DNA. *Nature (London)* **313**: 495–498.

Myers, R. M., S. G. Fischer, T. Maniatis, and L. S. Lerman. 1985b. Modification or the melting properties of duplex DNA by attachment of a GC-rich DNA sequence as determined by denaturing gradient gel electrophoresis. *Nucleic Acids Res.* **13**: 3111–3129.

Myers, R. M., S. G. Fischer, L. S. Lerman, and T. Maniatis. 1985c. Nearly all single base substitutions in DNA fragments joined to a GC-clamp can be detected by denaturing gradient gel electrophoresis. *Nucleic Acids Res.* **13**: 3131–3145.

Myers, R. M., and T. Maniatis. 1986. Recent advances in the development of methods for detecting single-base substitutions associated with human genetic diseases. *Cold Spring Harbor Symp. Quant. Biol.* **51**: 275–284.

Myers, R. M., T. Maniatis, and L. S. Lerman. 1987. Detection and localization of single base changes by denaturing gradient gel electrophoresis. *Methods Enzymol.* **155**: 501–527.

Myers, R. M., V. C. Sheffield, and D. R. Cox. 1988. In *Genome analysis: a practical approach* (ed. K. Davies), p. 95. IRL Press, Oxford.

Saiki, R. K., S. Scharf, F. Faloona, K. B. Mullis, G. T. Horn, H. A. Erlich, and N. Arnheim. 1985. Enzymatic amplification of β-globin genomic sequences and restriction site analysis for diagnosis of sickle cell anemia. *Science* **230**: 1350–1354.

Saiki, R. K., T. L. Bugawan, G. T. Horn, K. B. Mullis, and H. A. Erlich. 1986. Analysis of enzymatically amplified β-globin and HLA-DQ α DNA with allele-specific oligonucleotide probes. *Nature (London)* **324**: 163–166.

Sheffield, V. C., D. R. Cox, L. R. Lerman, and R. M. Myers. 1989. Attachment of a 40-base-pair G+C-rich sequence (GC-clamp) to genomic DNA fragments by the polymerase chain reaction results in improved detection of single-base changes. *Proc. Natl. Acad. Sci. USA* **86**: 232–236.

27

AMPLIFICATION OF FLANKING SEQUENCES BY INVERSE PCR

Howard Ochman, Meetha M. Medhora, Dan Garza, and Daniel L. Hartl

A major limitation of conventional PCR is that DNA sequences situated immediately outside the primers are inaccessible because an oligonucleotide that primes synthesis into a flanking region produces only a linear increase in the number of copies since there is no primer in the reverse direction. A method has been developed independently by three groups (Ochman *et al.* 1988; Silver and Keerikatte 1989; Triglia *et al.* 1988) that allows the *in vitro* amplification of DNA *flanking* a region of known sequence. This technique ("inverse PCR") is based on the simple procedures of digestion of source DNA with restriction enzymes and circularization of cleavage products before amplification using primers synthesized in the opposite orientations to those normally employed for PCR. In general, inverse PCR permits the amplification of the upstream and/or downstream regions flanking a specified segment of DNA without resorting to conventional cloning procedures and can also be used to rapidly produce hybridization probes for identifying and orienting the adjacent or overlapping clones from a DNA library.

In this chapter we present a detailed protocol for inverse PCR along with several modifications that have proven to be useful in

PCR Protocols: A Guide to Methods and Applications
219

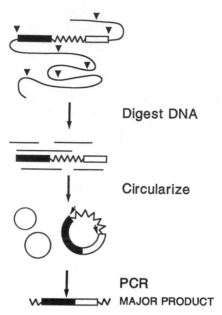

Figure 1   Application of inverse PCR. The core region is depicted as a jagged line. Filled and open boxes represent the upstream and downstream flanking regions, respectively, and restriction enzyme recognition sites are denoted by triangles. Oligonucleotide primers constructed to anneal to the core region and the direction of DNA synthesis are shown by arrows. See text for protocols and applications.

certain applications. As an example of the technique, we employ inverse PCR to generate end-specific DNA fragments for chromosome walking within a yeast artificial chromosome (YAC) library constructed from *Drosophila melanogaster* (Burke *et al*. 1987; Garza *et al*. 1989; in press).

Protocol for Inverse PCR

The basic methodology is shown schematically in Fig. 1. In the following discussions we refer to the segment of DNA that serves as the template for the oligonucleotide primers as the "core" region.

DNA Digestions

There are several factors that influence the selection of restriction endonucleases for inverse PCR, including the restriction map of the core and flanking regions and the size of the resulting DNA fragments to be circularized. In most cases, however, the choice of restriction enzymes can be readily established by examining the sequence of the core region or by determining cleavage sites and fragment lengths by Southern blots. For example, to obtain sequences both upstream and downstream from the core region in a single amplification, the DNA is initially cleaved by a restriction enzyme (or enzymes) having no recognition sites within the core region, as shown in Fig. 1. (If the restriction enzymes generate ends that are not compatible for ligations, DNA fragments are end-repaired with Klenow or T4 DNA polymerase before ligation.) Alternatively, by selecting enzymes that cleave *within* the core region, these procedures can, at this point, be modified to obtain only upstream or downstream flanking regions. The released fragment should be no more than 2 to 3 kb longer than the core region, a limitation imposed by the size of a region that can be efficiently amplified by using PCR.

After the restriction enzyme digestion of source DNA, one can usually proceed directly to the ligation steps detailed as follows. In some applications, as in the case where there are multiple copies of the core region or when the size of the desired product is not known, it is advisable to size-fractionate the DNA fragments containing the core region to enrich for a specific product. In any case, a total of about 1 μg of cleaved genomic DNA is sufficient for subsequent procedures.

Circularization

DNA fragments are ligated under conditions that favor the formation of monomeric circles (Collins and Weissman 1984). Samples are extracted once with phenol and once with chloroform to inactivate and remove restriction endonucleases. The aqueous phase is saved, one-tenth volume of cold 2.5 M ammonium acetate is added, and the DNA is then precipitated with 2 volumes of ethanol and collected by centrifugation. The DNA pellet is washed twice with 70% ethanol to remove salts that might inhibit ligase activity and dried under vacuum. DNA is then resuspended in ligation buffer [50 mM Tris–HCl (pH 7.4), 10 mM MgCl$_2$, 10 mM dithiothreitol, 1 mM adenosine triphosphate] to a concentration of less than 2.0 μg/ml. The reaction is initiated by the addition of T4 DNA ligase to 0.02 Weiss units per microliter and allowed to proceed 16 hours at 15°C.

The procedure is often simplified by initially digesting source DNA in a small volume with heat-labile restriction endonucleases. After digestion, the sample is heated to 68°C to inactivate restriction enzymes, whereby subsequent deproteinization and precipitation steps are eliminated. An aliquot is then diluted at least fivefold in ligation buffer to assure that both salt and DNA concentrations promote efficient ligations.

Polymerase Chain Reaction

Circularized DNA molecules are precipitated by the addition of salt and ethanol and resuspended in 10 μl of distilled water. Typical PCR mixtures and conditions are used (30 cycles of denaturation at 94°C for 30 seconds, primer annealing at 50°C for 30 seconds, and extension by *Taq* polymerase at 70°C for 2 minutes) that can be modified to accommodate certain primers and amplification products (Saiki *et al*. 1985, 1988). Unlike the oligonucleotide primers normally employed in PCR, those used for inverse PCR are complementary to the opposite strand and, therefore, oriented such that extension proceeds outward from the core region. We should also note that it is not necessary to reopen the circularized molecule at a site within the core region to promote efficient amplifications.

Using Inverse PCR for Chromosome Walking

Since regions of rather limited size can be enzymatically amplified by PCR, and primers must be constructed from characterized sequences, inverse PCR is apparently not amenable to proceeding long distances in the genome. We have, however, adapted inverse PCR to generate end-specific hybridization probes to be used for orienting and aligning large DNA fragments cloned as YACs.

Recovery of Yeast Artificial Chromosomes

Burke *et al.* (1987) have described YAC cloning vectors and methods for the construction of YACs containing inserts on the order of hundreds of kilobases in length. Chromosomal DNA is prepared in agarose plugs to minimize breakage, and intact yeast chromosomes are resolved by pulsed-field gel electrophoresis (Carle and Olson 1984; Schwartz and Cantor 1984; Chu *et al.* 1986). Yeast chromosomes range from 240 kb to well over a megabase, and YACs appear as bands in addition to the normal complement of the host.

To generate a DNA fragment corresponding to the 3' end of a 140-kb YAC clone (DY2, mapped to *D. melanogaster* chromosome 3 by *in situ* hybridization), the chromosomes of a yeast transformant are separated on a 1% low-melting agarose (SeaKem) gel according to the conditions described by Carle *et al.* (1986). After electrophoresis, the region of the gel containing the YAC (DY2) is excised with a razor blade and equilibrated in 50 volumes of TE [10 m*M* Tris–HCl (pH 8.0), 1 m*M* EDTA] to elute borate from the gel slice.

Preparation of Restriction Fragments

After two washes (each for 2 hours) with TE, the gel slice is placed in a microfuge tube and heated to 68°C for 20 minutes to melt the agarose. On the basis of the number of cells cultured for DNA preparations, the gel slice contains roughly 5 to 10 ng of DY2 DNA. One-fourth of this sample (25 μl) is mixed with an equal volume of distilled water and cleaved with five units each of *Eco*RV and *Hinc*II according to the specifications of the supplier (New England Bio-Labs, Inc., Beverly, Massachusetts). These enzymes were selected by examining the sequence of the vector arms and, because of their recognition sequences, should generate fragments of convenient lengths for PCR. Moreover, *Eco*RV and *Hinc*II both yield blunt ends and can be denatured by heat. (If the resulting fragments are too large to amplify by PCR, one can introduce a third enzyme that is heat labile and produces flush ends.) The reaction proceeds for 2 to 4 hours at 37°C and is terminated by heating to 68°C for 30 minutes.

Circularization

Ligations are performed in a total reaction volume of 50 μl containing 10 μl of the molten restriction enzyme reaction mix, 5 μl of a 10× ligation buffer, 34 μl of distilled water, and 1 Weiss unit of T4 DNA ligase. The reaction is left overnight at 15°C and terminated by heating to 68°C for 15 minutes.

Amplification of End-Specific Fragments by PCR

A 10-μl sample of the ligation mix is used for PCR following the standard reaction conditions previously described. To generate a fragment including only the 3' end of the YAC, we employed the following primers in our PCR amplifications:

5'-AGGAGTCGCATAAGGGAG-3'

which corresponds to a sequence near the HincII site in the non-centromeric arm of the YAC vector; and

5'-GGGAAGTGAATGGAGAC-3'

a sequence adjacent to the SmaI cloning site in the sup4 gene of the vector (see diagram in Fig. 2). A single pair of primers is used to amplify each end in separate reactions.

The results of these procedures are presented in Fig. 2. In Panel (A), EcoRV- and HincII-cut genomic DNAs from D. melanogaster strain Oregon-R and from two yeast transformants containing artificial chromosomes are probed with a segment of pBR322 present on the YAC vector. This fragment hybridizes to sequences present on the YACs (DY2 and DY20) but not to any in the D. melanogaster genome. After amplifying an end-specific DNA fragment from DY2 by the procedures described above, the PCR product is radiolabeled and hybridized to identical DNA samples (Panel B). This probe detects the fragments in the D. melanogaster genome and DY2, but

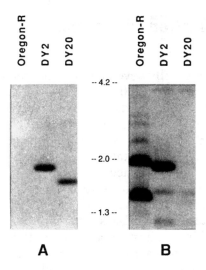

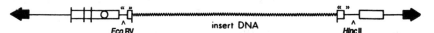

Figure 2 Generating end-specific hybridization probes from YACs. The diagram shows relevant features of the YAC: (1) *Eco*RV and *Hinc*II recognition sites nearest the cloning site in the vector, (2) positions of oligonucleotide primers and direction of DNA synthesis (« »), and (3) regions of the vector derived from yeast and pBR322 are depicted as boxes and narrow lines, respectively. Ten micrograms of Oregon-R *D. melanogaster* genomic DNA (lane 1), one microgram of genomic DNA from a yeast transformant containing a YAC with the 150-kb *D. melanogaster* DNA insert (DY2) used to generate an end-specific probe (lane 2), and one microgram of genomic DNA from a yeast strain containing a YAC with a different *D. melanogaster* DNA insert (DY20) (lane 3) are all digested with *Eco*RV and *Hinc*II. Samples in Panel A are probed with a ^{32}P-labeled, *Bam*HI-*Sal*I fragment derived from pBR322. Panel B shows the same samples probed with the end-specific inverse PCR product from DY2. Sizes are in kilobases.

none in DY20, a YAC-containing *D. melanogaster* DNA from another location in the *D. melanogaster* genome. As expected, both probes detect the same 1.8-kb fragment in DY2. The probe generated by this method can therefore be used to screen an entire YAC library to detect overlapping clones.

Further Applications of Inverse PCR

Inverse PCR was originally developed as a method to rapidly amplify regions of unknown sequence flanking any characterized segment of the genome. The technique eliminates the need to construct and screen DNA libraries to walk hundreds, if not thousands, of base pairs into flanking regions and is particularly useful for determining the insertion sites of translocatable genetic elements and other moderately repetitive DNA sequences.

In the present study we demonstrate another application of inverse PCR; that is, to enzymatically amplify end-specific DNA fragments of a specific orientation from YACs. Such probes can be employed for chromosome walking in any library containing overlapping DNA fragments and also offer a method for eliminating repetitive elements from certain YACs, cosmids, or lambda clones without fabricating and transforming a new construct prior to library screening.

Inverse PCR can be adapted to generate linking fragments—pieces of DNA that identify adjacent restriction fragments (Collins and Weissman 1984)—for creating an ordered library and large-scale physical map of complex genomes. In the usual case, linking fragments connect large genomic DNA fragments produced by rare-cutting restriction enzymes such as *Not*I or *Sfi*I. These large *Not*I or *Sfi*I fragments are either separated by pulsed-field gel electrophoresis or cloned in the form of YACs . To obtain linking fragments, DNA (usually chromosome-specific or from a very limited portion of the genome) is first digested with an enzyme that cleaves frequently, and the resulting fragments are ligated under dilute conditions. The monomeric circles are then treated with the rare-cutting restriction enzyme that will linearize only those molecules containing the rare restriction site and the regions adjacent to them (Poustka and Lehrach 1986). Finally, small pieces of DNA of known sequence are ligated onto the ends of the linearized fragment, which, in effect, places a core region onto the anonymous DNA fragments at the rare-cutting restriction enzyme cleavage site. Primers constructed to this core region are used to amplify the linking fragments as in the inverse PCR procedure.

In addition, Helmsley *et al.* (1989) have devised a method for site-directed mutagenesis using inverse PCR. In this scheme, one of the oligonucleotide primers is synthesized with an alternate base reflecting the desired modification. In addition, the primers are designed such that (1) their 5' ends hybridize to adjacent nucleotides on opposite strands of a circular double-stranded molecule containing the region of interest and (2) their 3' ends prime synthesis in opposite directions around this circular template. Amplification is followed by phosphorylation with T4 polynucleotide kinase to provide a 5' phosphate for the primer termini, intramolecular ligation, and transformation into the appropriate host.

Literature Cited

Burke, D., G. F. Carle, and M. V. Olson. 1987. Cloning of large segments of exogenous DNA into yeast by means of artificial chromosome vectors. *Science* **236**:806–812.

Carle, G., and M. V. Olson. 1984. Separation of chromosomal DNA molecules from yeast by orthogonal-field-alternation gel electrophoresis. *Nucleic Acids Res.* **12**:5647–5664.

Carle, G., M. Frank, and M. V. Olson. 1986. Electrophoretic separations of large DNA molecules by periodic inversion of the electric field. *Science* **232**:65–68.

Chu, G., D. Vollrath, and R. W. Davis. 1986. Separation of large DNA molecules by contour-clamped homogeneous electric fields. *Science* **234**:1582–1585.

Collins, F. S., and S. M. Weissman. 1984. Directional cloning of DNA fragments at a large distance from an initial probe: A circularization method. *Proc. Natl. Acad. Sci. USA* **81**:6812–6816.

Garza, D., J. W. Ajioka, D. T. Burke, and D. L. Hartl. *Science*, in press.

Garza, D., J. W. Ajioka, J. P. Carulli, R. W. Jones, D. H. Johnson, and D. L. Hartl. 1989. Physical mapping of complex genomes. *Nature (London)* **340** : 577–578.

Helmsley, A., N. Arnheim, M. D. Toney, G. Cortopassi, and D. J. Galas. A simple method for site-directed mutagenesis using the polymerase chain reaction. *Nucleic Acids Res.* **17**:6545–6551.

Ochman, H., A. S. Gerber, and D. L. Hartl. 1988. Genetic applications of an inverse polymer chain reaction. *Genetics* **120**:621–625.

Poustka, A., and H. Lehrach. 1986. Jumping libraries and linking libraries: The next generation of molecular tools in mammalian genetics. *Trends in Genetics.* **2**:174–179.

Saiki, R., S. J. Scharf, F. Faloona, K. B. Mullis, G. T. Horn, H. A. Erlich, and N. Arnheim. 1985. Enzymatic amplification of β-globin genomic sequences and restriction site analysis for diagnosis of sickle cell anemia. *Science* **230**:1350–1354.

Saiki, R., D. H. Gelfand, S. Stoffel, S. J. Scharf, R. G. Higuchi, G. T. Horn, K. B. Mullis, and H. A. Erlich. 1988. Primer-directed enzymatic amplification of DNA with a thermostable DNA polymerase. *Science* **239**:487–491.

Schwartz, D. C., and C. R. Cantor. 1984. Separation of yeast chromosome-sized DNAs by pulsed field gradient gel electrophoresis. *Cell* **37**:67–75.

Silver, J., and V. Keerikatte. 1989. *J. Cell. Biochem.* (suppl.) **13E**:306.

Triglia, T., M. G. Peterson, and D. J. Kemp. 1988. A procedure for *in vitro* amplification of DNA segments that lie otside the boundaries of known sequences. *Nucleic Acids Res.* **16**:8186.

28

DETECTION OF HOMOLOGOUS RECOMBINANTS

Michael A. Frohman and Gail R. Martin

The stable incorporation of cloned copies of genes into eukaryotic cells occurs when the incoming DNA integrates into random sites in the genome or when it undergoes homologous recombination (HR), a process by which it replaces its normal chromosomal homolog (Smithies *et al.* 1985; Thomas *et al.* 1986). The latter can be exploited to replace endogenous genes with "HR constructs" containing sequence changes that result in subtle or dramatic changes in the expression or function of the gene. For example, interrupting a coding sequence can result in a null mutation. Such direct manipulation of genes in their chromosomal sites provides a valuable means of exploring gene function. The fact that it may soon be feasible to use this approach to create heritable changes in mice is especially exciting (reviewed in Frohman and Martin 1989).

In yeast the HR frequency, defined here as the ratio of homologous recombination events to random insertions, is extremely high, and there is no difficulty in isolating homologous recombinants (cells in which homologous recombination has taken place). In mammalian cells, however, the HR frequency varies greatly and in general is extremely low. Several strategies have been developed to enrich for homologous recombinants (Jasin and Berg 1988; Mansour *et al.* 1988),

PCR Protocols: A Guide to Methods and Applications

but these selection methods may ultimately prove to be inadequate for many genes. In such cases, the very rare homologous recombinants can be found by using a screening method known as "sib-selection," in which homologous recombinants are isolated by repeatedly subdividing cell pools and testing for the presence of appropriately modified cells. This method depends on the ability to determine whether a pool contains cells in which the intended genetic change has occurred, even when a phenotypic change does not result from the homologous recombination. PCR is an obvious method of choice for such testing.

Basic Strategy for Screening for Homologous Recombinants

The basic strategy is to select primers such that amplification can occur only when homologous recombination has taken place. As illustrated in Fig. 1, primer 1 is chosen from the region unique to the HR construct and primer 2 from genomic DNA sequences in the gene of interest but outside of the region used for the HR construct (Zimmer and Gruss 1989; Joyner *et al.* 1989; Kim and Smithies 1988). In theory, amplification products should be produced only when an HR construct has replaced its normal chromosomal homolog and the primer 1 sequence has thus been integrated in contiguity with the primer 2 genomic sequence; no amplification products should appear when the HR construct has failed to integrate or has integrated into a random genomic site. In practice, however, a number of artifacts can complicate the detection of homologous recombinants.

Potential Artifacts

False Negatives

Some regions are difficult to amplify using PCR (probably because of hairpin structures), and primer pair combinations occasionally do fail (because of mismatching or self annealing). To be absolutely sure

homologous recombination construct

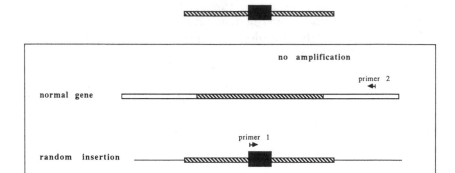

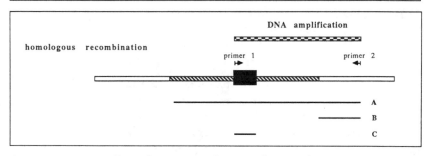

Figure 1 Detection of homologous recombinants. This is a schematic representation of the homologous recombination (HR) construct to be introduced into cells. The cross-hatched horizontal bar represents genomic DNA sequences of the gene of interest. The black box represents sequences that interrupt the normal gene sequence and are thus "unique" to the HR construct. The upper box contains a diagram of a chromosomal copy of the gene of interest (horizontal bar) with the relative positions of the sequences included in the HR construct (cross-hatched) and of primer 2 indicated. Also shown is a schematic representation of the HR construct integrated at a random site in genomic DNA (thin horizontal line), with the relative position of primer 1, which is unique to the HR construct, illustrated. Note that since the sequences of primers 1 and 2 are not adjacent in either of these cases, amplification should not occur. The lower box contains a schematic representation of a chromosomal copy of the gene of interest in a homologous recombinant. Since the sequences of primers 1 and 2 are now adjacent, amplification of the fragment illustrated by the checkered horizontal bar should occur. The fragments designated by thick horizontal lines and labeled A, B, and C are discussed in the text.

that an amplification will take place as expected, one can construct a replica of the genomic fragment that would be amplified in homologous recombinants and determine the appropriate amplification conditions. Be warned, however, that this is a particularly dangerous construction; contamination of reagents with this DNA will prove disastrous. A sensible precaution would be to create a new restriction site somewhere in the test construct, such that it can be differentiated from the experimental amplification product when necessary. A safer but less definitive alternative is to test whether the region of interest (fragment A, Fig. 1) can be successfully amplified by using normal genomic DNA as a substrate.

False Positives Due to Nonspecific Primer Annealing

When genomic DNA from normal cells is used as a substrate, artifactual amplification products can be generated as a consequence of nonspecific annealing of one primer to a sequence contiguous with the normal genomic sequence of primer 2 (Fig. 1) or of nonspecific annealing of two primers to adjacent sites anywhere in the genomic DNA. When DNA from stable transfectants is used as the substrate, additional artifactual products can be generated as a consequence of nonspecific annealing of a primer to a genomic sequence contiguous with the randomly integrated HR construct or of nonspecific annealing of two primers, one complementary to a sequence unique to the construct and one to a sequence in the adjacent genomic DNA. Most artifactual products are easily distinguished from the *bona fide* HR amplification product by size, but in some cases further scrutiny is required. This is most easily accomplished by Southern blot analysis of the amplification products using a probe derived from the part of the amplified region not included in the HR construct (fragment B, Fig. 1).

False Positives Due to Polymerase Halt-Mediated Linkage of Primers (PHLOP)

The most troubling artifact, which we have termed PHLOP, is one that can be created during the PCR process (see also Chapter 20 for discussion of jumping PCR). Although the mechanism by which this artifact is produced (Fig. 2) is unproven, it is likely that it occurs after annealing of complementary single strands generated by normal extension of the amplification primers. If one of the two strands has its 3' terminus in the region of complementarity, it can be further extended by using the other strand as a template; the DNA thus synthesized will have its original primer at its 5' end and a sequence complementary to the primer of the other strand at its 3' end. This

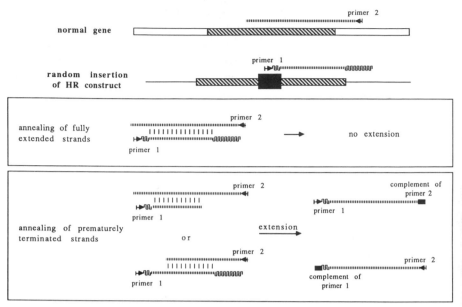

Figure 2   Proposed mechanism of the PHLOP artifact. Illustrated here are the strands generated after specific annealing and extension of primer 1 in all cells and primer 2 in cells containing the HR construct integrated at a random site (see Fig. 1). The portion of each extension product that contains genomic DNA sequences of the gene of interest is illustrated as a broken line. The portions of the extension product of primer 1 that contain sequences not found in the gene of interest are illustrated as wavy lines. The upper box illustrates the region of complementarity and specific annealing of these two extension products. Note that neither strand can serve as a template for extension of the other because their 3′ termini are in regions of noncomplementarity. The lower box illustrates the annealing of the two extension products when either strand 1 (generated by extension of primer 1) or strand 2 (generated by extension of primer 2) has been truncated because of stalling of the *Taq* polymerase in the region of complementarity between the two strands. Note that in either case the fully extended strand can now serve as a template for extension of the truncated strand. Note also that such extension of strand 1 results in a product that contains the complement of primer 2 at its 3′ end, and the extension of strand 2 results in a product that contains the complement of primer 1 at its 3′ end. In both cases the product is a substrate for further rounds of amplification by primers 1 and 2, and the product of such amplification is identical to that generated in a *bona fide* homologous recombinant (see Fig. 1).

DNA will thus serve as a substrate in future rounds of amplification. Such artificial joining of primers could theoretically occur during any PCR, even when there is no specific complementarity between the two strands (in a G + C-rich region, for example) but is much more likely to occur in cases in which the two strands do have regions of perfect complementarity.

In experiments aimed at detecting homologous recombinants, the single strands generated by the PCR have regions of perfect complementarity, but not at their 3' ends (see Fig. 2). If, however, during the synthesis of one of these strands the *Taq* polymerase should stall within the region of complementarity, and if this truncated strand anneals to a complementary strand and extends, the primers will effectively become linked and a product created that is identical in every way to the amplification product that would be produced in a *bona fide* homologous recombinant. As a consequence of this artifact the desired amplification product is observed despite the fact that no homologous recombinants are present in the cell pool being tested.

The factors that determine the frequency at which PHLOP occurs are unknown. With some constructs, it does not ever seem to occur (Kim and Smithies 1988), whereas with others the PHLOP product arises whenever genomic DNA and the HR construct are present in the same amplification mixture (J. Tuttleman, personal communication). Finally, in some cases PHLOP appears to occur only when the HR construct has integrated in an unfavorable random site. In general, however, PHLOP products only appear after more than 30 rounds of amplification, and they are more frequently observed when the substrate DNA has been isolated from a cell pool derived from a relatively large number of individual transfectants.

One possible way to avoid being misled by the PHLOP artifact in the detection of homologous recombinants might be to replace the amplification primer that is unique to the HR construct with a primer sequence that is contained within the HR construct but not unique to it, such that DNA from every cell in the population would be amplified, but DNA from homologous recombinants would include the unique portion of the HR construct in the amplified product (fragment A, Fig. 1). The high efficiency at which this product is amplified should prevent the accumulation of significant amounts of the PHLOP artifact product. The presence of homologous recombinants would then be detected by Southern blot analysis by using a construct-specific probe (fragment C, Fig. 1) or by restriction analysis using sites specific to the modified gene fragment. One limitation of this approach is that it requires that homologous recombinants are sufficiently frequent in the population for the *bona fide* HR amplification product to be produced at levels that are detectable by restriction or Southern blot hybridization analysis.

If PHLOP is a persistent problem, it may be worthwhile investing some effort in defining conditions under which the amount of artifactual product is minimized and homologous recombinants can be reliably detected. This could involve the isolation of cells stably transfected with a replica of the genomic fragment that would be amplified in homologous recombinants. DNA from such cells could be mixed with normal genomic DNA in experiments aimed at defining the best conditions for detection of a homologous recombinant.

Protocols

Sample Collection

The first step is to obtain an estimate of the HR frequency. In the absence of the PHLOP artifact this can be achieved by assaying cultures of serially diluted transfectants using the basic strategy described above. The protocol described below is best used when the HR frequency is $\geq 10^{-3}$. In cases where it is lower, efforts should be made to raise it by repeated subdivision of the population that proves to contain homologous recombinants.

Conditions of cell culture should then be chosen such that each plate contains 10 to 50 clonal colonies of ~500 cells each. Colonies to be sampled and/or pooled should be marked on the underside of each plate. Wash the cells and flood the dish with PBS. Under a dissecting microscope use a drawn-out Pasteur pipette with a very fine tip to aspirate a fragment of each colony (50 to 100 cells) and transfer it with a small amount of PBS to a 0.6-ml Eppendorf tube. When pooling colony fragments it is generally best to transfer them one at a time rather than collecting them all in the pipette at one time, particularly if the cells tend to adhere to the pipette wall. (If a colony detaches when a fragment is being collected, transfer the detached portion to a drop of trypsin in a tissue culture dish, incubate for 10 minutes at room temperature, and then add culture medium.) After sampling all the designated colonies, replace the PBS with fresh culture medium and return the plate to the incubator.

Preparation of DNA

Pellet the samples and remove all but 10 μl of the PBS. Add 0.5 μl of Proteinase K (10 mg/ml) and resuspend the pellet. Add 1 μl of SDS (1%) to the tube and then incubate at 55°C for 30 minutes. Add 20 μl of water and heat-inactivate the Proteinase K at 98°C for 5 minutes.

Quench on ice and spin down moisture. Use half of this mixture (15 μl) for a 50-μl PCR amplification mixture. To ensure that the DNA preparations are suitable substrates for PCR amplification, use the other half of one or more samples to perform a control amplification reaction with primers that will amplify normal genomic DNA fragments. The remaining half-samples can be used for duplicates of the experimental samples or for controls.

PCR

Perform a standard 50-cycle amplification reaction, using appropriate annealing temperatures and extension times. A typical profile is 94°C, 40 seconds; 55°C, 1 minute; 72°C, 1 minute. Follow the cycle program with a 15-minute extension at 72°C. It is important to note that the SDS present in the lysis buffer will inhibit amplification unless detergents are also included. Specifically, the PCR buffer should contain (final concentration) 0.3% NP-40 and 0.3% Tween 20. Negative controls should include samples to which (1) no DNA, (2) culture medium from an experimental plate but no DNA, and (3) genomic DNA from normal cells have been added.

Separate the amplification products (10 μl of each PCR sample) in a minigel. The HR product should be visible after EtBr staining. Perform a Southern blot analysis of the products with an appropriate probe (see above) to confirm that the positive samples do indeed contain HR products and to detect HR products that may not have been visible with EtBr staining. If a large proportion of the samples are positive, this could signify that the PHLOP artifact is a problem.

Follow-up

Sample and assay individual colonies that contributed to positive pools as previously described. Colonies of presumed homologous recombinants should be subcultured, and when there is a sufficient number of cells, genomic DNA should be isolated and assayed by Southern blot analysis to confirm that the cells are homologous recombinants.

Literature Cited

Frohman, M. A., and G. R. Martin. 1989. Cut, paste, and save: new approaches to altering specific genes in mice. *Cell* **56**: 145–147.

Jasin, M., and P. Berg. 1988. Homologous integration in mammalian cells without target gene selection. *Genes and Develop.* **2**:1353–1363.

Joyner, A. L., W. C. Skarnes, and J. Rossant. 1989. Production of a mutation in mouse *En-2* gene by homologous recombination in embryonic stem cells. *Nature (London)* **338**:153–156.

Kim, H. S., and O. Smithies. 1988. Recombinant fragment assay for gene targeting based on the polymerase chain reaction. *Nucleic Acids Res.* **16**:8887–8903.

Mansour, S. L., K. R. Thomas, and M. R. Capecchi. 1988. Disruption of the proto-oncogene *int-2* in mouse embryo-derived stem cells: a general strategy for targeting mutations to non-selectable genes. *Nature (London)* **336**:348–352.

Smithies, O., R. G. Gregg, S. S. Boggs, M. A. Koralewski, and R. S. Kucherlapati. 1985. Insertion of DNA sequences into the human chromosomal beta-globin locus by homologous recombination. *Nature (London)* **317**:230–234.

Thomas, K. R., K. R. Folger, and M. R. Capecchi. 1986. High frequency targeting of genes to specific sites in the mammalian genome. *Cell* **44**:419–428.

Zimmer, A., and P. Gruss. 1989. Production of chimaeric mice containing embryonic stem (ES) cells carrying a homoeobox *Hox 1.1* allele mutated by homologous recombination. *Nature (London)* **338**:150–153.

RNA PROCESSING: Apo-B

Lyn M. Powell

Uptake of dietary cholesterol and lipid and their transport in mammalian serum is dependent upon apolipoproteins. These proteins contain hydrophobic domains to interact with the immiscible lipid, and many act as ligands for receptors to clear the lipid and direct different lipoprotein particles to particular tissues. One of the largest apolipoproteins is apolipoprotein (apo-) B, which is the sole protein component of low-density lipoprotein (LDL). High levels of LDL and, therefore, apo-B are associated with increased risk of coronary artery disease in humans. Two different sizes of apo-B are synthesized in mammals (Kane 1983). The larger form, known as apo-B100, is produced in the liver and is the form found in LDL. The smaller protein is called apo-B48 and is approximately half the size of apo-B100. In humans, Apo-B48 is produced only in the gut and is essential for the uptake of dietary fat. The cloning and sequencing of apo-B100 (Knott *et al.* 1986; Yang *et al.* 1986) made it possible to address the question of how these two different-sized forms are produced. Apo-B100 is produced as a 4163-amino-acid protein, 512 kD, from a 14,121-bp mRNA. There is an identical-sized message in human small intestine, which produces apo-B48. Sequencing of this message showed a single difference that could be responsible for produc-

PCR Protocols: A Guide to Methods and Applications
Copyright © 1990 by Academic Press, Inc. All rights of reproduction in any form reserved.

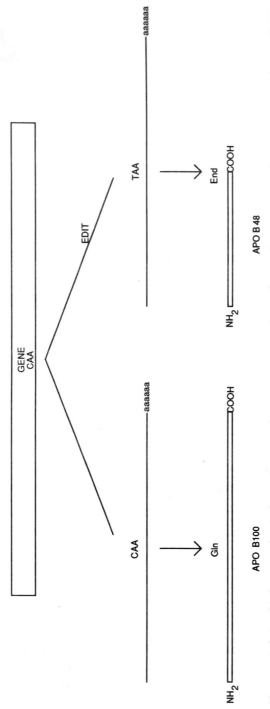

Figure 1 Relationship between the single apo-B gene and its two protein products. The apo-B gene contains the sequence encoding apo-B100. Under certain circumstances detailed in the text, the mRNA is modified to produce an in-frame stop codon 48% of the way through the apo-B100 protein coding sequence, resulting in the synthesis of apo-B48.

ing the smaller protein. This difference was the first base in codon 2153, which changed it from CAA (Gln) to TAA, a stop codon. Theoretically, this difference could have resulted from differential splicing of two nearly identical exons in this region. However, the exon containing this base is >7.5 kbp in size, and no evidence could be found for duplication or internal splicing. It was also shown that there was only one apo-B gene. PCR followed by differential hybridization of sequence-specific oligonucleotides was used to demonstrate that intestinal DNA contains a CAA and encodes apo-B100. This suggested that the single base difference producing apo-B48 was due to a post-transcriptional modification of the mRNA in the small intestine (Fig. 1). This was proved by performing PCR amplification on the intestinal mRNA, which contained a UAA at this position (Powell *et al.* 1987).

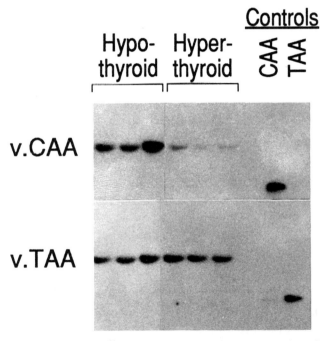

Figure 2 PCR and differential hybridization on liver apo-B RNA from hypothyroid and hyperthyroid rats. PCR-amplified cDNAs from three hypothyroid and three hyperthyroid rat livers were fractionated and transferred to nylon membranes along with plasmid DNAs containing, respectively, CAA or TAA codons and hybridized with discriminating oligonucleotides. The amounts of apo-B48 sequences present were 39 ± 6.1% (mean ± SD) for hypothyroid samples and 93 ± 2% for hyperthyroid samples.

One interesting aspect of apo-B48 protein production in human small intestine is that it is developmentally regulated. Early fetal gut expresses predominantly apo-B100; by 16-weeks gestation it synthesizes both apo-B100 and apo-B48 protein. Adult gut tissue produces only apo-B48 (Glickman et al. 1986). PCR is being used to determine when this switch takes place at the mRNA level. An analogous system to this developmental regulation is found in rodents. Rats and mice also produce two different-sized apo-B proteins. However, they produce significant amounts of apo-B48 in the liver. Moreover, the relative amounts of apo-B100 and apo-B48 produced vary with the amount of thyroid hormone produced in the animal. Hypothyroid rats produce predominantly apo-B100, while hyperthyroid animals produce almost exclusively apo-B48. Having determined that the region surrounding the C-U transition is conserved between rat and human, it was possible to follow the degree of mRNA modification in these animals and use PCR coupled with differential hybridization to quantitate the relative amounts of apo-B100 and apo-B48 mRNAs (Davidson et al. 1988) (Fig. 2).

Protocols

mRNA Production

RNA can be prepared by any of the published procedures. As the message for apo-B is 14.1 kb, I routinely use the method of Cathala et al. (1983). PCR amplification is always performed on RNA populations before cDNA synthesis to check that genomic contamination is minimal.

cDNA Synthesis

PCR amplification for RNA analysis requires synthesis of first-strand cDNA by using reverse transcriptase. This can be performed using a commercially available kit (e.g., Amersham Corp. PLC Cat. No. RPN 1256) or as follows: 1 to 5 μg total RNA or 50 to 200 ng poly(A)$^+$ RNA is reverse transcribed in 50 mM Tris-HCL (pH 8.8) at room temperature, 6 mM MgCl$_2$, 10 mM DTT. All four dNTPs are present at 1 mM concentration, and 200 ng of specific primers is used. Reverse transcriptase (2 units) (Life Sciences, Inc.) is used in a total volume of 20 μl and incubated at 42°C for 40 minutes.

After the first-strand synthesis, the sample may be ethanol precipitated or aliquots may be diluted directly into PCR buffer. Between 10 and 100% of cDNA reactions are used as templates for PCR amplification.

PCR Amplification

The optimal conditions for PCR amplification of the region surrounding the C-U modification are determined empirically for each set of oligonucleotides and each species, using cloned sequences at 1 pmol concentration. Typically the following regime gives good results using the Perkin-Elmer Cetus DNA Thermal Cycler, Cetus *Taq* polymerase, and the conditions recommended for polymerase activity.

DNA/RNA hybrids are melted at 95°C for 12 minutes, followed by 20 to 30 cycles of 1 minute at 50°C to anneal, 2 minutes at 72°C to extend, and 1 minute at 95°C to denature. After cycling, the reactions are allowed to extend for 12 minutes at 72°C, and one-tenth of the reaction mix is fractionated on agarose gels.

Discriminating Hybridization

Hybridization of discriminating oligonucleotides is performed on PCR amplification products bound to nylon membranes (ZetaProbe, Bio-Rad Laboratories), after either Southern transfer in 0.4 *M* NaOH or dot blotting in H$_2$O followed by denaturation in 0.4 *M* NaOH.

1. Labeling of oligos. Oligonucleotides are labeled to high specific activity using 150 μCi [^{32}P]ATP and 20 units of T4 polynucleotide kinase (Pharmacia PL). Aliquots (20 ng) of 19-base oligonucleotides are labeled in the following conditions: 50 m*M* Tris–HCl (pH 7.5), 10 m*M* MgCl$_2$, 5 m*M* DTT, 50 μg/ml bovine serum albumin, in a volume of 30 μl at 37°C for >1 hour.

   Oligonucleotides as short as 15 bp can be used to discriminate between single-base differences after PCR amplification; however, the 19-base antisense oligonucleotides

   BSTOP (TACTGATCAAATTATATCA)

   and

   BGLN (TACTGATCAAATTGTATCA)

   have been found to be effective for examining the C-U modification of apo-B mRNA in humans, rabbit, rat, and mouse.

2. Hybridization and washing conditions are determined em-

pirically on cloned or amplified sequences known to contain the two different sequences. For the C-U modification in apo-B, the conditions used are hybridization with 10^6 cpm ml^{-1} in 6× SSC, 1 to 2% SDS, 0.5% "Blotto" at 42°C for > 4 hours. Washes are performed in 6× SSC, 1% SDS. Wash once at room temperature for 15 to 30 minutes and then at either 46°C (for the BSTOP oligo) or 48°C (for the BGLN oligo) until no hybridization to the noncomplementary control sequences is detected (usually 15 minutes is sufficient).

Quantitation

The relative amounts of the two sequences present in the hybridized DNA are measured by exposing the filters to preflashed X-ray film at −80°C. The signals are measured by using integrated laser densitometry, and the proportion of total signal due to hybridization of each oligonucleotide is calculated.

Discussion

The technique of PCR coupled with discriminating hybridization of oligonucleotides was indispensable in discovering the post-translational modification of apo-B mRNA. Further studies on the mechanism and regulation of this modification make use of this powerful technique. During the course of these investigations, protocols have been modified, and important controls and caveats have been discovered. One of the essential controls when attempting to quantitate post-transcriptional modification of RNA is to ensure that the RNA sample is entirely free from DNA contamination. The sensitivity of PCR amplification means that control amplification of RNA without prior cDNA synthesis may detect very low levels of contaminating genomic DNA that will give a signal upon hybridization, even in the absence of ethidium-stainable PCR product. Minimizing signal from genomic sequences may entail treating the RNA preparations with DNase, digesting with restriction enzymes before cDNA preparation, amplifying from poly(A)$^+$ mRNA, or a combination of these treatments. The choice of treatment is determined by the amount of sample available and the degree of message integrity. For example, selection of poly(A)$^+$ mRNA can only be performed on samples in which there is no degradation of the message. cDNA can be primed using oligo(dT) or a primer specific for the mRNA of interest. This specific primer can be located further 3' of the region to be amplified or may be the same as the 3' PCR primer. One of the advantages of using specific primers close to the site of modification is that cDNA synthesis and subsequent amplification may be performed on less than intact preparations of total RNA. The position of the apo-B C-U site in the middle of a very large exon precludes the possibility of amplifying across the site of an intron, which would separate PCR products from genomic and RNA templates by size and, therefore, may be more useful in the study of different RNA modifications that remain to be discovered.

While many samples can be screened using dot blots of amplified material, comparison of amounts of ethidium bromide-stainable material with signals obtained upon subsequent Southern blotting and hybridization may be more convincing, to both worker and reviewer. Hybridization of control sequences containing the two alternative nucleotides should be included in every experiment to exclude the background hybridization of oligonucleotides, especially when low levels of one signal are seen in samples. Quantitation of amounts of PCR product should be performed using both oligos together or a third oligo to a different sequence within the amplified product.

Literature Cited

Cathala, G., J.-F. Savouret, B. Mendez, B. L. West, M. Karin, J. A. Martial, and J. D. Baxter. 1983. Laboratory methods: a method for isolation of intact, translationally active ribonucleic acid. *DNA* **2**:329–335.

Davidson, N. O., L. M. Powell, S. C. Wallis, and J. Scott. 1988. Thyroid hormone modulates the introduction of a stop codon in rat liver apolipoprotein B messenger RNA. *J. Biol. Chem.* **263**:13482–13485.

Glickman, R. M., M. Rogers, and J. N. Glickman. 1986. Apolipoprotein B synthesis by human liver and intestine *in vitro*. *Proc. Natl. Acad. Sci. USA* **83**:5296–5300.

Kane, J. P. 1983. Apolipoprotein B: structural and metabolic heterogeneity. *Annu. Rev. Physiol.* **45**:637–650.

Knott, T. J., R. J. Pease, L. M. Powell, S. C. Wallis, S. C. Rall, Jr., T. L. Innerarity, B. Blackhart, W. H. Taylor, Y. Marcel, R. Milne, D. Johnson, M. Fuller, A. J. Lusis, B. J. McCarthy, R. W. Mahley, B. Levy-Wilson, and J. Scott. 1986. Complete protein sequence and identification of structural domains of human apolipoprotein B. *Nature (London)* **323**:734–738.

Powell, L. M., S. C. Wallis, R. J. Pease, Y. H. Edwards, T. J. Knott, and J. Scott. 1987. A novel form of tissue-specific RNA processing produces apolipoprotein-B48 in intestine. *Cell* **50**:831–840.

Yang, C.-Y., S.-H. Chen, S. H. Gianturco, W. A. Bradley, J. T. Sparrow, M. Tanimura, W.-H. Li, D. A. Sparrow, H. DeLoof, M. Rosseneu, F.-S. Lee, Z.-W. Gu, A. M. Gotto, Jr., and L. Chan. 1986. Sequence, structure, receptor-binding domains and internal repeats of human apolipoprotein B-100. *Nature (London)* **323**:738–742.

A TRANSCRIPTION-BASED AMPLIFICATION SYSTEM

T. R. Gingeras, G. R. Davis, K. M. Whitfield, H. L. Chappelle, L. J. DiMichele, and D. Y. Kwoh

Biologically important nucleic acids are frequently present within cells at very low levels. The isolation and detection of this genetic information have often been a difficult problem for clinicians and molecular biologists. *In vitro* nucleic acid amplification was developed to respond to this need. The first reported *in vitro* amplification protocol proceeds by a strategy of DNA replication and is called PCR (Saiki *et al.* 1985; Mullis and Faloona 1987). Enhancements of the PCR protocol have been described and include the use of thermostable DNA polymerase (Saiki *et al.* 1988; Erlich *et al.* 1988), the use of reverse transcriptase to copy both DNA and RNA target molecules (Byrne *et al.* 1988), and the addition of the phage T7 RNA polymerase recognition sequences to PCR primers such that after multiple cycles of PCR, RNA transcripts can be produced from the PCR-amplified DNA (Mullis and Faloona 1987; Murakawa *et al.* 1988; Stoflet *et al.* 1988). In this chapter, an RNA transcription-based amplification system (TAS) is described, and its application to the detection of HIV-1 present in cultured lymphocytes illustrates both its sensitivity and specificity (Kwoh *et al.* 1989). The specificity

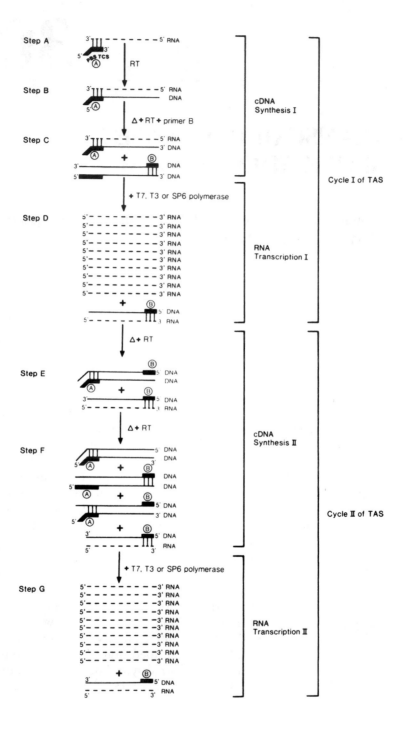

of the TAS protocol can be enhanced by the use of a bead-based sandwich hybridization system (BBSHS). This sandwich hybridization approach fits well with the single-stranded nature of the TAS products and permits direct analysis of the results of the TAS reaction.

The TAS strategy is outlined in Fig. 1. Each cycle of TAS copies a segment (100 to 500 bases) of an RNA or DNA target molecule into 20 to 100 copies of RNA. A single cycle of TAS is composed of a cDNA step (to convert a target nucleic acid sequence to a cDNA template containing an RNA polymerase binding site) and an RNA transcription step (to increase the copy number of the cDNA template). Consequently, relatively few cycles are required to achieve high levels of sequence-specific amplifications (10^5- to 10^6-fold increases).

← **Figure 1**   Scheme depicting a transcription-based amplification system (TAS). A two-cycle scheme for amplifying an RNA target sequence by using sequential cDNA synthesis and RNA transcription is displayed as seven steps. A target nucleic acid molecule RNA (or denatured DNA) (___) is hybridized to a primer oligonucleotide (A) for the *vif* region of HIV-1

(88–77;5′-AATTTAATACGACTCACTATAGGGA
CACCTAGGGCTAACTATGTGTCCTAATAAGG-3′)

which contains an RNA polymerase (T7, T3, or SP6) binding sequence (PBS) and a target complementary sequence (TCS) (step A). Reverse transcriptase (RT) elongates primer A to yield a newly synthesized DNA strand complementary to the target RNA (step B). The RNA-DNA heteroduplex is denatured by heat (Δ), and oligonucleotide B for the *vif* region of HIV-1

(86–29; 5′-ACACCATATGTATGTTTCAGGGAAAGCTA-3′)

is annealed to the newly synthesized DNA strand containing the PBS. Reverse transcriptase is added to produce a double-stranded cDNA and a new DNA-RNA heteroduplex (step C). Incubation of the double-stranded cDNA with T7 (T3 or SP6) RNA polymerase results in the synthesis of multiple RNA transcripts from the PBS-containing, double-stranded DNA template (step D). Some of this RNA is immediately converted to RNA-DNA heteroduplex by RT (still in the reaction mixture from step C) with oligonucleotide B as a primer. Further amplification of target sequences can be obtained by a second cycle of cDNA synthesis (steps E and F) and RNA transcription (step G).

Protocols

Reagents

Tris-buffered saline	135 mM NaCl
	5 mM KCl
	25 mM Tris base
	(Adjust pH to 7.4 with 1 N HCl. Sterilize by autoclaving)
Lysis buffer	20 mM Tris (pH 7.5)
	150 mM NaCl
	10 mM EDTA
	0.2% SDS
	200 μg/ml of Proteinase K
5× transcription buffer	200 mM Tris (pH 8.1)
	40 mM MgCl$_2$
	125 mM NaCl
	10 mM spermidine
	25 mM dithiothreitol
	0.4 mg/ml of bovine serum albumin (nuclease free)
	(Filter sterilize)

Isolation of Total Nucleic Acids from HIV-1-Infected and Uninfected CEM Cells

1. Pellet 10^5 to 10^6 CEM cells from 1 ml Tris-buffered saline in an Eppendorf tube at 5000 rpm in a variable-speed microfuge for 10 minutes. Draw off and discard the supernatant, leaving 50 μl with pellet.
2. Resuspend pellet in 600 μl lysis buffer.
3. Vortex vigorously and incubate at 50°C for 45 minutes, vortexing for 10 to 15 seconds every 10 minutes.
4. Add 600 μl phenol : chloroform : isoamyl alcohol (50 : 48 : 2). Shake and vortex to emulsify mixture. Centrifuge at 14,000 rpm for 2 minutes to separate phases.

5. Draw off 575 μl from aqueous (top) phase. Add 600 μl phenol:chloroform:isoamyl alcohol. Shake and vortex to emulsify mixture. Centrifuge for 2 minutes to separate phases.

6. Draw off 525 μl from the aqueous phase. Add 600 μl chloroform : isoamyl alcohol (24 : 1). Shake and vortex to emulsify the mixture. Centrifuge for 2 minutes to separate the phases.

7. Draw off 400 μl from the aqueous phase. (Do not transfer any cell debris that may be at the interface.)

8. Add 1/10 volume (40 μl) 8 M LiCl. At this step, samples may be split for parallel processing.

9. Add 3 volumes ice-cold 100% ethanol to samples. Mix well and precipitate at −20°C overnight or in a dry ice/ethanol bath for 15 minutes.

Transcription-Based Amplification

Nucleic acids in ethanol are collected by 20-minute centrifugation at 14,000 rpm in a microfuge at 4°C. The supernatant is drawn off and the pellet is air dried. The nucleic acids are redissolved in the volume of H_2O used in the TAS reaction (follows).

For a 100-μl amplification reaction, combine 20.0 μl of 5× transcription buffer, 8.0 μl of 2.5 mM dNTP mix, 8.0 μl of 25 mM rNTP mix, 5.0 μl of A oligo (88–77; see Fig. 1 for sequence) (50 ng/μl), 5.0 μl of B oligo (86–29; see Fig. 1 for sequence) (50 ng/μl), and 54.0 μl of H_2O.

1. 100°C for DNA or 65°C for RNA, 1 minute.

2. 42°C, 1 minute.

3. Add reverse transcriptase (10 units).

4. 42°C, 12 minutes.

5. 100°C, 1 minute.

6. 37°C, 1 minute.

7. Add reverse transcriptase (10 units).

8. Add T7 RNA polymerase (100 units).

9. 37°C, 30 minutes.

Steps 1–9 comprise one cycle of TAS. Repeat steps 1–9 for additional cycles. Replace step 1 with a 100°C, 1-minute denaturation step (all target types) for cycles 2–6. Add enzymes of steps 7 and 8 as a mix.

After step 9 of cycle 3, add 0.25 μg of each oligo from a 0.25 μg/μl concentrated stock of each. A 55°C annealing step may be used.

Bead-Based Sandwich Hybridization System

Sephacryl beads containing an oligonucleotide for the *vif* region of HIV-1

(87–83; 5'-ATGCTAGATTGGTAATAACAACATATT-3')

were prepared as described previously (Ghosh and Musso 1987). This bead-bound oligonucleotide was used to capture the TAS-amplified HIV-1 sequences after these targets had hybridized to a ^{32}P-labeled detection oligonucleotide (86–31; for the *vif* region of HIV-1;

5'-GCACACAAGTAGACCCTGAACTAGCAGACCA-3')

In the BBSHS, the TAS-amplified target and ^{32}P-labeled oligonucleotide are denatured in 10 μl of TE containing 0.2% SDS at 65°C for 5 minutes in an Eppendorf tube. To this, 10 μl of 2× solution hybridization mix (10× SSPE, 10% dextran sulfate) is added. The solution is mixed and incubated at 42°C for 2 hours.

Approximately 50 mg (wet weight) of Sephacryl beads are prehybridized in 250 μl of hybridization solution (5x SSPE, 10% dextran sulfate, 0.1% SDS) for 30 minutes at 37°C. Immediately before the capture step, the beads are centrifuged for 10 seconds, the prehybridization solution is removed, and 80 μl of fresh hybridization solution is added, after which the solution hybridization is transferred to the beads. The beads are then incubated at 37°C for 1 hour with occasional mixing.

After the bead hybridization step, the beads are centrifuged and the hybridization solution is transferred to a scintillation counter vial. The beads are then washed five times with 2× SSC at 37°C. The first three washes are rapid; 1 ml of wash is added, and the beads are mixed well and centrifuged 10 seconds. For the final two washes, 1 ml of wash is added, and the beads are mixed and incubated at 37°C for 5 minutes before being centrifuged. Cerenkov counts of the hybridization solution, each of the five washes, and beads are measured for 5 to 10 minutes. The scintillation counter background is subtracted from all samples (approximately 20 to 40 cpm). The fmol of target detected is calculated as follows:

(cpm on beads/total cpm) × fmol probe oligonucleotide

The total cpm is the sum of the cpm for the hybridization solution, five washes, and beads.

Results and Conclusions

The ability to perform quasi-homogeneous hybridizations (i.e., near solution-like hybridization conditions) by using bead-bound oligonucleotides as a hybridization matrix permits the rapid detection of the TAS-amplified HIV-1 RNA product. RNA produced by TAS amplification from a dilution series of HIV-1-infected cells was captured and detected with a BBSHS format (Table 1). As indicated in Table 1, this hybridization format is capable of quantitatively detecting the presence of HIV-1 sequences in infected cells over a two-log range while using only a small portion of the TAS amplification reaction mixture. Both the consistency and sensitivity of these results are noticeable. Thus, using a combination of the TAS amplification and BBSHS protocols, one can distinguish HIV-1 sequences present in one-quarter of a sample derived from one infected cell in a population of 10^6 uninfected cells.

Table 1

Bead-Based Sandwich Hybridization Detection of TAS-Amplified
HIV-1 Sequences

No. infected cells (in 10^6 cells)	Amount of reaction used (μl)	Total cpm ($\times 10^{-2}$)	Bead cpm ($\times 10^{-2}$)	% of probe	HIV-1 detected (fmol/μl)
10^3	0.016	857.3	41.1	4.8	53
		869.9	40.1	4.7	
10^2	0.033	876.4	23.9	2.7	13.6
		761.9	21.6	2.8	
10	0.33	770.1	19.1	2.5	1.2
		776.4	18.3	2.4	
1	0.33	790.1	9.2	1.2	0.4
		898.7	10.1	1.1	
0	0.33	939.0	4.9	0.5	—
		913.4	5.0	0.5	

Literature Cited

Byrne, B. C., J. J. Li, J. Sninsky, and B. J. Poiesz. 1988. Detection of HIV-1 RNA sequences by *in vitro* DNA amplification. *Nucleic Acids Res.* **16**:4165.

Erlich, H. A., D. H. Gelfand, and R. K. Saiki. 1988. Specific DNA amplification. *Nature (London)* **331**:461–462.

Ghosh, S. S., and G. F. Musso. 1987. Covalent attachment of oligonucleotides to solid supports. *Nucleic Acids Res.* **15**:5353–5372.

Kwoh, D. Y., G. R. Davis, K. M. Whitfield, H. L. Chappelle, L. J. DiMichele, and T. R. Gingeras. 1989. Transcription-based amplification system and detection of amplified human immunodeficiency virus type I with a bead-based sandwich hybridization format. *Proc. Natl. Acad. Sci. USA* **86**:1173–1177.

Mullis, K. B., and F. Faloona. 1987. Specific synthesis of DNA *in vitro* via a polymerase catalyzed chain reaction. *Methods Enzymol.* **155**:335–350.

Murakawa, G. J., J. A. Zaia, P. A. Spallone, D. A. Stephens, B. E. Kaplan, R. B. Wallace, and J. J. Rossi. 1988. Direct detection of HIV-1 RNA from AIDS and ARC patient samples. *DNA* **7**:287–295.

Saiki, R. K., S. Scharf, F. Faloona, K. B. Mullis, G. T. Horn, H. A. Erlich, and N. Arnheim. 1985. Enzymatic amplification of β-globin genomic sequences and restriction site analysis for diagnosis of sickle cell anemia. *Science* **230**:1350–1354.

Saiki, R. K., D. H. Gelfand, S. Stoffel, S. Scharf, R. Higuchi, G. T. Horn, K. B. Mullis, and H. A. Erlich. 1988. Primer-directed enzymatic amplification of DNA with a thermostable DNA polymerase. *Science* **239**:487–491.

Stoflet, E. S., D. D. Koeberl, G. Sarkar, and S. S. Sommer. 1988. Genomic amplification with transcript sequencing. *Science* **239**:491–494.

31

SCREENING OF λgt11 LIBRARIES

Kenneth D. Friedman, Nancy L. Rosen,
Peter J. Newman, and Robert R. Montgomery

Isolation of specific cDNAs from bacteriophage libraries is a time-consuming process involving screening, plaque purification of positive clones, λ DNA preparation, and subcloning of relevant inserts into plasmids. cDNA inserts from individual bacteriophage clones can be enzymatically amplified by using PCR (Saiki *et al.* 1988), and this technique can be adapted to obtain specific target cDNA directly from a λgt11 library (Friedman *et al.* 1988). PCR can be effectively used to screen cDNA libraries for the presence of specific clones as well as for the generation of nucleic acid probes, thereby bypassing the laborious steps of library screening through phage purification and subcloning. Furthermore, the ability to use cDNA libraries instead of purified cellular RNA (Newman *et al.* 1988) is advantageous, because cDNA libraries in bacteriophage are regenerative, stable to high temperature and ribonuclease, and often commercially available.

Protocol

Materials

λgt11 cDNA library
Appropriate oligonucleotide primers
Taq polymerase, 5.0 units/μl
dNTP mix (2.5 mM each of dATP, dCTP, dGTP, and dTTP)
10× *Taq* polymerase buffer: 500 mM KCl, 100 mM Tris–HCl
   (pH 8.3), 15 mM MgCl₂, 0.1% (w/v) gelatin

Screening λgt11 Libraries

1. Library aliquots (1, 5, and 50 μl) are placed in 600-μl siliconized reaction tubes and then brought to a final volume of 74 μl with double-distilled sterile water.
2. Phage particles are disrupted by incubation at 70°C for 5 minutes and then cooled on wet ice.
3. PCR master mix is made up in a 600-μl reaction tube for three reaction tubes as follows:
      30 μl of 10× *Taq* polymerase buffer
      24 μl of dNTP mix
      Appropriate volume of each oligonucleotide primer to supply 300 pmol of each
      Appropriate volume of double-distilled water to bring the final volume to 78 μl
   The final solution is thus 192 mM KCl, 38.5 mM Tris–HCl (pH 8.3), 5.8 mM MgCl₂, 0.038% (w/v) gelatin, 0.77 mM in each dNTP, and 3.8 μM in each oligonucleotide primer. Each reaction tube will require 26 μl of master mix.
4. Overlay the library aliquots and the master mix with 100 μl of mineral oil and heat the tubes to 94°C for 5 minutes. Bring the tubes to the desired primer annealing temperature.
5. Add 1.5 μl of *Taq* polymerase to the master mix. Transfer 26 μl of master mix to each of the library aliquots and gently mix the reactants in each tube by pipetting up and down.

6. Commence thermal cycling:

>72°C for 3 minutes (primer extension time may vary)
>
>94°C for 1 minute 20 seconds (denaturation)
>
>37°C for 2 minutes (to anneal 21-mer primers; for longer primers, see cautionary note that follows)
>
>Repeat for a total of 25 to 30 cycles. At the end of the last cycle add a final primer extension step for 7 minutes.

We have used this technique to amplify the von Willebrand factor (vWf) sequence from an endothelial cell cDNA library (Fig. 1) and

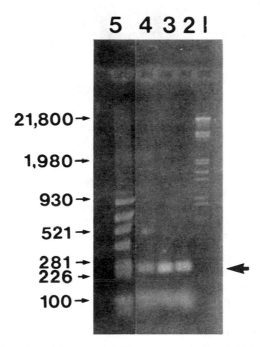

Figure 1 Ethidium bromide-stained 1.5% agarose gel. Von Willebrand factor (vWf) sequence was amplified from an endothelial cell cDNA library [gift of D. Ginsburg (Ginsburg et al. 1985), 1.3×10^7 recombinants/μl; approximately 4×10^6 independent recombinants] by using specific primers [nvWf 2275→2295 and nvWf 2502→2482 (Bonthron et al. 1986)]. Lanes 2, 3, and 4 were loaded with 15 μl of the 100-μl reaction (1-, 5-, and 50-μl aliquots of library used in the respective reactions). Molecular weight standards (noted in base pairs) were loaded in lanes 1 and 5. Arrow indicates the 227-bp PCR product. On the basis of the intensity of the band obtained and the known proportion of vWf clones present in the library (Ginsburg et al. 1985), we estimate that the target sequence could be amplified up to 10^7 times by using a 1-μl library aliquot. The identity of the product as an authentic vWf cDNA was confirmed by dideoxy sequence analysis.

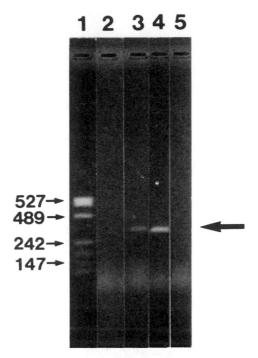

Figure 2   Ethidium bromide-stained 1.5% agarose gel. Platelet glycoprotein Ib (GpIb) sequence was amplified from a HEL cell cDNA library [gift of M. Poncz (Poncz *et al.* 1987) $>2 \times 10^{10}$ recombinants/μl], but not for an endothelial cell library (Ginsburg *et al.* 1985), using specific primers [nGpIb 748→780 and nGpIb 1047→1015 (Lopez *et al.* 1987)]. An anneal temperature of 51°C was used. Lane 1 was loaded with molecular weight markers, lanes 2, 3, and 4 were loaded with 15 μl of the 100-μl reaction (1-, 5-, and 50-μl aliquots of HEL cell library used in the respective reactions), and lane 5 was loaded with 15 μl from the reaction, using 5 μl of endothelial cell library as template. HEL cells are known to express GpIb, whereas evidence is accumulating to suggest that endothelial cells do not synthesize this protein (Montgomery *et al.* 1988).

the platelet glycoprotein Ib (GpIb) sequence from a HEL cell cDNA library (Fig. 2).

Variations

This technique combined with degenerate oligonucleotide-primed amplification of cDNA (see Chapter 5) allows for the isolation of cDNA sequences from a λgt11 library when only amino acid sequence information is available. In addition, Trahey and colleagues have shown that selected λgt11 clones can be screened by amplification between a λgt11 sequencing primer and an insert-specific oligonucleotide primer to generate 5' or 3' sequence information (Trahey *et al*. 1988). Briefly, to generate cDNA with more of the 5' sequence than is known, clones are screened by conventional hybridization techniques using an "antisense" oligonucleotide probe. Positive clones are then used as templates for PCR by using either a λgt11 forward or λgt11 reverse sequencing primer along with the "antisense" oligonucleotide probe. PCR products are then evaluated by gel electrophoresis, and the longer PCR products are then either subcloned or directly sequenced to reveal the 5' sequence.

Cautionary Notes

The degree of specificity of PCR is in part a function of the temperature at which primers are annealed (see Chapter 1). Using longer oligonucleotide probes and higher annealing temperatures will increase the specific product yield and decrease the background. We currently use primers that are 30 to 35 bases in length and anneal at 51°C. We have noticed that yields are low when large aliquots of library are used (see Figs. 1 and 2). This may be due in part to an inhibitory effect of contaminating phage material or because of the salt content of the buffer that libraries are frequently suspended in. For any individual library, optimization with various aliquots of library and, possibly, adjustment in the buffer conditions may be necessary.

Literature Cited

Bonthron, D., E. C. Orr, L. M. Mitsock, D. Ginsburg, R. I. Handin, and S. H. Orkin. 1986. Nucleotide sequence of pre-pro-von Willebrand factor cDNA. *Nucleic Acids Res.* **14**:7125–7127.

Friedman, K. D., N. L. Rosen, P. J. Newman, and R. R. Montgomery. 1988. Enzymatic amplification of specific cDNA inserts from λgt11 libraries. *Nucleic Acids Res.* **16**:8718.

Ginsburg, D., R. I. Handin, D. T. Bonthron, T. A. Donlon, G. A. P. Bruns, S. A. Latt, and S. H. Orkin. 1985. Human von Willebrand factor (vWF): isolation of complementary DNA (cDNA) clones and chromosomal localization. *Science* **228**: 1401–1406.

Lopez, J. A., D. W. Chung, K. Fujikawa, F. S. Hagen, T. Papayannopoulou, and G. J. Roth. 1987. Cloning of the α chain of human platelet glycoprotein Ib: a transmembrane protein with homology to leucine-rich α_2-glycoprotein. *Proc. Natl. Acad. Sci. USA* **84**:5615–5619.

Montgomery, R. R., E. A. Vokac, J. P. Scott, E. R. Reynolds, N. L. Rosen, P. J. Newman, and K. D. Friedman. 1988. Does von Willebrand factor bind to endothelial cells through a GPIb-like receptor? *Blood* **72**:333a.

Newman, P. J., J. Gorski, G. C. White II, S. Gidwitz, C. J. Cretney, and R. H. Aster. 1988. Enzymatic amplification of platelet-specific messenger RNA using the polymerase chain reaction. *J. Clin. Invest.* **82**:739–743.

Poncz, M., R. Eisman, R. Heidenreich, M. Silver, G. Vilaire, S. Surrey, E. Schwartz, and J. S. Bennett. 1987. Structure of the platelet membrane glycoprotein Ib: homology to the α subunits of the vitronectin and fibronectin membrane receptors. *J. Biol. Chem.* **262**:8476–8482.

Saiki, R. K., D. H. Gelfand, S. Stoffel, S. J. Scharf, R. Higuchi, G. T. Horn, K. B. Mullis, and H. A. Erlich. 1988. Primer-directed enzymatic amplification of DNA with a thermostable DNA polymerase. *Science* **239**:487–491.

Trahey, M., G. Wong, R. Halenbeck, B. Rubenfeld, G. A. Martin, M. Ladner, C. M. Long, W. J. Crosier, K. Watt, K. Koths, and F. McCormick. 1988. Molecular cloning of two types of GAP complementary DNA from human placenta. *Science* **242**:1697–1700.

GENETICS AND EVOLUTION

32

HLA DNA TYPING

Henry A. Erlich and Teodorica L. Bugawan

The HLA region, located on the short arm of chromosome 6, encodes a set of highly polymorphic integral membrane proteins that bind antigen peptide fragments. This complex of HLA proteins and antigen peptide is recognized by the T-cell receptor, leading to activation of the T lymphocyte and the initiation of a specific immune response. The HLA class II loci (e.g., HLA-DR, -DQ, and -DP) encode an α and a β glycopeptide chain; this cell surface heterodimer presents antigen to $CD4^+$ T lymphocytes. The HLA class I loci, (e.g., HLA-A, -B, and -C) encode a glycopeptide chain that, associated with β-2 microglobulin, binds antigen and is recognized by $CD8^+$ T lymphocytes (reviewed in Kappes and Strominger 1988).

The detection of genetic variation in the HLA region (or "HLA typing") is useful in tissue typing for transplantation to minimize graft rejection by selecting "HLA-matched" donor and recipient pairs. HLA typing has also proved highly informative in the analysis of genetic susceptibility to autoimmune diseases, since specific alleles at certain HLA loci have been associated with particular diseases (e.g., HLA-DR3 and DR4 with insulin-dependent diabetes mellitus). The degree of polymorphism at these loci has also made HLA typing valuable for individual identification in forensic analysis and paternity determination.

Traditionally, HLA typing has been carried out with serologic reagents or, in the case of the class II loci, by the mixed-lymphocyte culture (MLC) in which T lymphocytes from one sample will proliferate in response to different or "nonmatching" HLA class II gene products on the cell surface of cells from the other sample (reviewed in Bodmer 1984). More recently, with the availability of cloned HLA cDNA and genomic hybridization probes, HLA typing at the DNA level could be carried out by restriction fragment length polymorphism (RFLP). DNA typing offers a number of advantages over immunologic typing; these have been reviewed elsewhere (Erlich *et al.* 1986). RFLP analysis is based on the presence or absence of polymorphic restriction sites located primarily in noncoding regions that are in linkage disequilibrium (nonrandom association) with allelic variation in coding sequences. Until recently, the direct analysis of coding sequence polymorphism has been difficult. However, the enzymatic amplification of specific DNA sequences by PCR has provided a new approach to genetic typing.

PCR/Oligonucleotide Probe Typing

The capacity of the PCR to amplify a specific segment of genomic DNA has made it an invaluable tool in the study of polymorphism and evolution, as well as in the analysis of genetic susceptibility to disease. In all of these areas, a particular gene must be examined in a variety of individuals; either within a species, in different closely related species, or in patient and in healthy control populations. Here, we will focus on the use of PCR to perform HLA class II DNA typing and will use the analysis of allelic diversity at the HLA-DQα (now designated the DQA1) locus as an illustrative example. The polymorphism of class II genes is localized primarily to the NH_2 terminal outer domain encoded by the second exon. Using PCR primers to conserved regions, we have amplified and sequenced the second exon of these class II loci from many different individuals, revealing a remarkable degree of allelic diversity (Fig. 1). For some of these loci (e.g., DRβI), it is likely additional alleles will be revealed as samples from a variety of ethnic groups are analyzed. The sequences of the amplification primers used are shown in Table 1. These sequences

Class II Genes

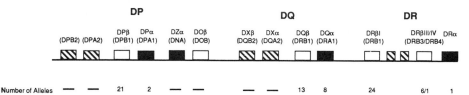

Figure 1   The expressed class II α loci are shown as filled-in boxes and the expressed β chain loci are shown as open boxes. Nonexpressed genes are represented by hatched boxes. The new nomenclature for these loci is shown in parentheses. In the DR region, some haplotypes (e.g., DRw8 and DR1) have only one expressed β locus. The number of alleles identified by us thus far are shown below the locus.

were determined either by M13 cloning of the DNA amplified with "linker-primers" (Scharf *et al.* 1986; see Chapter 4) followed by chain termination sequencing of the purified single-stranded phage DNA or by direct sequencing using the asymmetric primer method (Gyllensten and Erlich 1988; see Chapter 10) to generate single strands from the PCR.

Having determined the extent of allelic diversity by sequence analysis, one can then detect the presence of specific alleles in a PCR-amplified sample by dot blot hybridization with labeled oligonucleotide probes (Saiki *et al.* 1986). We have used this procedure with either ^{32}P, biotin, or horseradish peroxidase-labeled oligonucleotide probes for HLA-DQα, HLA-DQβ, HLA-DRβ, and HLA-DPβ typing (Bugawan *et al.* 1988; Horn *et al.* 1988; Scharf *et al.* 1988; Bugawan *et al.* 1989). This type of simple, rapid, and precise test is critical in typing the large number of patient and control samples necessary for the analysis of genetic susceptibility to disease. Unlike immunologic or RFLP approaches to HLA typing, the sequence based PCR/oligonucleotide probe analysis reveals not only that two alleles are different but *how* they differ, allowing the identification of individual polymorphic residues that are critical to disease susceptibility, graft rejection, T-cell recognition, antigen presentation, or whatever biological function is being examined.

The dot blot typing approach involves the PCR amplification of a specific region (i.e., the second exon of the DQα locus) and the subsequent immobilization of the amplified DNA to replicate filters (nylon membranes). The protein sequence alignment for the DQα alleles is shown in Fig. 2, and the nucleotide sequence alignment and the location of the oligonucleotide probes are shown in Fig. 3. Each

Table 1

Sequence of Oligonucleotide PCR Primers for HLA Class II Gene Amplification

HLA gene	Name	Sequence 5' to 3'	PCR amplification conditions		
			Denature	Anneal	Extend
DRβ	GH46	CCGGATCCTTCGTGTCCCCACAGCACG	96°C	55°C	72°C
	GH50	CTCCCCAACCCGTAGTTGTGTCTGCA			
DQα	GH26	GTGCTGCAGGTGTAAACTTGTACCAG	96°C	65°C	≥65°C
	GH27	CACGGATCCGGTAGCAGCGGTAGAGTTG			
DQβ	GH28	CTCGGATCCGCATGTGCTACTTCACCAACG	96°C	55°C	72°C
	GH29	GAGCTGCAGGTAGTTGTGTCTGCACAC			
DPα	GH98	CGCGGATCCTGTGTCAACTTATGCCGC	96°C	55°C	72°C
	GH99	GTGGCTGCCAGTGTGGTTGGAACGC			
DPβ	UG21	CGGATCCGCCCAAAGCCCTCACTC	96°C	65°C	≥65°C
	UG19	GCTGCAGGAGAGTGGCGCCTCCGCTCAT			

The denaturation, annealing, and extension times are 30 seconds, and ramping from 96° to 55° to 72°C is programmed in the PECI ThermoCycler for 1 sec. In some cases (DQα and DPβ), the temperature profile is a 2-step cycle from 65°C to 96°C. With these conditions, the DQα primers (GH26 and GH27) and the DQβ primers (GH28 and GH29) co-amplify to a limited extent the homologous sequence from the linked but nonexpressed loci DXα and DXβ. The DRβ primers GH46 and GH50 amplify *all* DRβ loci not simply the DRβ1 locus.

264

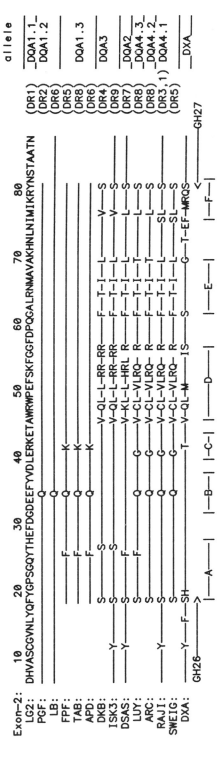

Figure 2 HLA-DQα protein sequences. The DQα allele designations are shown to the left, and their DR serotypes are shown to the right of the sequences. Polymorphic segments corresponding to specific oligonucleotide probes are designated A through F.

265

```
                    gh88*              rh54            gh76*              rh83
                          gh69                                              I
1.1: TTTGATGGAGATGAGGAGTTCTACGTGGACCTGGAGAGGAAGGAGACTGCCTGGCCGTGGCCTGAGTTCAGCAAATTTGGAGGTTTTGACCCGCAGGG
                          _____                                          _____
1.2: ------------------------C----------------------------------------------------------------------------
                                        gh77
1.3: ------------------------C------------A---------------------------------------------------------------
                                        ____                                  rh71
2:   -------------------------------C------------------------T-----AA---T----CT-----CA--G-C--  A-A-------ATT
                                                                                        ___gh67
3:   -------------------------------C------------------------T-----A--T-----CT----C---G-----A-A-A---------ATT
                                              gh66                        _____
4:   -----------------------------------------------G------------------T-----T-T--T---TTC----AC------  A--A-------ATT

X:   ------------------------------------------------------C---A------------T-----A--T---AT---T--------------AT-A----A--
```

* = probe sequence is other strand

probe	specificity	length	sequence, 5' to 3'
RH54	all	33-MER	CTACGTGGACCTGGAGAGGAAGGAGACTGCCTG
RH83	1	19-MER	GAGTTCAGCAAATTTGGAG
GH88	1.1	19-MER	CGTAGAACTCCTCATCTCC
GH76	all but 1.3	17-MER	GTCTCCTTCCTCTCCAG
GH89	1.2, 1.3, 4	19-MER	GATGAGCAGTTCTACGTGG
GH77	1.3	17-MER	CTGGAGAAGGAGGAGAC
RH71	2	21-MER	TTCCACAGACTTAGATTTGAC
GH67	3	19-MER	TTCCGCAGATTTAGAAGAT
GH66	4	19-MER	TGTTTGCCTGTCTCCAGAC

Figure 3 HLA-DQα DNA sequence. The DQα DNA sequence and location of allele-specific oligonucleotide probe are shown. The probe name, sequence, and its specificity are shown on the bottom of the figure.

filter is then hybridized with a labeled oligonucleotide probe, and the bound probe is detected by the enzymatic conversion of a colorless soluble substrate to a colored precipitate. A detailed description of the protocol for HLA-DQα typing is given below. For a locus with n alleles, each amplified sample must be immobilized on n membranes and each membrane hybridized to one of n labeled probes. Thus, the procedural complexity of this approach is a function of the number of oligonucleotide probes required for complete genetic analysis. To address this problem, we have recently developed a "reverse dot-blot" procedure in which the oligonucleotide probe is immobilized on a membrane and hybridized to a labeled PCR product (Saiki *et al.* 1989). In this method, a panel of oligonucleotide probes is tailed with poly(dT) using terminal transferase and UV-crosslinked to a nylon membrane. The PCR product, labeled during amplification by using biotinylated primers, is then hybridized to the immobilized array of oligonucleotide probes. The presence of the specifically bound PCR product is detected using a streptavidin-horseradish peroxidase conjugate. Both the dot blot and the reverse dot blot method represent rapid and precise approaches for typing HLA class II polymorphism.

Protocols

DQα-PCR Amplification

1. Prepare PCR mixture containing the following components (see Chapter 1).

DNA	10 pg to 1 μg
10× *Taq* salts	10 μl (500 mM KCl, 15 mM MgCl$_2$)
100 mM dNTPs (25 mM each)	0.7 μl
10 μM GH26	2.5 μl
10 μM GH27	2.5 μl

   5 units/μl *Taq* Polymerase 0.6 μl
   Bring up to 100 μl with glass-distilled water.
2. Include a positive (DNA that is known to successfully amplify for DQα) and negative (no DNA) control to check for PCR efficiency, specific oligonucleotide probe hybridization, and contamination.

3. Mix reaction by vortexing lightly.

4. Amplify for 25 cycles (more if less than 0.1 μg starting DNA) in Perkin-Elmer Cetus Thermal cycler. Denature at 94°C for 30 seconds and anneal and extend at 65°C for 30 seconds.

5. Load 3 μl of PCR into 3% NuSieve plus 1% agarose gel to monitor amplification efficiency.

Preparation of Dot Blots

Prepare four replicate dot blots by spotting 5 μl of denatured PCR product per dot onto Genatran (or ZetaProbe) membrane. For detailed protocol for dot blot preparation, see Chapter 15. The immobilized amplification products are each hybridized with one of four DQα-ASO probes to determine the four allelic major types (DQA1, DQA2, DQA3, DQA4). If subtyping is necessary, membranes can be stripped and rehybridized with DQA1 subtyping probes, or additional membrane can be prepared.

Allele-Specific Oligonucleotide Probe Hybridization

1. Pre-wet the dot blot membranes in water or 2× SSPE.

2. Prehybridize the membranes in Seal-a-Meal bags containing just enough hybridization solution (5× SSPE, 5× Denhardt's and 0.5% Triton X-100) to cover the membrane (for a full-size dot blot use 8 ml solution). Incubate the membranes for 5 to 10 minutes at 55°C.

3. Add the HRP-labeled Allele Specific Oligonucleotide probe at 1 pmol/ml hybridization solution into the bag in the liquid. Hybridize for 30 minutes at 55°C.

4. Wash the membranes in 2× SSPE, 0.1% Triton X-100 for 5 minutes at 55°C.

Detection

The presence of the probes is detected by using a colorless soluble substrate that is converted by HRP, in presence of H_2O_2, to a colored precipitate. The detection procedure was carried out at room temperature with moderate shaking (see Chapter 15).

1. Incubate the membranes for 5 minutes with Buffer B [137 mM NaCL, 2.7 mM KCl, 1.5 mM KH_2PO_4, 8.0 mM Na_2HPO_4 (pH 7.4), 5% (v/v) Triton X-100, 1 M Urea, and 1% Dextran Sulfate].

2. Wash the membranes for 5 minutes in Buffer C [100 mM Sodium Citrate (pH 5)].
3. Incubate the membranes for 10 minutes in Buffer C plus 0.1 mg/ml of TMB (3,3' 5,5'- tetramethlybenzidine in 100% Ethanol) under light exclusion.
4. Add 0.0015% hydrogen peroxide for blue positive signal to appear, which usually takes 1 to 5 minutes.
5. Stop the color development by washing the membranes in Buffer C or water for 20 minutes. If filter background is high, additional washing may be necessary. The membranes can be stored in Buffer C in the absence of light at 4°C with little fading of signals for at least 2 months.

Rehybridization

If subtyping for the DQA1 or DQA4 allelic type is required, the signals on the membranes can be decolorized by washing in 0.18% Na_2SO_3 at room temperature with shaking; this can take from 5 minutes to 1 hour depending on signal intensity. The annealed probes are then stripped by incubating the membrane in water and 0.5% SDS at 65°C for at least 1 hour with shaking. The membranes can now be hybridized with subtyping probes.

Reverse Dot Blots

1. Pre-wet Genatran or ZetaProbe nylon membrane in water or 2× SSPE.
2. Spot 4 pmol each of oligonucleotide probe tailed with poly(dT) using terminal transferase (see Saiki *et al.* 1989 for details).
3. Fix the probe to the membrane by UV-cross linking at 50 mjules using Stratagene Stratalinker apparatus.
4. Incubate the membrane for at least 30 minutes at 55°C in 5× SSPE, 0.5% SDS to remove the unbound probe.
5. The membrane can be stored dry at room temperature for future use.

Hybridization

1. Pre-warm hybridization solution (5× SSPE, 0.5% SDS) to 55°C before use.

2. Denature biotinylated PCR product at 95°C for 5 minutes and place on ice.
3. Place the membrane in Seal-a-Meal bags or troughs, add 2 to 3 ml hybridization solution, 20 μl denatured PCR product, and 15 μl of 20 mg/ml HRP-SA stock solution. Remove air bubbles and seal the bag or cover trough tray with a glass plate and place a lead doughnut weight on top of the tray.
4. Hybridize for 20 minutes at 55°C in a shaking water bath.
5. Wash the membrane in a bowl containing 200 to 300 ml of pre-warmed 2× SSPE, 0.1% SDS for 10 minutes in a shaking water bath at 55°C.
6. Proceed to the detection procedure.

Potential Problems

Contamination or carry-over from a previous PCR amplification can lead to problems in interpreting typing results. In general, this can be minimized by careful laboratory procedure (see Chapter 17) and monitored by the use of negative control (no template DNA) samples. In most cases of genetic typing, a contaminant can be detected as the presence of more than two alleles at a polymorphic locus.

For some loci, in particular at the HLA-DPβ locus (Bugawan *et al.* 1989), specific alleles cannot be defined by the hybridization of a particular oligonucleotide probe but by the pattern of probes that hybridize to the amplified DNA sample. The absence of HLA-DP allele-specific sequences and the requirement for multiple probes defining each individual allele reflects the patchwork pattern of polymorphism at this locus. Occasionally, an ambiguous typing result can be obtained when the sequence detected by a given probe could be assigned to either of the two alleles. In this case, the ambiguity can usually be resolved by using an allele-specific primer for amplification. This issue is discussed for HLA-DRβ typing with oligonucleotides in Scharf *et al.* (1989).

Another potential problem is that, on occasion, some nylon membranes appear to produce more "cross-hybridization" of the sequence-specific oligonucleotide typing probes, even when the hybridization and wash conditions remain constant. If this occurs, some minor modification (e.g., increasing stringency by lowering the salt concentration or increasing the temperature) may be required to address the variability in membrane properties.

Literature Cited

Bodmer, W. F. 1984. The HLA system. *Histocompatibility testing* (ed. E. D. Albert, M. P. Baur, and W. R. Mayr). Springer-Verlag KG, Berlin.

Bugawan, T. L., R. K. Saiki, C. H. Levenson, R. M. Watson, and H. A. Erlich. 1988. The use of non-radioactive oligonucleotide probes to analyze enzymatically amplified DNA for prenatal diagnosis and forensic HLA typing. *Bio/Technology* **6**:943–947.

Bugawan, T. L., G. Angelini, J. Larrick, S. Auricchio, G. B. Ferrara, and H. A. Erlich. 1989. HLA-DPβ sequence polymorphism and celiac disease: a combination of a particular DPβ allele and an HLA-DQ heterodimer confers susceptibility. *Nature (London)* **339**:470–472.

Erlich, H. A., E. L. Sheldon, and G. Horn. 1986. HLA typing using DNA probes. *Bio/Technology* **4**:975–981.

Gyllensten, U. B., and H. A. Erlich. 1988. Generation of single-stranded DNA by the polymerase chain reaction and its application to direct sequencing of the HLA-DQα locus. *Proc. Natl. Acad. Sci. USA* **85**:7652–7658.

Horn, G. T., T. L. Bugawan, C. Long, and H. A. Erlich. 1988. Allelic sequence variation of the HLA-DQ loci: relationship to serology and insulin-dependent diabetes susceptibility. *Proc. Natl. Acad. Sci. USA* **85**:6012–6016.

Kappes, D., and J. L. Strominger. (1988) Human class II major histocompatibility complex genes and proteins. *Ann. Rev. Biochem.* **57**:991-1028.

Saiki, R. K., T. L. Bugawan, G. T. Horn, K. B. Mullis, and H. A. Erlich. 1986. Analysis of enzymatically amplified β-globin and HLA-DQα DNA with allele-specific oligonucleotide probes. *Nature (London)* **324**:163–166.

Saiki, R. K., P. S. Walsh, C. H. Levenson, and H. A. Erlich. 1989. Genetic analysis of amplified DNA with immobilized sequence-specific oligonucleotide probes. *Proc. Natl. Acad. Sci. USA* **86**:6230–6234.

Scharf, S. J., G. Horn, H. Erlich. 1986. Direct cloning and sequence analysis of enzymatically amplified genomic sequences. *Science* **223**:1076–1078.

Scharf, S. J., A. Friedmann, C. Brautbar, F. Szafer, L. Steinman, G. Horn, U. Gyllensten, and H. A. Erlich. 1988. HLA class II allelic variation and susceptibility to *Pemphigus vulgaris*. *Proc. Natl. Acad. Sci. USA* **85**:3504–3508.

Scharf, S. J., A. Friedmann, L. Steinman, C. Brautbar, and H. A. Erlich. 1989. Specific HLA-DQβ and DRβI alleles confer susceptibility to *Pemphigus vulgaris*. *Proc. Natl. Acad. Sci. USA*, **86**:6215–6219.

33

MULTIPLEX PCR FOR THE DIAGNOSIS OF DUCHENNE MUSCULAR DYSTROPHY

Jeffrey S. Chamberlain, Richard A. Gibbs, Joel E. Ranier, and C.Thomas Caskey

PCR has achieved widespread use in the analysis of genetic diseases. Typically, a region of interest is amplified from genomic DNA or cDNA and examined for mutations or polymorphisms by sequencing, hybridization with allele-specific oligonucleotides, digestion with a restriction endonuclease, or cleavage of heteroduplexes either chemically or enzymatically (Erlich, 1989). PCR can also be used to rapidly identify deletions in genomic DNA by determining whether a specific region of DNA can be amplified (Chamberlain *et al.* 1989). In cases of small genes or transcripts, or in cases where the mutant allele or the polymorphism is previously known, amplification of single, small regions of DNA suffices for an analysis. However, when analyzing large genes and transcripts where the target region of interest is unknown, multiple PCRs may be required to identify the specific change of bases or deletion of interest.

Analysis of mutations at the Duchenne muscular dystrophy (DMD) locus is one example in which an enormous gene and transcript necessitates PCR amplification of many large regions of DNA. DMD is among the most common human genetic diseases, affecting approximately 1 in 3500 male births, and one-third of all cases arise via new mutations (Emery 1987). This gene is greater than 2 million

PCR Protocols: A Guide to Methods and Applications

base pairs in size and contains at least 70 exons separated by an average intron size of 35 kb (Koenig *et al.* 1987; van Ommen *et al.* 1987). Partial intragenic deletions account for up to 60% of all cases of this disease, and although these deletions arise in a heterogeneous manner, they are generally large (typically several hundred kilobases) and are concentrated around two specific regions of the gene (Koenig *et al.* 1987; Baumbach *et al.* 1989). These observations have led to the development of conditions under which at least nine separate regions of the dystrophin gene are co-amplified in a single PCR. These multiplex reactions permit the rapid identification of 80 to 90% of all dystrophin gene deletions (Chamberlain *et al.* 1988; Chamberlain, manuscript in preparation), can detect duplications of the gene (Fenwick *et al.*, manuscript in preparation), and should enable multiplex sequencing or heteroduplex mapping to identify point mutations as has been done at the HPRT locus (Gibbs *et al.* 1989a; Gibbs *et al.* 1989b; Gibbs and Caskey 1987).

This chapter describes the protocol that we use for multiplex amplification of the dystrophin gene, addresses potential difficulties and ways to avoid problems, and discusses general strategies for the development of multiplex reactions to amplify large numbers of separate sequences for the analysis of any given gene or sets of nucleic acid sequences.

Multiplex Amplification Protocol

Reagents

5× *Taq* polymerase buffer for this procedure is 83 mM $(NH_4)_2SO_4$, 335 mM Tris–HCl (pH 8.8), 33.5 mM $MgCl_2$, 50 mM β-mercaptoethanol, 850 μg/ml of bovine serum albumin, and 34 μM EDTA. dNTPs. We have obtained the most consistent results using dNTPs purchased from Pharmacia.

Procedure

1. Prepare template DNA from lymphoblasts, amniotic fluid cells, or chorionic villi specimens dissected of decidual tissue by using standard protocols (Ward *et al.* 1989).

2. Add the following to a 0.5-ml microfuge tube:

   H_2O —(to 45 μl final volume)
   5× *Taq* buffer 10 μl
   25 m*M* each dNTP 3 μl
   oligonucleotide primers —(25 pmol of each)
   DMSO 5 μl

   Reaction mixes prepared in this manner are stable at −70°C for least 3 months.

3. Mix gently; add 250 ng of template DNA (dilute DNA to a final concentration of between 50 and 250 ng/μl so that the DNA may be added to the reaction in a volume of 5 μl or less).

4. Add H_2O to a final volume of 50 μl.

5. Add 5 units of *Taq* polymerase (AmpliTaq, available from Perkin-Elmer Cetus), mix gently.

6. Add 25 μl of paraffin oil and centrifuge 5 seconds.

7. Place the sample in an automatic thermocycler. Cycle as follows:

   94°C for 6 minutes (once)
   94°C for 30 seconds; 53°C for 30 seconds; 65°C for 4 minutes (repeat 23 times)
   65°C for 7 minutes (once)
   4°C until analysis (up to 2 months)

8. Electrophorese 15 μl of the reaction products on a 1.4% agarose or a 3% NuSieve agarose gel (FMC Bioproducts) containing 0.5 μg/ml ethidium bromide for 2 hours at 3.7 V/cm in 90 m*M* Tris base, 90 m*M* boric acid, 1 m*M* EDTA (TBE).

9. Photograph the gel or otherwise record the results.

Discussion

The multiplex DNA amplification procedure described in this chapter provides a rapid and simple method to screen large numbers of patient samples for deletions leading to DMD. Amplification of the nine regions currently included in the assay (Fig. 1) would have detected at least 80% of the DMD gene deletions that have been reported by our laboratory and others (Koenig *et al.* 1987; Baumbach *et*

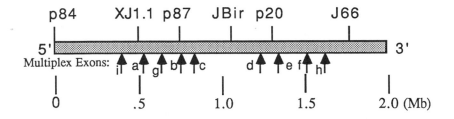

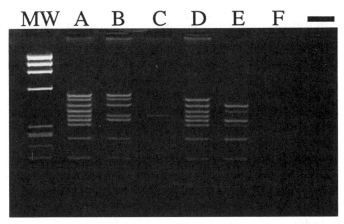

Figure 1   Detection of DNA deletions at the DMD locus by multiplex DNA amplification. (Top). Schematic illustration of the DMD gene indicating the relative location of the nine exon-containing DNA segments amplified with this procedure (arrows, a–i). Also shown are the location of several RFLP-detecting genomic probes. The exon contained within each amplified region is indicated in Table 1. (Bottom). Detection of deletions in the DNA of DMD males. (MW) *Hae*III-digested ΦX174 DNA molecular weight standard. (A–F) show the results of multiplex amplification from the DNA of six unrelated male DMD patients. The sample in (A) does not display a deletion, the samples in (B–E) are deleted for one or more of the nine regions, and the sample in (F) was deleted for all nine regions. (−) shows a negative control in which no template DNA was added to the reaction. Shown is a 3% NuSieve agarose gel through which 15 μl of each reaction was electrophoresed. The fragments correspond (top to bottom) to regions e, f, c, b, h, a, g, d, and i, respectively.

al. 1989; Chamberlain *et al.* 1988), thus directly identifying the molecular defect in almost 50% of all cases. This assay is uninformative for the approximately 40% of all cases that do not display intragenic deletions, and those cases need to be further examined via Southern blot analysis. However, several features of the method make it ideally suited for preliminary screening of all patients. The assay is extremely rapid and requires only about 5 hours from start-up to photography of results. Large quantities of premade reaction mixes can be prepared in advance and stored at $-70°C$. For analysis a tube is thawed, DNA and *Taq* polymerase are added, and the reaction is performed on an automatic thermocycler. The results are easily determined by electrophoresing a small aliquot of the reaction products on agarose gels and visually identifying which of the nine fragments have been amplified. Figure 1 displays an example of the use of the assay. The location of the nine amplified regions is indicated at the top, while the bottom of the figure shows a photograph of an agarose gel through which reaction products obtained with six patient samples were electrophoresed. Table 1 lists the sequences of

Table 1

Summary of DMD Gene Multiplex Amplification Primer Sets

Exon and Size[a]	Primer Sequence[b]	Amplified[c]
a. Exon8 182 bp; (probe 1b)	F-GTCCTTTACACACTTTACCTGTTGAG R-GGCCTCATTCTCATGTTCTAATTAG	360 bp
b. Exon17 178 bp; (probe 3)	F-GACTTTCGATGTTGAGATTACTTTCCC R-AAGCTTGAGATGCTCTCACCTTTTCC	416 bp
c. Exon 19 88 bp; (probe 3)	F-TTCTACCACATCCCATTTTCTTCCA R-GATGGCAAAAGTGTTGAGAAAAAGTC	459 bp
d. 4.1 Kb <u>HindIII</u> 148 bp; (probe 7)	F-CTTGATCCATATGCTTTTACCTGCA R-TCCATCACCCTTCAGAACCTGATCT	268 bp
e. 0.5 Kb <u>HindIII</u> 176 bp; (probe 7)	F-AAACATGGAACATCCTTGTGGGGAC R-CATTCCTATTAGATCTGTCGCCCTAC	547 bp
f. 1.2/3.8 Kb <u>HindIII</u> 186 bp; (probe 8)	F-TTGAATACATTGGTTAAATCCCAACATG R-CCTGAATAAAGTCTTCCTTACCACAC	506 bp
g. Exon 12 151 bp; (probe 2)	F-GATAGTGGGCTTTACTTACATCCTTC R-GAAAGCACGCAACATAAGATACACCT	331 bp
h. 3.1 Kb <u>Hind</u> III 233 bp; (probe 8)	F-GAAATTGGCTCTTTAGCTTGTGTTTC R-GGAGAGTAAAGTGATTGGTGGAAAATC	388 bp
i. Exon 4 78 bp; (probe 1a)	F-TTGTCGGTCTCCTGCTGGTCAGTG R-CAAAGCCCTCACTCAAACATGAAGC	196 bp

[a] Each exon is designated by a letter. When known, the exon number is listed; when not known the size of the genomic <u>Hind</u> III fragment that the exon is located on is listed. Also indicated is the human DMD cDNA probe that hybridizes with each exon (Koenig *et al.* 1987), as well as the size of the exon in base pairs (bp).

[b] Shown is the sequence in 5′-3′ orientation for the PCR primers used to amplify each region. F: forward primer, hybridizes 5′ of the exon; R: reverse primer, hybridizes 3′ of the exon.

[c] The size of the amplified fragment obtained with each primer set.

the 18 oligonucleotide primers that have been combined for these reactions.

We have analyzed more than 350 patient samples by this method, and in each case the results have been subsequently confirmed via Southern analysis. Cases in which all nine of the regions are deleted should be confirmed via Southern or dot blot analysis with the appropriate cDNA probe. Alternatively, an additional set of primers derived from elsewhere in the genome could be added to the reactions to serve as a positive control for amplification.

Potential Problems and Solutions

Through the development and testing of the method, we have observed that almost all problems that may be encountered can be traced to one of the following causes.

Amount of DNA

The reactions are designed to produce visibly detectable results after 23 cycles of PCR amplification. Adding too little DNA to the reactions produces weak or undetectable bands. In such cases reactions can be returned to the thermocycler and amplified for several additional cycles. Reamplifications are generally successful when performed up to a week after the initial PCR, and no additional enzyme is required. Alternatively, adding too much DNA to a reaction disrupts the relative intensity of the nine bands and can complicate interpretation of the results. This also greatly increases the hazard of false-positive amplification (see following).

Contamination

There are many ways that exogenous DNA can be introduced into a PCR assay, and the problem is particularly acute with this method, as failure to achieve amplification is the only indication that a mutation has been identified. Serious errors could result if false-positive amplification from contaminating exogenous or maternal DNA were to occur. Reagent contamination is minimized in two ways. First, the preparation of large quantities of individually aliquoted reaction mixes ("kits") allows batches of reagents to be checked for contamination before use. Second, the preparation and analysis stages of the reactions should be physically separated; i.e., amplified reactions are opened and aliquots are removed for analysis at a separate location to where the reactions are initiated. In addition, separate pipettors are used to sample amplified reactions from those that are used to mix together the initial ingredients. These precautions are critical to prevent minute quantities of prior reaction products from serving as efficient template for future reactions. To eliminate or preclude the possibility of contamination, pipettors may be effectively cleaned of DNA as follows. Soak the barrel of a pipetman in a beaker containing 0.25 N HCl for 30 minutes, repeat with 0.5 N NaOH for 30 minutes, rinse well with distilled water, and dry. Sample contamination must also be kept to a minimum. Equipment that has been in contact with either prior reaction products or cloned DNA complementary to any of the regions targeted for amplification must not be used to prepare template DNA. We prepare all DNA samples for PCR analysis on an Applied Biosystems model 340A DNA extractor. Preparation of several hundred samples on this machine has not yet led to any detectable contamination from either exogenous or prior sample sources. However, standard methods should work well as long as the previously stated care is taken.

Maternal DNA contamination of amniotic fluid cells or chorionic villi specimens is a problem less easily controlled by laboratory personnel. Chorionic villi specimens should be routinely dissected of maternal decidual tissue before extraction of the DNA. Beyond this, care must be taken to ensure that amplification is kept to the absolute minimum required to make a diagnosis. By performing reconstitution experiments with normal and partially deleted DNA samples, we have observed that levels of maternal DNA at up to 5% of the total will not lead to false-positive amplification as long as the reactions do not approach saturation (Chamberlain *et al.* 1988).

Specificity

Strict adherence to the recommended temperatures and times (particularly for the annealing step) of the PCR cycles (step 7, Protocol) is necessary to prevent the generation of spurious amplification products. For DMD PCRs, the specificity of reaction mixes and the quality of all ingredients are easily checked by using template DNA containing partial dystrophin gene deletions that have been previously delineated via Southern analysis with full-length cDNA clones.

Future Developments

Development of multiplex PCR for other regions of the genome will require both appropriate DNA sequence data for synthesis of primers and modification of reaction conditions to permit reproducible and specific co-amplification of each targeted DNA fragment. The rationales behind the choices of primers and reaction conditions used in this assay were as follows. (1) The location of PCR priming sites needs to be chosen while considering the entire multiplex reaction. Flexibility in choosing the size of the regions to be amplified facilitates obtaining multiple reaction products that can be resolved on agarose gels. (2) Oligonucleotide PCR primers 23 to 28 bases long have worked well for multiplex amplification. Primers of this length permit high-stringency annealing, thus providing greater specificity during multiplex amplification reactions. Combinations of sets of shorter primers that may work well individually can lead to artifactual bands during multiplex amplification. Often such bands can be traced to only one or two of the primers, which can then be replaced by more specific oligonucleotides. The G + C content of primers generally has varied between 35 and 50%. Lower G + C contents have lead to poor amplification, while higher contents have occasionally led to spurious amplification products. (3) The reaction buffer system that has permitted multiplex amplification contains 10% (V/V) dimethylsulfoxide (Kogan *et al.* 1987). Although the presence of organic solvents is not optimal for most uses of *Taq* polymerase, we have been unable to achieve efficient and reproducible multiplex amplification of nine regions in an aqueous buffer. Additional parameters that have been varied include the amount of enzyme used, the amount of template DNA, the concentrations of Mg^{2+}, dNTPs, and primers, and the time and temperature of annealing and DNA polymerization steps. Different combinations of primers require different conditions for optimal multiplex amplification, and it will probably be necessary to optimize reaction conditions for any given set of oligonucleotides. It is advisable to limit the annealing step to 30 seconds at the highest possible temperature.

Finally, as more primer sets are added, the permissive reaction conditions have been observed to become increasingly less flexible.

Acknowledgments

We gratefully acknowledge the expert technical assistance of Andrew Civitello, Nancy Farwell, Donna Muzny, and Phi-Nga Nguyen. This work was supported by a Task Force on Genetics grant from the Muscular Dystrophy Association. JSC and this work were supported by the Texas Advanced Technology Program under grant #3034.

Literature Cited

Baumbach, L. L., J. S. Chamberlain, P. A. Ward, N. J. Farwell, and C. T. Caskey. 1989. Molecular and clinical correlations of deletions leading to Duchenne and Becker muscular dystrophies. *Neurology* **39**:465–474.

Chamberlain, J. S., R. A. Gibbs, J. E. Ranier, P. N. Nguyen, and C. T. Caskey. 1988. Deletion screening of the Duchenne muscular dystrophy locus via multiplex DNA amplification. *Nucleic Acids Res.* **16**:11141–11156.

Chamberlain, J. S., J. Ranier, J. A. Pearlman, P. N. Nguyen, N. J. Farwell, R. Gibbs, D. M. Muzny, and T. Caskey. 1989. Analysis of Duchenne muscular dystrophy gene mutations in mice and humans. *UCLA Symp. Cell. Mol. Biol. New Ser.* **93**:951–962.

Emery, A. E. H. 1987. In *Oxford monographs on medical genetics*, No. 15. Oxford University Press, Oxford, England.

Erlich, H. A. (ed.). 1989. *PCR technology: principles and applications for DNA amplification.* Stockton Press, New York.

Gibbs, R. A., and C. T. Caskey. 1987. Identification and localization of mutations at the Lesch-Nyhan locus by ribonuclease A cleavage. *Science* **236**:303–305.

Gibbs, R. A., P. N. Nguyen, L. J. McBride, S. M. Keopf, and C. T. Caskey. 1989a. Identification of mutations leading to the Lesch-Nyhan syndrome by automated direct DNA sequencing of *in vitro* amplified cDNA. *Proc. Natl. Acad. Sci. USA* **86**:1919–1923.

Gibbs, R. A., J. S. Chamberlain, and C. T. Caskey. 1989b. Diagnosis of new mutation diseases using the polymerase chain reaciton. In *PCR technology: principles and applications for DNA amplification.* Chapter 15. (ed. H. A. Erlich). Stockton Press, New York.

Koenig, M., E. P. Hoffman, C. J. Bertelson, A. P. Monaco, C. Feener, and L. M. Kunkel. 1987. Complete cloning of the Duchenne muscular dystrophy (DMD) cDNA and preliminary genomic organization of the DMD Gene in normal and affected individuals. *Cell* **50**:509–517.

Kogan, S. C., M. Doherty, and J. Gitschier. 1987. An improved method for prenatal diagnosis of genetic diseases by analysis of amplified DNA sequences. *N. Engl. J. Med.* **317**:985–990.

van Ommen, G. J. B., C. Bertelsen, H. B. Ginjaar, J. T. den Dunnen, E. Bakker, J. Chelly, M. Matton, A. J. van Essen, J. Bartley, L. M. Kunkel, and P. L. Pearson. 1987. Long-range genomic map of the Duchenne muscular dystrophy gene: isolation and use of J66 (dxs268), a distal intragenic marker. *Genomics* **1**:329–336.

Ward, P. A., J. F. Hejtmancik, J. S. Witkowski, L. L. Baumbach, S. Gunnell, J. Speer, P. Hawley, U. Tantravahi, and C. T. Caskey. Prenatal diagnosis of Duchenne muscular dystrophy: prospective linkage analysis and retrospective dystrophin cDNA analysis. 1989. *Am. J. Hum. Genet.* **44**:270–281.

34

ISOLATION OF DNA FROM FUNGAL MYCELIA AND SINGLE SPORES

Steven B. Lee and John W. Taylor

Previous methods for rapid isolation of total DNA for the comparative study of many fungi have focused on obtaining a high yield of DNA for restriction enzyme analysis (Lee *et al.* 1988; Taylor and Natvig 1987; Biel and Parrish 1986; Zolan and Pukkila 1986). Fungal genetic studies have also required tedious methods to separate nuclear, mitochondrial, and ribosomal DNA fractions as well as other extranuclear fractions (Lambowitz 1979; Cramer *et al.* 1983; Hudspeth *et al.* 1980) to study specific DNA molecules by cloning or use in DNA hybridization as probes (Kwok *et al.* 1986; Anderson *et al.* 1987; Bruns *et al.* 1988; Lee and Taylor, manuscript in preparation).

The introduction and application of PCR to fungal genetic studies has changed these requirements for isolating DNA (Bruns *et al.* 1989). Our previous concern for a high yield is of little consequence since single-copy genomic sequences can be routinely amplified from 1 μg of total DNA (Saiki *et al.* 1988). If the target sequence is present in multiple copies, as with nuclear ribosomal DNA or mitochondrial DNA, even less starting material is required (0.1 to 10.0 ng of total DNA). Moreover, the separation of nuclear and organelle

PCR Protocols: A Guide to Methods and Applications

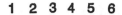

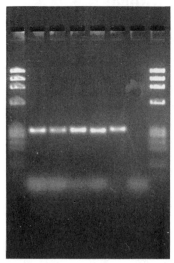

Figure 1 PCR-amplified 16s rDNA from ten thousandfold dilutions of miniprepped total DNA isolated from fresh *Thraustotheca, Dictyuchus, Apodachlya, Aqualinde-rella,* and *Pythium* mycelium using the DNA protocol given in this chapter, lanes 1–5. Distilled water control, lane 6. Primers NS5 and NS6 are listed in Chapter 38. For each sample, 5 μl from a total 100-μl reaction was run on a 2% NuSieve, 1% agarose gel at 75 V for 2 hours. The minigel was then stained in ethidium bromide (0.5 μg/ml), and DNA was visualized under UV.

genomes is unnecessary because target amplification is primer directed, and primers can be designed to amplify specific genomic or organellar fractions (Chapter 38).

We have simplified our previous method of total DNA isolation (Lee *et al.* 1988) and have used this method to amplify multicopy genes from small quantities of fresh and freeze-dried mycelia and herbarium samples (Bruns *et al.*, in press). Thousandfold to tenthousandfold dilutions of miniprep total DNAs have been used successfully in amplification reactions (see steps 10 and 11 in procedure) yielding double-stranded and single-stranded product (see Fig. 1). This method has worked for members of fungi in every group examined to date, including the *Chytridiomycetes, Oomycetes, Zygomycetes, Deuteromycetes, Ascomycetes* and *Basidiomycetes.*

DNA Isolation Protocol

Solutions

Lysis buffer   50 mM Tris–HCl (pH 7.2)
50 mM EDTA
3% SDS
1% 2-mercaptoethanol

Chloroform : TE-saturated phenol (1 : 1, v : v)
[TE: 10 mM Tris–HCl (pH 8.0), 1 mM EDTA]

3 M NaOAc (pH 8.0)

Isopropanol

Ethanol (70%)

TE (for resuspending pellet): 10 mM Tris-HCl, 0.1 mM EDTA

Procedure

1. Fill a 1.5-ml Eppendorf microcentrifuge tube one-third up the conical portion with ground lyophilized mycelium (20 to 60 mg dry) or fresh mycelium (0.1 to 0.3 g wet [see Note]) ground in liquid nitrogen. In either case the mycelium is ground by hand with a mortar and pestle.

2. Add 400 μl lysis buffer and stir with a dissecting needle and/or vortex so the mixture is homogeneous. If the mixture is too viscous, add more lysis buffer (up to 700 μl).

3. Incubate at 65°C for 1 hour.

4. Add 400 μl chloroform : phenol and vortex briefly, but be careful as the caps may loosen during vortexing. (If more than 400 μl of lysis buffer was used in step 2, add an equal amount of chloroform : phenol.)

5. Microcentrifuge at 10,000 x g for 15 minutes at room temperature or until aqueous (top) phase is clear.

6. Remove 300 to 350 μl of the aqueous phase containing the DNA to a new tube. Be careful not to take any cellular debris from the interface.

7. Add 10 μl of 3 M NaOAc to the aqueous phase followed by 0.54 volumes of isopropanol. Invert gently to mix. DNA clots that precipitate may or may not be visible depending on the amount of starting material.

8. Microcentrifuge as above (step 5) for 2 minutes at room temperature. Pour off the supernatant. Rinse the pellet once with 70% ethanol. Invert the tubes for 1 minute and drain them on a paper towel.

9. Place the tubes in a vacuum oven at 50°C for 15 minutes or until dry.

10. Resuspend the pellet in 100 to 500 μl TE (10 mM Tris–HCl, 0.1 mM EDTA) or distilled water. The amount of TE used depends on the amount of DNA in the pellet. The final concentration of total DNA should be in the range of 0.1 to 10 μg/μl.

11. For PCR amplification, 1 μl of the DNA sample should be diluted in distilled water or TE to a final concentration of 0.1 to 10 ng DNA in each PCR amplification. We have found a 1 : 1000 dilution works well for double-stranded amplification and a 1 : 10,000 dilution works for asymmetric amplification of single strands. Note: Do not despair if you are unable to get this much starting material. We have started with a colony of 2 cm in diameter scraped from an agar petri dish and have successfully amplified rDNA from the extracted DNA.

Notes

DNA isolated by this method from basidiocarps stored in herbaria for 30 years has been successfully amplified with PCR (Bruns *et al.*, in press). Amplification from such samples will allow comparisons of genetic variability present in samples collected many years ago with that of contemporary samples. Recent reports of PCR-amplified sequences from even older samples of human DNA (Pääbo *et al.* 1988; Chapter 20) and formalin-fixed samples (Impraim *et al.* 1987) indicate that extraction and amplification of DNA from ancient collections of fungi may be possible.

Gardes, Fortin, Bruns, Taylor, and White (unpublished) have used PCR to amplify fungal DNA from ectomycorrhizal roots by using fungal-specific primers. We anticipate that with taxon-specific primers, this method will be extended to other heterogenous DNA samples such as plant tissues containing obligate fungal pathogens and natural soil samples. Many fungi that cannot be cultured, e.g., endomycorrhizal fungi and some lichens, would be amenable to comprehensive molecular studies by isolating and amplifying DNA by using these techniques.

Amplification from Single Spores

Recent reports of amplification from a single human sperm (Li *et al.* 1988; see also Chapter 36) and a single human hair (Higuchi *et al.* 1988) have encouraged us to try to amplify sequences from single *Neurospora* ascospores. Using a modified version of our DNA isolation method, PCR amplification of rDNA sequences from a single spore of *Neurospora tetrasperma* was possible (Fig. 2). Amplification and analysis of specific DNA fragments from single spores

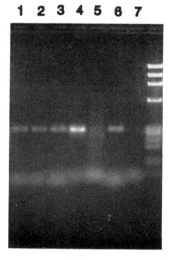

Figure 2 PCR-amplified rDNA from single ascospores of *Neurospora tetrasperma*. Single ascospores were crushed between two siliconized baked slides. Slides were siliconized in silicoat (CHCl₃, 5% dichlorodimethlysilane) and baked at 180°C for 2 hours. The spore and contents were then washed off the slides with sterile-distilled H₂O directly into the PCR mixture. PCR was performed with 70 cycles of amplification: 97°C denature / 51°C anneal / 73°C extension. For each sample, 5 µl from a total 100-µl reaction was run on a 2% NuSieve, 1% agarose gel at 75 V for 2 hours. The minigel was then stained in ethidium bromide (0.5 µg/ml) and DNA was visualized under UV. Lanes 1–3 each contained a single ascospore, lane 4 contained 10 ascospores, and lane 5 contained sterile H₂O.

would be especially useful for studies of recombination frequencies and genetic linkage with species whose spores cannot be germinated.

Literature Cited

Anderson, J. B., D. M. Petsche, and M. L. Smith. 1987. Restriction fragment polymorphisms in biological species of *Armillaria mellea. Mycologia* **79**:69–76.

Biel, S. W., and F. W. Parrish. 1986. Isolation of DNA from fungal mycelia and sclerotia without use of density gradient ultracentrifugation. *Anal. Biochem.* **154**:21–25.

Bruns, T. D., J. D. Palmer, D. S. Shumard, L. I. Grossman, and M. E. S. Hudspeth. 1988. Mitochondrial DNAs of *Suillus*: three fold size change in molecules that share a common gene order. *Curr. Genetics* **13**:49–56.

Bruns, T. D., R. Fogel, T. J. White, J. D. Palmer. 1989. Accelerated evolution of a false-truffle from a mushroom ancestor. *Nature (London)* **339**:140–142.

Bruns, T. D., R. Fogel, and J. W. Taylor. Amplification and sequencing of DNA from fungal herbarium specimens. *Mycologia*, in press.

Cramer, C. L., J. L. Ristow, T. J. Paulus, and R. H. Davis. 1983. Methods for mycelial breakage and isolation of mitochondria and vacuoles of Neurospora. *Anal. Biochem.* **128**:384–392.

Higuchi, R., C. H. von Beroldingen, G. F. Sensabaugh, and H. A. Erlich. 1988. DNA typing from single hairs. *Nature (London)* **332**:543–546.

Hudspeth, M. E. S., D. S. Shumard, C. J. R. Braford, and L. I. Grossman. 1980. Rapid purification of yeast mitochondrial DNA in high yield. *Biochim. Biophys. Acta* **610**:221–228.

Impraim, C. C., R. K. Saiki, H. A. Erlich, and R. L. Teplitz. 1987. Analysis of DNA extracted from formalin-fixed, paraffin embedded tissues by enzymatic amplification and hybridization with sequence-specific oligonucleotides. *Biochem. Biophys. Res. Commun.* **142**:710–716.

Kwok, S., T. J. White, and J. W. Taylor. 1986. Evolutionary relationships between fungi, red algae, and other simple eucaryotes inferred from total DNA hybridization to a cloned Basidiomycete ribosomal DNA. *Exper. Mycol.* **10**:196–204.

Lambowitz, A. M. 1979. Preparation and analysis of mitochondrial ribosomes. *Methods Enzymol.* **59**:421–433.

Lee, S. B., M. G. Milgroom, and J. W. Taylor. 1988. A rapid, high yield mini-prep method for isolation of total genomic DNA from fungi. *Fungal Gen. Newsl.* **35**:23–24.

Li, H., U. B. Gyllensten, X. Cui, R. K. Saiki, H. A. Erlich, and N. Arnheim. 1988. Amplification and analysis of DNA sequences in single human sperm and diploid cells. *Nature (London)* **335**:414–417.

Pääbo, S., and A. C. Wilson. 1988. PCR Reveals cloning artefacts: the case of the extinct Quagga. *Nature (London)* **334**:387–388.

Saiki, R. K., D. H. Gelfand, S. Stoffel, S. J. Scharf, R. Higuchi, G. T. Horn, K. B. Mullis, and H. A. Erlich. 1988. Primer-directed enzymatic amplification with a thermostable DNA Polymerase. *Science* **239**:487–491.

Taylor, J. W., and D. Natvig. 1987. Isolation of fungal DNA. In *Zoosporic fungi in teaching and research* (ed. M. S. Fuller and A. Jaworski), p. 252–258. Southeastern Publishing Corporation, Athens, Georgia.

Zolan, M. E., and P. J. Pukkila. 1986. Inheritance of DNA methylation in *Coprinus cinereus. Mol. Cell. Biol.* **6**:195–200.

35

GENETIC PREDICTION OF HEMOPHILIA A

Scott C. Kogan and Jane Gitschier

Hemophilia A is an X-linked bleeding disorder affecting one in 5000 males throughout the world. The disease is due to a defect in coagulation factor VIII, a protein serving as a cofactor in the coagulation cascade. Unlike diseases such as sickle cell anemia, thalassemias in certain populations, or phenylketonuria, hemophilia A stems from a wide variety of mutations. Generally, the mutation causing the defect is unknown. Thus, genetic prediction of hemophilia A, which includes the detection of female carriers and prenatal diagnosis, depends upon the analysis of DNA polymorphisms in and near the factor VIII gene.

At least eight two-allele polymorphisms have been discovered in the factor VIII gene. These are of two types, those that affect a restriction site and those that do not. Restriction site polymorphisms include *Bcl*I (Gitschier *et al.* 1985), *Xba*I (Wion *et al.* 1986), *Bgl*I (Antonarakis *et al.* 1985), *Msp*I (Youssoufian *et al.* 1987), and *Hin*dIII (Ahrens *et al.* 1987). Two sequence polymorphisms were found in exons 14 and 26 by comparing the factor VIII sequences of two different individuals (Wood *et al.* 1984). We have observed an additional sequence polymorphism in intron 7, just 5' to exon 8 (Kogan and Gitschier 1988). In the past, restriction site polymorphisms

PCR Protocols: A Guide to Methods and Applications

were analyzed with Southern blots of digested genomic DNA and sequence polymorphisms by hybridization of allele-specific oligonucleotides to genomic Southern blots. Analysis of both types of polymorphism is now made far easier by the application of PCR.

Unfortunately, we and others have observed that many of the factor VIII gene polymorphisms are in linkage disequilibrium with one another. For example, the exon 14 and exon 26 polymorphisms appear to be in complete linkage disequilibrium with the *Bcl*I polymorphism, on the basis of our study of 25 independent chromosomes. Moreover, the *Bgl*I, *Hind*III, and *Msp*I polymorphisms are also in linkage disequilibrium with the *Bcl*I site. Thus we have chosen to concentrate on the *Bcl*I, *Xba*I, and intron 7 polymorphisms, which together should provide useful information in close to 70% of hemophilia pedigrees.

For the prenatal diagnosis of hemophilia, as well as for other X-linked disorders, the first question to be answered is whether the fetus is male. We have developed a protocol for amplification of Y chromosome-specific sequences that allows rapid fetal sexing (Kogan *et al.* 1987).

What follows is a set of procedures for amplification and analysis of the factor VIII gene polymorphisms and for fetal sexing. These procedures routinely work in our hands for genetic prediction of hemophilia; however, only a limited attempt has been made to optimize them.

Protocols and Examples

DNA Amplification with *Taq* Polymerase

Table I describes the oligonucleotide primers used for the amplification of short regions of the factor VIII gene containing the *Bcl*I, *Xba*I, and intron 7 polymorphisms, as well as Y chromosome repetitive sequences. A set of general conditions for amplification follows. Also presented are some troubleshooting notes and procedures for amplification directly from blood, chorionic villi, and amniocytes.

Table 1

Oligonucleotide Primers and Probes for Analysis of Factor VIII Gene Polymorphisms

Polymorphism		Primers and ASO Probes[a]	Expected fragment sizes (bp)
BclI (intron 18)			
Primers	8.1	5'-TAAAAGCTTTAAATGGTCTAGGC	142 vs. 99 & 43
	8.2	5'-TTCGAATTCTGAAATTATCTTGTTC	
ASO Probes	8.3	5'-CAATCAGTGATCAAAGCAG	
	8.4	5'-CAATCAGTGAACAAAGCAG	
XbaI (intron 22)			
Primers	7.1	5'-CACGAGCTCTCCATCTGAACATG	96 vs. 68 & 28[b]
	7.10	5'-GGGCTGCAGGGGGGGGACAACAG	
ASO Probes	7.3	5'-GCGCATTTCTAGACTGTTG	
	7.4	5'-GCGCATTTCTGGACTGTTG	
Intron 7			
Primers	11.6	5'-TGCAGAACATGAGCCAATTC	314
	11.2	5'-TAATGTACCCAAGTTTTAGG	
ASO Probes	11.9	5'-GCAAGACACTCTGACATTG	
	11.10	5'-GCAAGACACTCTAACATTG	
Y repeat			
Primers	Y1.1	5'-TCCACTTTATTCCAGGCCTGTCC	154
	Y1.2	5'-TTGAATGGAATGGGAACGAATGG	

[a] ASO probes = allele-specific oligonucleotide probes.
[b] See note on page 297.

Reagents

Amplification mix per sample	10× "Cetus" buffer
2.5 μl of 10× "Cetus" buffer	100 mM Tris–HCl
	(pH 8.3) at 25°C
2.5 μl of 2 mM each dNTPs	500 mM KCl
10 pmol each primer	15 mM MgCl$_2$
Bring final volume to 25 μl with dH$_2$O.	0.1% gelatin

Note: Include 2.5 μl of 10 mM MgCl$_2$ in amplification mix if the *Bcl*I region is being amplified alone or in combination for a final MgCl$_2$ concentration of 2.5 mM.

1. Prepare a large-volume amplification mix on the basis of the number of samples to be amplified.
2. Place 25 to 250 ng DNA to be amplified (generally 1 μl) into 0.5-ml microfuge tubes. Include one tube without any DNA as a negative control.
3. Add 24 μl of amplification mix.
4. Dilute enzyme to 1 unit per 2 μl in 1× buffer.
5. Add 1 unit of enzyme to each sample.
6. Cover with 25 μl of mineral oil.

To amplify by hand, heat sample 5 minutes at 95°C before adding enzyme and mineral oil. Amplify 30 rounds as follows: 30 seconds at 92°C, 30 seconds at 40°C, and 30 seconds at 70°C.

To amplify using a Perkin-Elmer Cetus Thermal Cycler, prewarm the machine to 95°C. Denature DNA with Time-Delay File: 2 minutes at 94°C. Amplify 30 rounds with Step-Cycle File: 10 seconds at 94°C, 1 second at 55°C, and 10 seconds at 72°C. Do final extension with Time-Delay File: 3 minutes at 72°C.

Amplification Directly from Chorionic Villi

Spin down villi. Wash 2× with 250 ml of STE (Maniatis *et al.* 1982). Add 20 μl of 0.1 N NaOH, 2 M NaCl, 0.5% SDS. Vortex to disrupt and pipet up and down. Heat 2 minutes at 100°C. Spin 10 minutes. Take supernatant. Dilute 10× by adding 2 μl of supernatant to 18 μl of dH$_2$O. Use 1 μl to amplify in a 25-μl amplification reaction.

Amplification Directly from Amniocytes

Take 1 ml of amniotic fluid. Spin 2 minutes. Discard supernatant. Spin again and discard supernatant. Take up pellet in 10 μl of 0.1 N

NaOH, 2 M NaCl. Heat 2 minutes at 100°C. Spin 10 minutes. Take supernatant. Dilute 10× by adding 2 μl of supernatant to 18 μl of dH$_2$O. Use 1 μl to amplify in a 25-μl amplification reaction.

Amplification Directly from Blood

Collect blood into EDTA. Freeze. Thaw. Take 200 μl of freeze-thawed blood. Heat 10 minutes at 95°C. Spin 10 minutes. Take supernatant (don't be greedy). Dilute 10× by adding 2 μl of supernatant to 18 μl of dH$_2$O. Use 1 μl to amplify in a 25-μl amplification reaction. Some samples amplified directly from blood may be refractory to restriction enzyme digestion.

> Notes: Occasionally, genomic DNA samples are difficult to amplify. We have found that diluting these 10- or 20-fold solves the problem. Another approach has been suggested by de Franchis et al. (1988). For the Bcl I region, an alternative set of amplification primers can be used to take advantage of a nearby Bcl I site as an internal digestion control (Sarkar et al. in press).

Gel Analysis

Run 5 μl of amplified DNA samples on 6% polyacrylamide gel for inspection. To analyze the Bcl I polymorphism, digest 2.5 μl in 30 μl with Bcl I 2 hours at 50°C and run on a 6 or 12% polyacrylamide gel (see Note at end of Discussion). For the Xba I polymorphism, digest 5 μl in 30 μl with Xba I 2 hours at 37°C and run on a 12% polyacrylamide gel. These minipolyacrylamide gels measure 10 × 10 cm with 1-mm spacers. Photographs of UV fluorescent ethidium bromide-stained digest fragments may require >1-second exposure.

Example: Carrier Detection

The Bcl I and Xba I polymorphisms can be analyzed by digesting the amplified DNA with the appropriate enzyme. Figure 1A demonstrates carrier detection of hemophilia A using the Bcl I polymorphism. The mother of the hemophilic son is known by familial history to be a carrier of hemophilia. Thus the question is whether the two daughters have inherited the hemophilia mutation. In this fam-

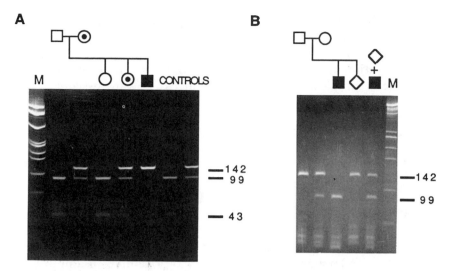

Figure 1 Genetic prediction of hemophilia A by DNA amplification and digestion of the *Bcl*I polymorphism. In (A), a carrier detection is performed by amplification of the *Bcl*I region by using primers 8.1 and 8.2. The amplified DNAs are digested with *Bcl*I and run on a polyacrylamide gel that has been stained with ethidium bromide. Control samples include digested amplified DNAs from a male known to have the *Bcl*I site and from a known heterozygote. Marker lane (M) (here and in [B]) contains *Hae*III-digested ΦX174 DNA. Similarly, in (B), a prenatal diagnosis is made by *Bcl*I digestion of amplified DNA. For the fetal sample, indicated by a diamond, the amplification was performed directly on chorionic villi, without DNA extraction, as described in protocols and examples. The lane indicated by the open diamond plus the solid square contains the digestion product of the mix of fetal and hemophiliac amplified DNAs, to control for the possibility that the fetal sample inhibits *Bcl*I.

ily, the hemophilia mutation is associated with absence of the *Bcl*I site, as evidenced by the 142-bp amplified DNA fragment. The mother is heterozygous for the *Bcl*I polymorphism. Thus we can predict that one daughter has inherited the mutation, whereas the other has not.

Example: Prenatal Diagnosis

A prenatal diagnosis of hemophilia by using the *Bcl*I polymorphism is demonstrated in Fig. 1B. In contrast to the case shown in Fig. 1A, the hemophilia mutation in the patient is associated with the presence of the *Bcl*I site. The mother was found to be heterozygous for the *Bcl*I alleles, and the fetus to be male by DNA amplification of Y-specific sequences (see following example). Since the fetal sample is lacking the *Bcl*I site, the fetus is predicted to be unaffected by hemophilia. Interestingly, the hemophilia in this family is a "sporadic" occurrence, that is, there was no prior genetic history of hemophilia. The mutation may have arisen de novo in the patient. Consequently, if the fetal sample had shown the presence of the site, no conclusion could be drawn as to whether or not the fetus is affected. In this circumstance, fetal blood sampling would have been offered to directly measure the factor VIII level. Note that in this diagnosis the fetal sample was amplified directly from chorionic villi, as previously described, without prior DNA extraction.

Fetal Sexing

The sex of a fetus can be determined by amplification of Y-specific sequences. As a control for the ability of a sample to amplify, Y-specific sequences are co-amplified with the *Xba*I region. In the female DNA sample, only the *Xba*I sequence is amplified, whereas in the male DNA sample, amplification of the Y-repeat sequence is so abundant as to exclude amplification of the *Xba*I sequence. As a general practice, it is a good idea to also amplify the DNAs of the mother and father when fetal sexing, to control for the unlikely event that the mother has some Y-specific sequences or that the father is deleted for them. (See Fig. 2.)

Analysis of Polymorphisms by Hybridization of Allele-Specific Oligonucleotides

Another approach for analyzing polymorphisms is discriminant hybridization of the amplified DNA to allele-specific oligonucleotides. This approach is, of course, the method of choice for analyzing sequence polymorphisms that do not affect restriction sites. Additionally, this method allows all polymorphic regions to be amplified and analyzed simultaneously by using allele-specific oligonucleotide probes of identical length and the salt tetramethylammonium chloride (TMACl). Washing in 3 *M* TMACl allows all perfectly matched oligonucleotides of a given length to dissociate from the DNA sample at the identical temperature (Wood *et al*. 1985). Thus, for example, all 19-mers that hybridize to the amplified DNA with-

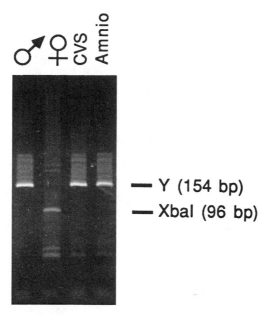

Figure 2 Fetal sexing by amplification of Y-specific sequences. Adult male and female DNAs and fetal DNAs from two separate pregnancies were amplified with the Y.1, Y.2, 7.1, and 7.10 primers. The amplification products were run on a gel and stained with ethidium bromide. Amplification was performed on chorionic villi and amniocytes directly, as described in the protocols and examples.

out any mismatches will remain bound at 60°C, but any 19-mers that are mismatched by only a single base will dissociate at this temperature. Examples of discriminant hybridization of allele-specific oligonucleotides can be seen in Kogan *et al.* (1987) and in other chapters of this book.

Slot or Dot Blotting

Use a vacuum blotting apparatus. Pre-wet the nylon membrane first with water and then with 20× SSPE (Maniatis *et al.* 1982) and assemble the blotting apparatus. Take 5 to 20 μl of amplified sample and add 0.4 M NaOH, 25 mM EDTA, to 200 μl. Heat the amplified DNA 2 minutes at 95°C to be sure the DNA is denatured. Put on ice. Load sample into wells. Load all samples before applying vacuum. Shut off vacuum. Rinse wells with 400 μl 20× SSPE. Apply vacuum. Let run 3 minutes. Take off filter and bake 1 hour at 80°C. Rinse in prehybridization solution before prehybridizing.

Hybridization

Prehybridize 1 hour at 37°C in 5× SSPE, 5× Denhardt's, 0.5% SDS. Add probe and hybridize 1 hour to overnight at 37 to 42°C. We routinely use ^{32}P-end-labeled oligonucleotides, but ^{35}S-labeling or biotin labeling (as described in Chapters 12, 13, 14, and 15) should work equally well.

Tetramethylammonium Chloride Wash (TMACl)

Rinse posthybridization blot with 6× SSC three times. Wash blot in 6× SSC for 30 minutes at room temperature. Rinse blot with room temperature TMACl solution to exchange salts. Wash twice 20 minutes in TMACl solution at an appropriate temperature (60°C for 19-mer probes). Blot the filter to Whatman filter paper to remove moisture, wrap in plastic wrap, and expose with screens at −80°C.

TMACl solution: 3.0 M TMACl, 50 mM Tris–HCl (pH 8.0), 2 mM EDTA (pH 8.0), 0.1% SDS.

TMACl is available through Aldrich (T1, 952−6) in 500-g quantity. Add 400 ml H_2O to get 5 M stock, roughly. Heat at 68°C to dissolve. Use a refractometer to determine exact molarity by formula:

$$\text{molarity} = \frac{\text{(reading of solution—reading of dH}_2\text{O)}}{0.018}$$

Discussion

Many different diagnostic situations might be encountered for hemophilia, the most important and urgent being prenatal diagnosis for a pregnant carrier. In this case, the issue of informativeness is equivalent to whether or not the mother is heterozygous for any given polymorphism. In this discussion, we will consider strategies for making a rapid prenatal diagnosis.

How Informative Are the Polymorphisms, Alone or in Combination?

We have found the $BclI$ polymorphism to be informative in 41% of cases. When $BclI$ is uninformative and +/+ (51% of total cases),

XbaI is informative in 41% of cases (or 21% of total). The intron 7 polymorphism is never informative when BclI is +/+. In cases where BclI is uninformative and −/− (8% of total), XbaI is virtually never informative, but for 50% of these cases (or 4% of total) the intron 7 polymorphism is informative. The intragenic polymorphisms together are informative in 66% of cases. The tightly linked marker St14 is informative in at least 90% of cases but has a 4% risk of recombination.

One of the problems with using the XbaI polymorphism is that this DNA sequence is present not only in the factor VIII gene but also at another X-chromosomal location where the XbaI site is routinely missing. The two sequences are so similar that both amplify with primers 7.1 and 7.10, as well as with a variety of primers (alone and in a nested configuration) under stringent conditions. Therefore, while amplification and digestion or dot blot allows the presence or absence of the XbaI site within the factor VIII gene to be unequivocally determined in males, if one of a female's X chromosomes contains the XbaI site, this approach cannot directly detect the presence or absence of the XbaI site on her other chromosome (Kogan *et al.* 1987).

What Is a Good Diagnostic Strategy?

When samples arrive, amplify with the BclI, XbaI, and intron 7 primers in separate tubes. Digest the appropriate samples with BclI and XbaI (see Note at the end of Discussion) and analyze them on a polyacrylamide gel. Table 2 lists a set of results that might be obtained for a carrier mother as well as explanations and suggestions for the next step.

A simple alternate strategy to the above would be to amplify the BclI, XbaI, and intron 7 polymorphic regions together in a single tube and to analyze them individually by allele-specific hybridization.

We are looking for more polymorphisms in and near the factor VIII gene and are continuing to improve the amplification protocols and strategies.

> NOTE: For many samples, analysis of the XbaI site is not possible by digestion because of the presence of background bands that interfere with interpretation. These bands are due to two other non-factor VIII regions that are homologous to the XbaI region within the factor VIII gene. It has become apparent that these other regions are also

Table 2

Example of Diagnostic Strategy

Results		Explanations and Suggestions
BclI	XbaI	
+/−	Anything	Diagnosis can be made by using the BclI digests.
+/+	+/?	Depending on family members available, the XbaI site on the second chromosome may often be inferred. If so, and the mother is +/+, proceed to the St14 Southern blot. If the mother is +/−, the diagnosis can be made with XbaI digest. If the XbaI genotype cannot be inferred, proceed to XbaI and St14 Southern blots.
+/+	−/−	Proceed to the St14 Southern blot.
−/−	+/?	This will happen very rarely.
−/−	−/−	Proceed to the intron 7 slot blot, and, if that is uninformative, the St14 Southern blot.

polymorphic for *Xba*I and are located near the factor VIII gene. We are attempting to develop PCR conditions for analyzing the three polymorphisms separately; at the moment we recommend analysis of the *Xba*I polymorphisms by Southern blot.

Acknowledgments

We are extremely grateful to Marie Doherty for oligonucleotide synthesis, Shiping Cai for contributing Fig. 1A, and Barbara Levinson for many helpful suggestions.

Literature Cited

Ahrens, P., T. A. Kruse, M. Schwartz, P. B. Rasmussen, and N. Din. 1987. A new *Hind*III restriction fragment length polymorphism in the hemophilia A locus. *Hum. Genetics* **76**:127–128.

Antonarakis, S. E., P. G. Waber, S. D. Kittur, A. S. Patel, H. H. Kazazian, Jr., M. A. Mellis, R. B. Counts, G. Stamatoyannopoulos, E. J. W. Bowie, D. N. Fass, D. D. Pittman, J. M. Wozney, and J. J. Toole. Hemophilia A: detection of molecular defects and of carriers by DNA analysis. 1985. *N. Engl. J. Med.* **313**:842–848.

de Franchis, R., N. C. P. Cross, N. S. Foulkes, and T. M. Cox. 1988. A potent inhibitor of *Taq* polymerase copurifies with human genomic DNA. *Nucleic Acids Res.* **16**:10355.

Gitschier, J., D. Drayna, E. G. D. Tuddenham, R. L. White, and R. M. Lawn. 1985. Genetic mapping and diagnosis of hemophilia A achieved through a *Bcl*I polymorphism in the factor VIII gene. *Nature* **314**: 738–740.

Kogan, S. C., M. Doherty, and J. Gitschier. 1987. An improved method for prenatal diagnosis of genetic diseases by analysis of amplified DNA sequences. Application to hemophilia A. *N. Engl. J. Med.* **317**: 985–990.

Kogan, S. C., and J. Gitschier. 1988. Detection of hemophilia A mutations near the acidic region of factor VIII by DNA amplification and denaturing gradient gel electrophoresis. *Blood* **72**: 300a.

Maniatis, T., E. F. Fritsch, and J. Sambrook. 1982. In *Molecular cloning: a laboratory manual*. Cold Spring Harbor Laboratory, Cold Spring Harbor, New York.

Sarkar, G., M. I. Evans, S. Kogan, J. Lusher, and S. S. Sommer. *Obstet. Gynecol.*, in press.

Wion, K. L., E. G. D. Tuddenham, and R. M. Lawn. 1986. A new polymorphism in the factor VIII gene for prenatal diagnosis of hemophilia A. *Nucleic Acids Res.* **14**: 4535–4542.

Wood, W. I., D. J. Capon, C. C. Simonsen, D. L. Eaton, J. Gitschier, B. Keyt, P. H. Seeburg, D. H. Smith, P. Hollingshead, K. L. Wion, E. Delwart, E. G. D. Tuddenham, G. A. Vehar, and R. M. Lawn. 1984. Expression of active human factor VIII from recombinant DNA clones. *Nature (London)* **312**: 330–337.

Wood, W. I., J. Gitschier, L. A. Lasky, and R. M. Lawn. 1985. Base composition-independent hybridization in tetramethylammonium chloride: a method for oligonucleotide screening of highly complex gene libraries. *Proc. Natl. Acad. Sci. USA* **82**: 1585–1588.

Youssoufian, H., D. G. Phillips, H. H. Kazazian, Jr., and S. E. Antonarakis. 1987. MspI polymorphism in the 3′ flanking region of the human factor VIII gene. *Nucleic Acids Res.* **15**: 6312.

36

HAPLOTYPE ANALYSIS FROM SINGLE SPERM OR DIPLOID CELLS

Ulf Gyllensten

A haplotype is a combination of alleles at two or more loci on the same chromosome. Because of selective forces or short physical distances between loci, combinations of alleles at different loci on the same chromosome may be inherited as a unit. Previously, analysis of the combinations of alleles on individual chromosomes has only been feasible by reconstruction of parental chromosomes from the segregation of alleles in pedigrees. An example of this method for haplotype determination is shown in Fig. 1. To be efficient this approach requires access to large, and preferably complete, pedigrees. In higher vertebrates, where complete families may not be available, generation times are long, and family sizes are limited, haplotype determinations may not be possible using this approach. These limitations could be bypassed if the combination of alleles along a chromosome could be examined directly in individual gametes (sperm or egg) or diploid somatic cells (white blood cells, epithelial cells).

The development of PCR (Mullis *et al.* 1986; Mullis and Faloona 1987; Saiki *et al.* 1985) has for the first time made possible the genetic analysis of individual chromosomes without resorting to pedigree analysis (Saiki *et al.* 1988; Li *et al.* 1988). The combination of alleles along a chromosome can now be determined from a small

PCR Protocols: A Guide to Methods and Applications

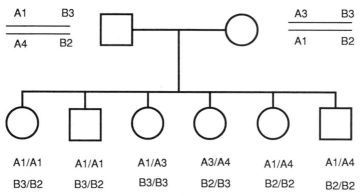

Figure 1   Schematic representation of the traditional method for reconstruction of parental haplotypes by pedigree analysis. The distribution of combinations of alleles at the A locus (with the alleles A1–A4) and the B locus (with the alleles B1–B3) in the six offspring makes it possible to deduce the haplotypes of the parents.

number of sperm or single diploid cells by using DNA-based typing of regions amplified *in vitro* by the PCR method. This method can also be used to study the recombination frequency between loci that are too close for pedigree analysis to result in statistically reliable recombination rates.

Method

The principle for haplotype analysis based on single chromosomes is shown in Fig. 2. The single chromosomes used in the analysis can be derived from single sperm or egg, single diploid cells, or single flow-sorted chromosomes. Single sperm and single diploid cells can be obtained by micromanipulation or by low-speed flow-sorting. Single diploid cells can be used as a source for single chromosomes if a serial dilution is applied subsequent to cell lysis and prior to PCR amplification to separate the two homologous chromosomes. For instance, the theoretical probability that the two homologous chromosomes in a cell are found in the same PCR after five serial twofold dilutions is only about 3%. However, if the loci studied are not physically very close, a high frequency of erroneous combinations of alleles due to chromosome breakage may be found. This artifact can only be overcome by increasing the number of cells analyzed. Alter-

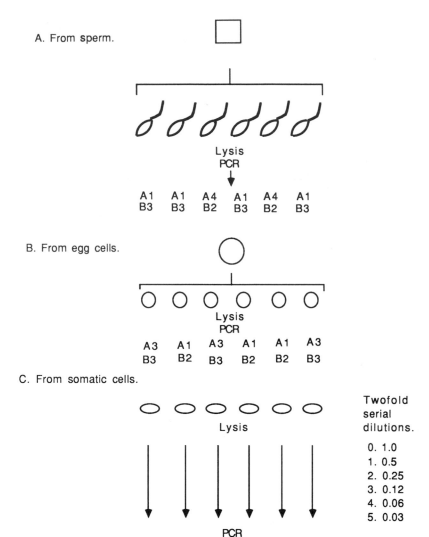

Figure 2 The principle for haplotype analysis by PCR. The individual germ cells are lysed directly and analyzed by PCR, while the somatic cells are lysed and subjected to serial dilutions prior to PCR to separate to two homologous chromosomes. The haplotypes can be identified directly from a small number of individual PCRs.

natively, if possible, a high-speed flow-sorter may be used to prepare single chromosomes.

Before analysis the cells or sperm are lysed in a mixture of Proteinase K, SDS, and (in the case of sperm) DTT. The Proteinase K is then heat-inactivated, and the PCR reagents are added. For the haplotype analysis two or more loci have to be co-amplified from the same molecule. Co-amplification of multiple single-copy targets (loci) is usually possible when starting from 10 to 100 ng of genomic DNA (3000 to 30,000 target molecules). By contrast, co-amplification of several loci from a *single* target DNA molecule usually results in very dissimilar yields of PCR product. Amplification from a single target molecule appears to enhance differences in the specificity, or efficiency, of the different primer pairs and results in the preferential amplification of only one locus. To counteract this and allow for typing of several loci, the first PCR is for 20 cycles, and aliquots of this reaction are then diluted 1 : 50 to 1 : 100 into new reactions, each containing PCR primers for a single locus. This second PCR is then continued for 20 to 30 cycles. The splitting of a PCR from a single target will allow for the enrichment of a sufficient amount of product from each locus to permit detection.

Haplotype analysis of single DNA molecules by the PCR method can be complicated by several artifacts: (1) the presence of more than one original target molecule (representing another haplotype) in the reaction; (2) the presence of artifact DNA molecules composed of one locus from each of two individual chromosomes; and (3) the presence of PCR product (or target) from previous experiments in the reagents. Since there is initially only one starting target molecule, any contamination will be at least as abundant as the true target. *Therefore, the purity of the reagents is crucial!*

To minimize these artifacts (1) designate one specific area in the laboratory (e.g., sterile hood) for preparing the PCR and another for analyzing the PCR products, (2) use, if possible, specific sets of pipettors with disposable positive-displacement tips for pipetting water, buffers, nucleotides, enzyme, and oligos, (3) autoclave the buffers and prepare small aliquots for one-time use or, alternatively, treat the appropriate reagents with DNase that is later inactivated by heat, (4) test all reaction components for purity by performing PCR without target DNA, (5) never open a tube with PCR product in an area where reactions are prepared, and (6) include several negative controls for each solution. For additional precautions, see Chapters 17 and 54.

After the PCR, the presence or absence of product of a specific type can be detected by hybridization isotopically or nonisotopically

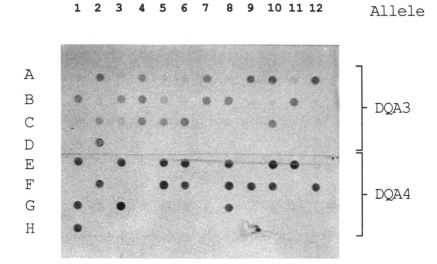

Figure 3 Genetic analysis of PCR-amplified DNA from single sperm, using a dot blot format. Individual sperm were lysed as described, and the second exon of the HLA-DQA locus was amplified for 45 cycles [using the primers GH26 and GH27 (Gyllensten and Erlich 1988)]. Five percent of the 100-μl PCR was applied to a Genatrans membrane and probed with HRP-labeled oligonucleotides made specific to the different alleles [A3 and A4 (Gyllensten and Erlich 1988)]. The products of 38 PCRs are shown repeated in the panels A–D and E–H. The reactions A(E)1–C(G)6 contain sperm, C(G)7 and C(G)6 are negative controls for the lysis solution, and C(G)11 and 12 are negative controls for PCR mixture. The PCR A(E)10 and B(F)8 are examples of reactions that contained more than one haplotype, presumably because of the presence of double sperm. D(H)1 is a positive control for the DQA4 allele, and D(H)2 is a positive control for the DQA3 allele.

labeled oligonucleotides. An example of nonisotopic detection of PCR-amplified alleles at the HLA-DQA locus from individual sperm is shown in Fig. 3. If individual target molecules have been prepared and there is no product contamination from the previous PCR, it is only necessary to examine a small number of reactions to establish the haplotypes.

Protocols

Preparation of Sperm for Flow-Sorting or Micromanipulation

1. Let the sperm sit for 30 minutes to liquefy.
2. Wash the sperm in 5× PBS three times.

3. Layer the sperm on a cushion of 25% Percoll and spin them at 5000 rpm for 5 minutes. Resuspend the sperm in 1× PBS and visually inspect the purity. If necessary the centrifugation can be repeated several times.
4. Resuspend the sperm to the appropriate concentration for the subsequent analysis in 1× PBS or a solution of choice.

Lysis of Sperm and Diploid Cells

1. Lyse the sperm in a final volume of 20 μl of 1× PCR buffer (see following) containing 0.05 mg/ml of Proteinase K, 20 mM DTT, and 1.7 μM SDS.
2. After 1 hour at 37°C heat-inactivate the Proteinase K at 85°C and cool; add 100 μl of PCR mix (see following).

First PCR

The PCR conditions have to be optimized for each set of PCR primers as described in Chapter 1. A typical first PCR (100 μl) with two primer pairs contains:

10 μl of 10× buffer [500 mM KCl, 100 mM Tris–HCl (pH 8.3), 15 to 25 mM MgCl$_2$]
0.8 μl of dNTP mix (25 mM each dNTP)
5.0 μl of each PCR primer at 10 pmol/μl
2 units of Taq polymerase
distilled water to 100 μl

Second PCR

1 μl of first PCR (1 : 100 dilution)
5 μl each of PCR primers for an individual locus at 10 pmol/μl
0.8 μl of dNTP mix (25 mM each dNTP)
2 units of Taq polymerase,
distilled water to 100 μl

Detection of PCR Product

The protocol for the detection of specific PCR-amplified products by labeled oligonucleotides has been described elsewhere in this book.

Haplotype determination from individual germ cells may also be useful in forensic analyses. The frequency of a specific chromosomal combination of alleles at several polymorphic loci is much more informative than the allelic types alone. For example, by analysis of individual sperm from the crime scene (e.g., from the victim of a rape), as well as from the suspects, a much higher power of exclusion can be obtained. The procedure for analyzing the allelic combinations of individual germ cells can also be used to study the recombination frequency between loci. This application is of particular interest for loci that are too close for pedigree analysis to yield statistically reliable recombination values. It may also be used to determine recombinational hot- and coldspots between loci separated by a known physical distance. Finally, the construction of genetic maps for species that are endangered or hard to breed in captivity, or have limited population and family sizes in nature could greatly benefit from the analysis of individual germ cells.

Literature Cited

Gyllensten, U. B., and H. A. Erlich. 1988. Generation of single-stranded DNA by the polymerase chain reaction and its application to direct sequencing of the *HLA-DQA* locus. *Proc. Natl. Acad. Sci. USA* **85**:7652–7656.

Li, H., U. B. Gyllensten, X. Cui, X. F., R. K. Saiki, H. A. Erlich, and N. Arnheim. 1988. Amplification and analysis of DNA sequences in single human sperm and diploid cells. *Nature (London)* **335**:414–417.

Mullis, K. B., F. A Faloona, S. J. Scharf, R. K. Saiki, G. T. Horn, and H. A. Erlich. 1986. Specific enzymatic amplification of DNA in vitro: the polymerase chain reaction. *Cold Spring Harbor Symp. Quant. Biol.* **51**:263–273.

Mullis, K. B., and F. A. Faloona. 1987. Specific synthesis of DNa *in vitro* via a polymerase-catalyzed chain reaction. *Methods Enzymol.* **155**:335–350.

Saiki, R. K., S. Scharf, F. Faloona, K. B. Mullis, G. T. Horn, H. A. Erlich, and N. A. Arnheim. 1985. Enzymatic amplification of β-globin genomic sequences and restriction site analysis for diagnosis of sickle cell anemia. *Science* **230**:1350–1354.

Saiki, R. K., D. G. Gelfand, S Stoffel, S. J. Scharf, R. Higuchi, G. T. Horn, K. B. Mullis, and H. A. Erlich. 1988. Primer-directed enzymatic amplification of DNA with a thermostable DNA polymerase. *Science* **239**:487–491.

37

AMPLIFICATION OF RIBOSOMAL RNA GENES FOR MOLECULAR EVOLUTION STUDIES

Mitchell L. Sogin

Sequence comparisons of small subunit ribosomal RNAs (16S-like rRNAs) or their genes have revolutionized our perspectives on molecular and cellular evolution (Sogin *et al.* 1989; Gunderson *et al.* 1987a; Woese 1987). These macromolecular sequences contain sufficient evolutionary information to allow the measurement of both close and distant phylogenetic relationships (Sogin and Gunderson 1987). Strategies that take advantage of conserved sequence elements distributed along the length of rRNA genes (Elwood *et al.* 1985) permit the rapid analysis of both coding and noncoding DNA strands. Oligonucleotides (15 to 20 nucleotides in length) that are complementary to highly conserved elements can be used to initiate DNA synthesis in dideoxynucleotide sequencing protocols (Sanger and Coulson 1975), thus eliminating the requirement for constructing a nested set of overlapping fragments in the M13 cloning and sequencing system. The rate-limiting step is that required to construct genomic libraries and identify recombinant clones containing rRNA coding regions.

The cloning step can be circumvented by using PCR methods (Mullis and Faloona 1987; Saiki *et al.* 1988) to rapidly amplify 16S-like rRNA coding regions for sequence analysis (Medlin *et al.* 1988).

PCR Protocols: A Guide to Methods and Applications
Copyright © 1990 by Academic Press, Inc. All rights of reproduction in any form reserved.

Comparison of 60 eukaryotic 16S-like rRNA sequences reveals that there are conserved sequence elements proximal to the 5' and 3' termini that can be used as initiation sites for DNA synthesis in PCR experiments. Sequences between the conserved elements can be exponentially amplified within a few hours, and several micrograms of DNA encoding 16S-like rRNAs can be obtained from as little as 0.1 ng of bulk genomic DNA. The resulting product can be cloned into the single-stranded phage M13 or characterized by modifications of the dideoxynucleotide sequencing protocols for analyzing double-stranded DNA templates.

Protocols

Preparation of Genomic DNA for PCRs

Optimal results in PCR amplification of nuclear-encoded 16S-like rRNA genes are achieved if phenol-extracted DNA is purified on CsCl gradients as previously described (Maniatis *et al.* 1982). This optional cesium step removes RNA and in some cases carbohydrate fractions. The subfractionation of organellar DNA is not required because eukaryotic-specific primers (primers that recognize sequences in eukaryotic but not prokaryotic 16S-like rRNAs) are used in the PCRs. When preparing DNA for PCR amplification, it is not necessary to avoid mechanical shearing if the average DNA chain length can be maintained at 20 kb or larger. For some organisms (e.g., diatoms) it is very important to extract the bulk nucleic acids at 0°C.

Primers Used for PCR Amplification of 16S-like rRNA Genes

Amplification primers for the 16S-like rRNA genes can be designed with polylinkers at their 5' termini to facilitate the cloning of the PCR amplification products into the single-stranded phage M13. The synthetic oligos for amplifying eukaryotic 16S-like rRNA coding regions are presented in Fig. 1. The sequence of primer (A) is complementary to 21 nucleotides in the coding strand at the 5' terminus of eukaryotic 16S-like rRNAs and contains restriction sites for *Eco*RI and

PRIMER A

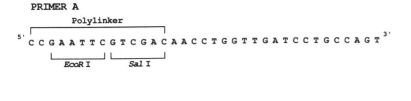

PRIMER B

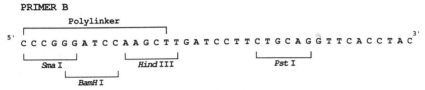

Figure 1 Synthetic oligos used for *in vitro* amplification of eukaryotic 16S-like rRNA coding regions. Primer (A) contains a polylinker plus a conserved 21-nucleotide sequence that is complementary to the coding strand at the 5' terminus of eukaryotic 16S-like rRNAs (nucleotides 1–21 in the *Saccharomyces cerevisiae* sequence). Primer (B) contains a polylinker plus a conserved sequence that is complementary to 24 nucleotides in the noncoding strand at the 3' terminus of eukaryotic 16S-like rRNAs (the complement of nucleotides 1772–1795 in the *S. cerevisiae* sequence).

*Sal*I. Primer (B) is complementary to 24 nucleotides in the noncoding strand at the 3' terminus of eukaryotic 16S-like rRNAs and includes restriction sites for *Sma*I, *Bam*HI, *Hind*III, and *Pst*I.

A modification of the PCR methods (Mullis and Faloona 1987; Saiki *et al.* 1988) is used to amplify the 16S-like rRNA coding regions. PCR mixtures (100 μl each) are prepared in 0.6-ml Eppendorf tubes containing 1× reaction buffer [10 m*M* Tris (pH 8.4), 50 m*M* KCl, 2 m*M* MgCl$_2$, and 0.1 mg/ml of gelatin] plus 0.1 ng to 1 μg of genomic DNA, 1 μ*M* amplification primer A, 1 μ*M* amplification primer B, and 200 μ*M* each dATP, dGTP, dCTP, and dTTP. The mixture is heated to 94°C for 5 minutes. After a 30-second microfuge spin, 0.1 to 0.5 units of *Taq* DNA Polymerase are added, and the reaction mixes are overlayed with 100 μl mineral oil.

The samples are cycled among three incubations: denaturation at 94°C for 2 minutes, cooling (or "ramping") over a 2-minute period to 37°C where it is maintained at annealing temperature for an additional 2 minutes, and ramping over a 3-minute period to 72°C where primer extension is allowed to continue for an additional 6 minutes. The sequential incubations can be repeated between 30 and 90 cycles by using a Perkin-Elmer Cetus DNA Thermal Cycler. After the final incubation at 72°C, the reaction mixture is incubated for an additional 9 minutes at 72°C and then held at 4°C until the samples can be processed. These cycle times and temperatures are optimized for the amplification of 16S-like rRNA genes by using the primers described above. "Ramping" between 37°C and 72°C significantly improves yields and reproducibility. During initial rounds of synthesis several mismatches can be tolerated in the primer/DNA annealing step. As the incubation slowly approaches 72°C, the primer is extended by the *Taq* polymerase to a length that allows it to form a more stable duplex. [Note: The amount of polymerase should probably be increased, and the number of cycles (90) are of concern. See Chapters 1 and 38 for alternate cycling parameters for large fragments.]

The PCR products are extracted with phenol and concentrated by precipitation with ethanol (Maniatis *et al.* 1982). Subsequent to purification on Quiagen columns, the products can be sequenced directly or subcloned into the replicative form of the single-stranded phage M13. [Note: also see Chapters 10 and 24.]

Results and Discussion

Figure 2 shows the electrophoretic separation of PCR amplification products by using genomic DNAs from the diatom *Skeletonema costatum* (lane 2) or the ascomycete *Kluveromyces lactis* (lane 3). The prominent reaction products are represented by single bands with chain lengths of 1.8 kb. PCR products produced from recombinant plasmids containing cloned rDNA genes are shown in lane 4 for *Plasmodium falciparum* or in lane 5 for *Tetrahymena tropicalis*. Yields from the *in vitro* amplification reactions vary with cloned rRNA genes from different organisms. For example, despite nearly perfect complementarity between the *P. falciparum* coding region and the amplification primers, the PCR yields are substantially less than those with any other tested clone. In most cases, with as little as 1 ng of genomic DNA, the yield for 40 PCR cycles is on the order of 5 μg of total nucleotides. Increasing the number of cycles beyond 40 does not significantly improve the yield and can affect the quality of the product. The *S. costatum* PCR products appear more heterogeneous (as judged by higher backgrounds) than the *K. lactis* PCR products. The *S. costatum* reaction cycle was repeated 90 times, which compares to only 40 cycles for *K. lactis* and *P. falciparum*. A

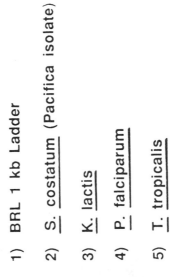

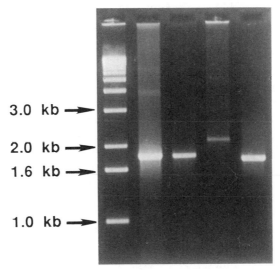

Figure 2 Agarose gel electrophoresis of amplified small subunit rRNA coding regions. PCRs directed by genomic DNA or cloned rDNA templates were primed with oligos complementary to conserved sequences in eukaryotic 16S-like rRNAs. Aliquots (1–5 μl) from each of the reactions were mixed with 10 μl LT buffer [10 mM Tris (pH 7.2), 10 mM NaCl, and 0.5 mM EDTA (pH 7.2)] plus 5 μl stop solution [25% Ficoll, 0.05% bromphenol blue, 40 mM Tris (pH 8.3), and 20 mM EDTA (pH 8.0)]. The samples were electrophoresed for 2 hours at 7 V/cm in 2% agarose gels built in E buffer [40 mM Tris/CH$_3$COOH (pH 7.0), 1 mM EDTA]. The bands are located by staining with 2% ethidium bromide. Lane 1 is a 1-kb ladder (Bethesda Research Laboratories). Products from 90 amplification cycles using 10 ng of genomic DNA from *S. costatum* are shown in lane 2, and products from 40 amplification cycles using 10 ng of genomic DNA from *K. lactis* are shown in lane 3. Lanes 4 and 5 show the products of 40 amplification cycles using 10 ng of cloned rDNA genes from *P. falciparum* and *T. tropicalis*, respectively.

lower number of cycles significantly reduces the background level in the *S. costatum* amplification reactions (data not shown).

The speed of the PCR amplification for rDNA structure analyses compares favorably with the primer extension methods employed in direct analyses of rRNAs (Qu *et al.* 1983; Lane *et al.* 1985). Microgram quantities of 16S-like rRNA coding regions can be produced from as little as 0.1 ng of genomic DNA even in the presence of high concentrations of mitochondrial or chloroplast DNA. Complementarity of the primers to regions that are conserved in all eukaryotic 16S-like coding regions assures their specific amplification. (Sequences that are conserved in prokaryotic rRNA genes may be useful for specifically amplifying mitochondrial or chloroplast 16S-like rRNA coding regions against a background of nuclear DNA.) The PCR amplification of a 16S-like rRNA coding region from a typical eukaryotic microorganism with a genome complexity of 1×10^8 bp may require fewer than 100 cells to produce 0.5 μg of product. Thus the analysis of rDNA genes from organisms that are difficult or impossible to propagate in the laboratory, including obligate parasites and symbionts, is now possible.

A major concern with PCR amplification of genes for subsequent sequence analysis is the introduction of errors during the amplification process. Comparison of sequences from amplified rRNA genes with genomic clones demonstrates differences at one to five positions per 1000 base pairs. These differences reflect heterogeneity in the rDNA sequence or errors incorporated during the amplification process. Microheterogeneity in multicopy rRNA genes has been demonstrated in several organisms, including *P. berghei* (Gunderson *et al.* 1987b), *P. falciparum* (McCutchan *et al.* 1988), and *Escherichia coli* (Sogin *et al.* 1972).

There are two strategies for guarding against potential errors in the amplification process. One strategy is to sequence the products of the amplification reaction directly by using procedures for analyzing double-stranded DNA (Wrischnik *et al.* 1987). Positions that are incorrectly copied during early rounds of amplification will appear to be ambiguous in the DNA sequencing gels, while errors in late rounds will not contribute to the sequence interpretation. The products can be sequenced directly if the amplification primers are dialyzed away or removed by chromatography before initiating DNA synthesis in the DNA sequencing reactions. As an alternative, single-stranded templates can be prepared from mixtures of 30 or 40 recombinant M13 clones. This provides a sampling of variation in the amplification products. Regardless of which method is employed,

the error rate appears to be sufficiently low so as not to confuse in-
terpretations of phylogenetic relationships on the basis of sequence
comparisons of PCR amplification products.

Conclusions

High-quality sequences on both DNA strands can be generated from
amplified rDNA coding regions. The number of cells required for the
analysis is much less than that necessary for primer extension analysis of rRNAs or for construction of genomic libraries. The specific
amplification of nuclear rRNA genes is possible by using complex
DNA populations containing rRNA genes from mitochondrial or
chloroplast genomes.

Acknowledgment

This work was partially supported by the National Institutes of Health grant
GM32964.

Literature Cited

Elwood, H. J., G. J. Olsen, and M. L. Sogin. 1985. The small subunit ribosomal RNA
    gene sequences from the hypotrichous ciliates *Oxytricha nova* and *Stylonychia
    pustulata. J. Mol. Biol. Evol.* **2**:399–410.
Gunderson, J. H., H. Elwood, A. Ingold, K. Kindle, and M. L. Sogin. 1987a. Phylo-
    genetic relationships between chlorophytes, chrysophytes, and oomycetes. *Proc.
    Natl. Acad. Sci. USA* **84**:5823–5827.
Gunderson, J. H., M. L. Sogin, G. Wollett, M. Hollingdale, V. F. de la Cruz, and T. F.
    McCutchan. 1987b. Structurally distinct, stage-specific ribosomes occur in *Plas-
    modium. Science* **238**:933–937.
Lane, D. J., B. Pace, G. J. Olsen, D. A. Stahl, M. L. Sogin, and N. R. Pace. 1985. Rapid
    determination of 16S ribosomal RNA sequences for phylogenetic analyses. *Proc.
    Natl. Acad. Sci. USA* **82**:6955–6959.
Maniatis, T., E. F. Fritsch, and J. Sambrook. 1982. In *Molecular cloning: a laboratory
    manual.* Cold Spring Harbor Laboratory, Cold Spring Harbor, New York.
McCutchan, T. F., V. F. de la Cruz, A. A. Lal, J. H. Gunderson, H. J. Elwood, and M. L.
    Sogin. 1988. The primary sequences of the small subunit ribosomal RNA genes
    from *Plasmodium falciparum. Mol. Biochem. Parasitol.* **28**:63–68.
Medlin, L., H. J. Elwood, S. Stickel, and M. L. Sogin. 1988. The characterization
    of enzymatically amplified eukaryotic 16S-like rRNA-coding regions. *Gene*
    71:491–499.

Mullis, K. B., and F. A. Faloona. 1987. Specific synthesis of DNA *in vitro* via a polymerase-catalyzed chain reaction. *Methods Enzymol.* **155**:335–350.

Qu, L. H., B. Michot, and J.-P. Bachellerie. 1983. Improved methods for structure probing in large RNAs: a rapid 'heterologous' sequencing approach is coupled to the direct mapping of nuclease accessible sites. Applications to the 5' terminal domain of eukaryotic 28S rRNA. *Nucleic Acids Res.* **11**:5903–5920.

Saiki, R., D. H. Gelfand, S. Stoffel, S. J. Scharf, R. Higuchi, G. T. Horn, K. B. Mullis, and H. A. Erlich. 1988. Primer-directed enzymatic amplification of DNA with a thermostable DNA polymerase. *Science* **239**:487–491.

Sanger, F. and A. R. Coulson. 1975. A rapid method for determining sequences in DNA by primed synthesis with DNA polymerase. *J. Mol. Biol.* **94**:441–448.

Sogin, M. L., K. J. Pechman, L. Zablen, B. J. Lewis, and C. R. Woese. 1972. Observations on the post-transcriptionally modified nucleotides in the 16S ribosomal ribonucleic acid. *J. Bacteriol.* **112**:13–22.

Sogin, M. L., and J. H. Gunderson. 1987. Structural diversity of eukaryotic small subunit ribosomal RNAs: evolutionary implications. Endocytobiology III. *Ann. N. Y. Acad. Sci.* **503**:125–139.

Sogin, M. L., J. H. Gunderson, H. J. Elwood, R. A. Alonso, and D. A. Peattie. 1989. Phylogenetic meaning of the kingdom concept: an unusual ribosomal RNA from *Giardia lamblia*. *Science* **243**:75–77.

Woese, C. R. 1987. Bacterial evolution. *Microbiol. Rev.* **51**:221.

Wrischnik, L. A., R. G. Higuchi, M. Stoneking, H. A. Erlich, N. Arnheim, and A. C. Wilson. 1987. Length mutations in human mitochondrial DNA: direct sequencing of enzymatically amplified DNA. *Nucleic Acids Res.* **15**:529–541.

38

AMPLIFICATION AND DIRECT SEQUENCING OF FUNGAL RIBOSOMAL RNA GENES FOR PHYLOGENETICS

T. J. White, T. Bruns, S. Lee, and J. Taylor

Comparative studies of the nucleotide sequences of ribosomal RNA (rRNA) genes provide a means for analyzing phylogenetic relationships over a wide range of taxonomic levels (Woese and Olsen 1986; Zimmer *et al.* 1988; Medlin *et al.* 1988; Jorgensen and Cluster 1989). The nuclear small-subunit rDNA sequences (16S-like) evolve relatively slowly and are useful for studying distantly related organisms, whereas the mitochondrial rRNA genes evolve more rapidly and can be useful at the ordinal or family level. The internal transcribed spacer region and intergenic spacer of the nuclear rRNA repeat units evolve fastest and may vary among species within a genus or among populations.

Numerous sequences of rRNA genes have been obtained primarily by isolating and sequencing individual cloned genes (Medlin *et al.* 1988). Direct rRNA sequencing (Lane *et al.* 1985) has also been used to rapidly obtain sequence data. However, this method requires relatively large amounts of RNA and is prone to errors since only one strand is sequenced.

The polymerase chain reaction (PCR) and direct sequencing offer several advantages over cloning and direct rRNA sequencing: (1) the method utilizes relatively crude preparations of total DNA such as

PCR Protocols: A Guide to Methods and Applications
Copyright © 1990 by Academic Press, Inc. All rights of reproduction in any form reserved.

those from minipreps (see Chapter 34); (2) only small amounts of DNA are required, about 0.1 to 10 ng per amplification; (3) both strands of the gene can be sequenced, which reduces errors; and (4) the method is compatible with automated DNA sequencing instruments that utilize fluorescently labeled sequencing primers or dideoxynucleotide triphosphates.

Figure 1 shows the location of the primers, and Table 1 describes the sequences of primers that we have designed for amplifying various segments of the nuclear and mitochondrial rDNA genes of fungi. The specificity of the primers for amplification of the target genes is excellent using the cycling parameters indicated (Fig. 2). The range of organisms that can be studied (as described below) can be extended or restricted to some extent by altering the annealing temperature.

Primers NS1 through NS8 were based on conserved nucleotide sequences from the 18S rRNA genes from *Saccharomyces cerevisiae*, *Dictyostelium discoideum*, and *Stylonicha pustulata* (Dams *et al.* 1988). Primers NS1 and NS2 have amplified rDNA from a wide variety of fungi, protists, and red and green algae. NS3 through NS6 have amplified all fungal DNAs tested. NS7 and NS8 also amplify some plant and vertebrate rDNAs.

The primers NS1 through NS8 will generally allow the sequenc-

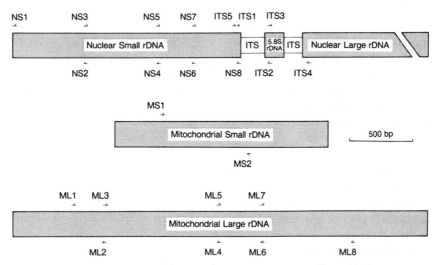

Figure 1 Locations on nuclear and mitochondrial rDNAs of PCR primers given in Table 1. The arrowheads represent the 3' end of each primer. The nuclear large rDNA is truncated in this figure.

Table 1

Primers for Amplification of Fungal Ribosomal RNA Genes

rRNA	GenePrimer[a]	Product Size (bp)[b]	T_m (°C)[c]
Nuclear, small			
NS1	GTAGTCATATGCTTGTCTC	555	56
NS2	GGCTGCTGGCACCAGACTTGC		68
NS3	GCAAGTCTGGTGCCAGCAGCC	597	68
NS4	CTTCCGTCAATTCCTTTAAG		56
NS5	AACTTAAAGGAATTGACGGAAG	310	57
NS6	GCATCACAGACCTGTTATTGCCTC		65
NS7	GAGGCAATAACAGGTCTGTGATGC	377	65
NS8	TCCGCAGGTTCACCTACGGA		65
Nuclear, ITS			
ITS1	TCCGTAGGTGAACCTGCGG	290	65
ITS5	GGAAGTAAAAGTCGTAACAAGG	315	63
ITS2	GCTGCGTTCTTCATCGATGC	290	62
ITS3	GCATCGATGAAGAACGCAGC	330	62
ITS4	TCCTCCGCTTATTGATATGC		58
Mitochondrial, small			
MS1	CAGCAGTCAAGAATATTAGTCAATG	716	65
MS2	GCGGATTATCGAATTAAATAAC		63
Mitochondrial, large			
ML1	GTACTTTTGCATAATGGGTCAGC	253	68
ML2	TATGTTTCGTAGAAAACCAGC		63
ML3	GCTGGTTTTCTACGAAACATATTTAAG	934	67
ML4	GAGGATAATTTGCCGAGTTCC		68
ML5	CTCGGCAAATTATCCTCATAAG	359	66
ML6	CAGTAGAAGCTGCATAGGGTC		65
ML7	GACCCTATGCAGCTTCTACTG	735	63
ML8	TTATCCCTAGCGTAACTTTTATC		57

[a]All odd-numbered primers are 5' primers; even numbers indicate 3' primers. Sequences are written 5'-3'.

[b]Product sizes are approximate based on the rRNA genes of *S. cerevisiae*; the size of the region amplified is the product size minus the primers. Primers NS3 and NS4 amplify mitochondrial and bacterial rDNA from some organisms; expected product sizes are approximately 365 bp and 425 bp, respectively (see Fig. 2, lane 5).

[c]T_m's were calculated by the method of Meinkoth and Wahl, 1984.

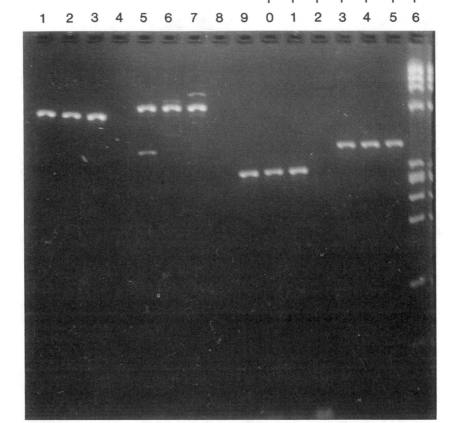

Figure 2 PCR amplification of adjacent segments of the nuclear small-subunit rRNA gene from ascomycetes: (lanes 1–4, NS1 and NS2; lanes 5–8, NS3 and NS4; lanes 9–12, NS5 and NS6; lanes 13–15, NS7 and NS8) from *Talaromyces flavus* (lanes 1, 5, 9, 13), *T. leycettanus* (lanes 2, 6, 10, 14), and *Byssochlamys nivea* (lanes 3, 7, 11, 15). Negative controls (no DNA) in lanes 4, 8, 12. Molecular weight standards: ΦX174RF *Hae*III digest in lane 16.

ing of most of the nuclear small rRNA gene excluding the areas of the primer sequences. NS2 and NS3, NS4 and NS5, and NS6 and NS7 are complementary, and a better design would have offset them to permit sequencing of the primer regions. NS1 and NS8 will amplify nearly the entire 18S gene in all fungi tested, but we have not yet devised optimal conditions for obtaining a good yield of single-strand template from an amplified product of this size.

The regions used for the mitochondrial primers were selected by comparison of sequences from the ascomycetes *S. cerevisiae* and

Aspergillus nidulans with the unpublished sequence of the basidiomycete *Suillus sinuspaulianus* (T. Bruns, personal communication). Potential primer sequences were then compared to nuclear 28S and 18S sequences to ensure that they were specific for mitochondrial genes. Where the three mitochondrial sequences differ within the primer regions, the *S. sinuspaulianus* sequences were chosen. Thus some of the primers are better matched to *Suillus* and related basidiomycetes than to ascomycetes. However, the primers were designed such that mismatches are few and central to the regions of homology. All primers have been tested with the ascomycete *Neurospora crassa* and found to amplify the correct fragment. Primers ML1, ML2, ML4, MS1, and MS2 will also amplify specific fragments from the oomycete *Phytophthora cinnamomi*. The region

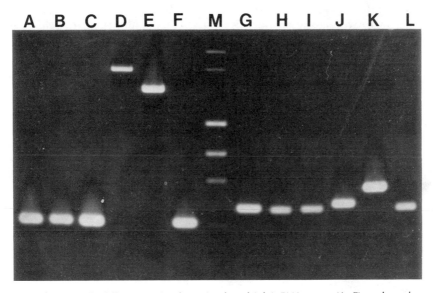

Figure 3   Length differences in the mitochondrial LrRNA gene (A–F) and nuclear ITS region (G–L). A portion of the mitochondrial LrRNA gene was amplified from six species of *Suillus* using the ML7 primer and one that is 3' to ML8 (MLID, CGGAGGATAGGATAAGTCG) and specific to the *Boletaceae* and related basidiomycetes. Four of the species yielded a fragment of approximately 610 bp, but two (D,E), which are known from previous mapping studies (Bruns and Palmer, in press) to contain an intron in the region, yielded fragments of approximately 2000 and 1700 bp. The ITS1 and ITS4 primers were used to amplify the ITS region from four species of *Suillus* and two species from the closely related genus *Rhizopogon*. Sizes of the region were found to be approximately 710 bp in the four *Suillus* species (G–I, L), but 740 and 850 bp in the two *Rhizopogon* species (J, K). The size markers (M, middle) are Phage λ *Hind*III and ΦX *Hae*III fragments of 2322, 2027, 1360, 1078, 878, 606, and 310 bp.

amplified by ML6 and ML5 contains an intron in some species of basidiomycetes (Fig. 3, Bruns and Palmer 1989), and in species in which this intron is large, the fragment is not amplified efficiently. Length mutations are common in these mitochondrial genes, and the length of the regions amplified may vary considerably from those listed in Table 1.

The ITS primers make use of conserved regions of the 18S, 5.8S, and 28S rRNA genes to amplify the noncoding regions between them. ITS1 is the complement of NS8 and was designed as described above. Comparisons among 5.8S rRNA sequences of *N. crassa*, *Schizosaccharomyces pombe*, *S. cerevisiae*, Broad bean (*Vicia faba*), and mouse (*Mus musculus*) were used to select ITS2 and ITS3. The 28S sequences of *S. pombe*, *S. cerevisiae*, and rice (*Oryza sativa*) were compared to select ITS4. The conserved region chosen overlaps the 28A primer of Zimmer *et al.* (1988), which has been previously used for direct sequencing of the 28S rRNA. ITS5 is identical in sequence to the *N. crassa* sequence in the 18S rDNA region (Kelly and Cox 1982) that is 25 base pairs 5' to ITS1.

Protocol

Reagents

10× amplification buffer
    15 mM MgCl$_2$
    500 mM KCl
    100 mM Tris HCl (pH 8.3) at 23°C
    0.1% gelatin
10×dNTP stock mixture
    2 mM each of dATP, dGTP, dTTP, and dCTP

Procedure

1. Using positive displacement pipets for steps 1–4, prepare a working reaction mixture sufficient for 10 amplifications, which consists of:
    275 μl   sterile-distilled water
    100 μl   10× amplification buffer

> 100 μl   dNTP stock mixture (2 mM each dNTP)
> 10 μl   excess primer (50 μM stock)
> 10 μl   limiting primer (1 μM stock)
> 5 μl   *Taq* polymerase (5 units/μl)

2. Add 50 μl of the working mixture to each tube.

3. Dilute the DNA samples in water or TE (10 mM Tris-HCl [pH 8.0], 0.1 mM EDTA) to 0.1 to 10 ng per 50 μl.

4. Add 50 μl of the diluted sample DNA to each tube followed by two drops of mineral oil.

5. Spin briefly in a microcentrifuge.

6. Cycling parameters

Initial denaturation	2 to 3 minutes at 95°C
Annealing	30 seconds at 50° to 60°C
Extension	0.5 to 2 minutes (depending on product size) at 72°C
Denaturation	30 seconds at 95°C
Final extension	10 minutes at 72°C

7. Number of cycles
   25 for double-stranded product, using 50 pmol of each primer. 35 for single-stranded product, using a primer ratio of 50 : 1 or 50 : 2.5 pmoles.

8. When the amplification is completed, briefly centrifuge the tubes and take 5-μl samples for analysis by minigels. Before sequencing the single-stranded PCR product, remove the oil by extraction in chloroform and remove the unincorporated nucleotides while concentrating the DNA by Centricon-30 centrifugal filtration (W. R. Grace, Danvers, Massachusetts).

9. We routinely use 7 μl of the Centricon-30 retentate for sequencing reactions. If this DNA proves too dilute, concentrate the retentate in a Speed Vac. Use 1 pmole of the limiting primer (or an internal primer) as the sequencing primer. Do not denature the double-stranded product prior to primer annealing, i.e., anneal at 65°C instead of 90 to 95°C.

10. Primers NS1 through NS8 and ITS1 through ITS4 have also been used with lower dNTP concentrations (32 μM each dNTP, instead of 200 μM, in the reaction).

11. These conditions have not been optimized for enzyme and magnesium ion concentrations.

Literature Cited

Bruns, T. D., and J. D. Palmer. 1989. Evolution of mushroom mitochondrial DNA: *Suillus* and related genera. *J. Mol. Evol.* **28**:349–362.

Dams, E., L. Hendriks, Y. Van de Peer, J.-M. Neefs, G. Smits, I. Vandenbempt, and R. De Wachter. 1988. Compilation of small ribosomal subunit RNA sequences. *Nucleic Acids Res.* **16** (Sup.):r87-r173.

Hamby, R. K., and E. A. Zimmer. 1989. Direct ribosomal RNA sequencing: optimization of extraction and sequencing methods for work with higher plants. *Plant Mol. Biol. Rep.* **6**:175–192.

Jorgensen, R. A., and P. D. Cluster. 1989. Modes and tempos in the evolution of nuclear ribosomal DNA: new characters for evolutionary studies and new markers for genetic and population studies. *Ann. Mo. Bot. Gard.* **75**:1238–1247.

Kelly, J. M., and R. A. Cox. 1982. The nucleotide sequence at the 3'-end of *Neurospora crassa* 18S-rRNA and studies on the interaction with 5S-rRNA. *Nucleic Acids Res.* **10**:6733–6745.

Lane, D. J., B. Pace, G. J. Olsen, D. A. Stahl, M. L. Sogin, and N. R. Pace. 1985. Rapid determination of 16S ribosomal RNA sequences for phylogenetic analyses. *Proc. Natl Acad. Sci. USA* **82**:6955–6959.

Medlin, L., H. J. Elwood, S. Stickel, and M. L. Sogin. 1988. The characterization of enzymatically amplified eukaryotic 16S-like rRNA-coding regions. *Gene* **71**:491–499.

Meinkoth, J., and G. Wahl. 1984. Hybridization of nucleic acids immobilized on solid supports. *Anal. Biochem.* **138**:267–284.

Woese, C. R., and G. J. Olsen. 1986. Archaebacterial phylogeny: perspectives on the urkingdoms. *Syst. Appl. Microbiol.* **7**:161–177.

DIAGNOSTICS AND FORENSICS

Part Four

DIAGNOSTICS
AND CONTROLS

39

DETECTION OF HUMAN T-CELL LYMPHOMA/LEUKEMIA VIRUSES

Garth D. Ehrlich, Steven Greenberg, and Mark A. Abbott

The utilization of the *in vitro* enzymatic gene amplification technique, polymerase chain reaction (PCR), has greatly facilitated the detection of the human oncogenic retroviruses (Kwok *et al.* 1988a; Duggan *et al.* 1988; Poiesz *et al.* 1988). The detection of human T-cell lymphoma/leukemia virus type I (HTLV-I) proviral sequences in patients with tropical spastic paraparesis was first accomplished by the use of PCR in our laboratories (Ehrlich and Poiesz 1988; Bhagavati *et al.* 1988; Kwok *et al.* 1988b). Since then PCR has been used extensively in epidemiological studies and diagnostic procedures to detect the presence of HTLV-I and -II in a wide spectrum of carriers and patients with lymphoproliferative and neurological disorders (Ehrlich *et al.* 1989a; Ehrlich *et al.* 1989b; Ehrlich *et al.*, in press; Greenberg *et al.* 1989). PCR has fostered the development of the discipline of molecular epidemiology by providing a facile means to unambiguously identify infected asymptomatic carrier individuals as well as patients with clinically expressed disease. Each population of at-risk individuals presents with their own set of technical and ethical problems that must be addressed before dispensing the results of PCR analyses (Abbott *et al.* 1988).

Inherent in all identification and detection programs that employ PCR is the threat of false-positive results caused by physical carry-over of amplified DNA into experimental samples (see Chapter 17). Carry-over usually occurs by aerosolization of amplified samples when tightly capped tubes are opened and by transfer of samples by pipetting devices that do not physically separate the tip of the device from the material being pipetted. If the amplified DNA is viewed as an infectious agent with the ability to replicate under the proper environmental conditions, i.e. the thermal cycler, it is relatively easy to establish certain physical precautions that should greatly reduce the specter of carry-over. Any laboratory using PCR for diagnostic or epidemiologic studies should separate its operations into three physically separate areas. One area should be used to receive and process blood and tissue samples. In most cases it is advisable to isolate the peripheral blood mononuclear cells (PBM) by a Ficoll-Hypaque step gradient prior to DNA extraction. Purified PBM can be stored indefinitely at −70°C while awaiting extraction. A second area is used for DNA extraction and purification. The latter is optional depending on the type of analyses to be performed. This area is also used for setting up the reaction cocktails prior to PCR; it is never used for the actual amplification or for the assessment of the products of amplification. The third area is used for the actual amplification and analysis of PCR products. The tightly capped tubes to be amplified are never opened until all amplification is complete. These tubes should never be opened and reclosed once amplification has begun. Likewise they should never be taken back into the first two areas after amplification has been allowed to proceed.

Each of the three areas should be completely equipped with its own set of appropriate pipetting and storage devices. Further, to reduce the chance of carry-over posed by aerosolized target DNA, we recommend the use of positive-displacement pipetting devices that have disposable tips and plungers, such as Rainin Microman, when pipetting the DNA. (Contamination with target DNA is much less likely since target DNA, as opposed to amplified DNA, is generally of high molecular weight.) Standard pipettors are appropriate for dispensing non-DNA-containing solutions and buffers.

In addition to physical precautions, a number of procedural safeguards should be employed. A diagnosis of HTLV infection should never be made on the result of a single PCR performed with a single primer pair in the absence of some corroborating evidence such as serological or antigenic data. It is recommended in all diagnostic cases that at least two sets of primer pairs produce a positive result, particularly if the patient is seronegative. Due to the biology of these viruses, a finite percentage of infected individuals will be truly seronegative. Many individuals infected neonatally do not sero-convert until their third decade of life (Ehrlich and Poiesz 1988; Ehrlich, in press), and a recent study of parenteral drug users re-vealed that over 30% of both HTLV-I and -II infected individuals were seronegative (Ehrlich *et al.*, in press). Further, it is suggested that all such results be confirmed from a second DNA sample. To expeditiously accomplish the latter, we routinely double or triple al-iquot the purified mononuclear cells from each patient during the initial processing of the sample. When DNA is to be extracted from PBMs, only a single aliquot is used for the initial screen. If a given DNA is positive, DNA from a second aliquot is extracted for confir-mational analyses.

If a DNA is to be analyzed generically for HTLVs, there are certain primer-probe systems that will amplify and detect both HTLV-I and -II (Kwok *et al.* 1988a and b). Likewise there are specific primer-probe systems that will discriminate between the two species (Kwok *et al.* 1988a and b; Greenberg *et al.* 1989). (See Table 1 for a represen-tative list of primer and probe oligonucleotides). For epidemiological screening we routinely utilize a generic system for an initial screen. Positive DNAs are then subsequently analyzed with type-specific systems. Some heterogeneity of HTLV-I strains has been reported. A useful way to determine if a given isolate represents a prototype strain or a variant is to do simultaneous amplification and subse-quent simultaneous liquid hybridizations for multiple regions of the proviral genome. We have successfully amplified and detected five regions at once by judiciously choosing our primers to produce am-plification fragments of varying size that can be discriminated on a polyacrylamide gel (Ehrlich *et al.*, in press).

The following methods developed over the last three years in our laboratories have proved useful in a wide range of applications. These procedures can be used to amplify DNA in simple cellular ly-sates or organically extracted DNA preparations. We have found that boiling and quenching on ice of the high-molecular-weight target DNA prior to PCR greatly enhances the amplification process (Ab-bott *et al.* 1988). Full denaturation of the chromosomal DNA does not take place at 94°C; therefore, the binding of primers is limited in the first critical rounds of amplification, which greatly reduces the quantity of the final product. This is of particular importance when

Table 1

Generic and Specific Primer Pairs and Probes for Amplification and Detection of the HTLV's

Primer Type & Orientation	Primer	Probe Type & Orientation	Probe	Region	Position	Sequence (5'-3')
Generic +	SK43			Tax	7358-7377 (I)	CGGATACCCAGTCTACGTGT
					7248-7267 (II)	TGGATACCCGTCTACGTGT
		Generic +	SK45	Tax	7447-7468 (I)	ACGCCCTACTGGCCACCTGTC-CAGAGCATCAGATCACCTG
Generic −	SK44			Tax	7516-7496 (I)	GAGCCGATAACGGGTCATCG
					7406-7386 (II)	GAGCTGACAACGCGTCATCG
Generic +	SK110	Specific (I) +	SK112	pol	4757-4778 (I)	CCCTACAATCCAACCAGCTCAG
		Specific (II) +	SK188	pol	4735-4756 (II)	CCATACAACCCACCAGCTCAG
				pol	4825-4850	GTACTTTACTGACAAACCGACCTAC
				pol	4880-4898	TCATGAACCCCAGTGGTAA
		Generic +	SK115	pol	4870-4895 (I)	CATAGCCCTATGGACAATCAACCACC
					4848-4873 (II)	CAAGCCCTTGGACTCTCAATCAGC
Generic −	SK111			pol	4942-4919 (I)	GTGGTGAAGCTGCCATCGGGTTTT
					4920-4897 (II)	GTGGTGGATTTGCCATCGGGTTTT
Specific + (I)	SG166			gag	1388-1411	CTGCAGTACCTTTGCTCCTCCCTC
		Specific (I) −	SG242	gag	1451-1412	ATATAAGGCTATCTAGCTGCTGGTG-ATGGAGGGAAGCCAC
Specific − (I)	SG296			gag	1660-1641	TTCTACGAAGGCGTGGTAAG
Specific + (I)	SG295			gag	1641-1660	CTTACCACGCCTTCGTAGAA
		Specific (I) −	SG169	gag	1700-1661	TGCCTTCTGGCAGCCCATTGTCAAG-AGCTATGTTGAGGCG
Specific − (I)	SG167			gag	1957-1934	CCCGGGGGGGGGACGAGGGCTGAGT

Specificity	Primer	Gene	Position	Sequence
Specific + [I]	SG231	pol	2801-2820	CCGGGCCCCCTGACTTGTC
Specific [I] +	SG232	pol	2821-2860	CAGCCTGCCAACCACTAGCCCAC-TTGCAAACTATAGAC
Specific [I] −	SG237	pol	3010-2971	GAACAGGGTGGGACTATTTTAAAC-CCTTGGGTAGTACT
Specific − [I]	SG238	pol	3037-3018	CTGCAGGATATGGGCCAGCT
Specific + [I]	SK54	pol	3365-3384	CTTCACAGTCTCTACTGTGC
Specific [I] −	SK56	pol	3460-3426	CCGCAGCTGCACTAATGATTGAACT-TGAGAAGGAT
Specific − [I]	SK55	pol	3483-3465	CGGCAGTTCTGTGACAGGG
Specific + [II]	SK58	pol	4198-4217	ATCTACCTCCACCATGTCCG
Specific [II] −	SK60	pol	4276-4237	TAAGGGAGTCTGTGTATTCATTGAAG-GTGAAATTGGGTC
Specific − [II]	SK59	pol	4300-4281	TCAGGGAACAAGGGGAGCT
Specific + [I]	SG219	env	5270-5292	CCCCAGCTGCTGACTCTCACAA
Specific [I] +	SG224	env	5301-5340	TCCTCATACCACTCTAAACCCTGCA-ATCCTGCCCAGCCAG
Specific − [I]	SG294	env	5540-5521	TGGGCACTTTAAGGAACAAG
Specific + [I]	SG293	env	5521-5540	CTTGTTCCTTAAAGTGCCCA
Specific [I] −	SG225	env	5650-5611	ATATTGAGGCGTGAAACTTCTTGAG-TAAAATTGACATCGT
Specific − [I]	SG220	env	5804-5785	CTCGAGGATGTGGTCTAGGT
Specific + [I]	SG452	env	5796-5818	ATCCTCGAGCCCTCTATACCATG
Specific [I] +	228	env	5841-5880	GTCCAGTAACCTACAAAGCACTA-ATTATACTTGCATTG
Specific [I] −	229	env	6050-6011	AGCTTGAATCTGGGGTCAAAGCAG-TGGGTCCAGTTAAAT
Specific − [I]	SG453	env	6128-6106	GCGGGATCCTAGGGTGGGAACAG

the target sequence is very rare, as is the case with many asymptomatic carriers. Boiling of samples prior to PCR also reduces the amount of primer oligonucleotide (and therefore the cost) that must be used to obtain a given signal.

Protocols

Stock Solutions and Reagents for PCR and Hybridization

Solution 1	1 M Tris–HCl (pH 8.3), autoclaved
Solution 2	2% Sigma gelatin (swine skin type), autoclaved
Solution 3	3.73 g KCl and 0.51 g MgCl$_2$ (hydrous) in 80 ml of water (1625 mM KCl; 31 mM MgCl$_2$), autoclaved
10× *Taq* Buffer	Using aseptic technique, add 10 ml each of solutions 1 and 2 to solution 3, bringing the total volume to 100 ml. Filter through Genescreen Plus (Dupont, Boston, Massachusetts), aliquot, and store frozen at −20°C. Use each aliquot for a single set of reactions only, and then discard. The concentrations of this final solution are 500 mM KCl, 25 mM MgCl$_2$, 100 mM Tris HCl (pH 8.3), and 0.2% gelatin.
20× SSPE	174 g NaCl 24 g NaH$_2$PO$_4$ (monobasic) 7.4 g Na$_2$EDTA pH to 7.4 with NaOH, bring to final volume of 1 liter
100× TE	60.5 g Tris 18.6 g EDTA pH to 7.5, bring to final volume of 500 ml with distilled water
1× TE	10 mM Tris–HCl (pH 7.5) 1 mM Na$_2$EDTA

	To make 1× TE, dilute 100× TE with dH₂O, filter through Genescreen Plus, aliquot, and store frozen at −20°C.

Nucleotides (dNTPs)
: Nucleotides are purchased from Pharmacia as 100 m*M* stocks. Add 38 μl of each of the four dNTPs to 848 μl of water. The final concentration is 3.75 m*M* of each nucleotide in 1 ml.

Amplification primers
: Dilute to a final concentration of 10 pmol/μl, aliquot, and store frozen in dH₂O at −20°C.

Denaturation Solution
: 0.5 *M* NaOH, 50 ml of 10 *M* stock
1.5 *M* NaCl, 300 ml of 5 *M* stock
dH₂O, bring to final volume of 1 liter

Neutralization Solution
: 0.5 *M* Tris–HCl (pH 7.5), 167 ml of 3 *M* stock
1.5 M NaCl, 300 ml of 5 *M* stock
dH₂O, bring to final volume of 1 liter

Prehybridization Solution
: 150 ml of 20× SSPE
0.1 g of Ficoll
0.1 g of PVP
0.1 g of bovine serum albumin
1.0 ml of 0.5 *M* EDTA (pH 8)
12.5 ml of 20% SDS
Bring to final volume of 500 ml with dH₂O

Hybridization
: Put 200 ml of prehybridization solution into a separate beaker and stir in 20 g of dextran sulfate while the beaker is on a hot plate (medium-low heat). After it has gone into solution, transfer to a bottle and store at 4°C. Store prehybridization solution likewise.

10× TBE
: 108 g of Tris
55 g boric acid
9.3 g EDTA
Bring to final volume of 1 liter

Gel Loading Dye
: 25 ml water
50 ml glycerol

> 20 ml 0.5 M EDTA (pH 8)
> 0.1 g Bromphenol blue
> 0.1 g Xylene cyanol
> Pass through 3MM filter paper before use

PCRs

Cautionary statement: You are amplifying sequences of interest more than six orders of magnitude. Use all due caution to avoid contamination, especially after the amplification, when the target sequences will be present as small molecules at high molar concentrations and relatively high vapor pressure. Always set up runs so that the positive controls are pipetted last to minimize the chance of contamination both before and after amplification. Use the appropriate pipetting device at all times, and always err on the side of throwing something out that might be contaminated. Never reuse disposable aliquots of reaction components. Use the appropriate sterilized tips, plungers, and tube cap openers at all times. Do not bring amplified DNAs to the sample processing and PCR set-up areas.

1. In a 500-μl Eppendorf tube, add 1 μg of DNA. Bring to 50 μl with 1× TE.
2. Boil DNA for 5 minutes. Place on ice ~1 minute. Spin down condensation in microfuge.
3. While DNA is boiling, prepare cocktail as below. Each reaction should contain:

> 10 μl of 10× *Taq* buffer
> 10 pmol each primer
> 6 μl of 15 mM dNTP
> 2.5 units of *Taq* polymerase
> glass-distilled H$_2$O to a final volume of 50 μl

Prepare enough mix to allow for losses due to multiple pipetting (i.e., to PCR 20 samples calculate for 22 samples).

> 220 μl of 10x *Taq* buffer
> 132 μl of dNTPS
> 22 μl of each primer (10 pmol/μl)
> 11 μl of *Taq* polymerase
> *693 μl TE*
> 1100 μl Total cocktail volume

4. Add 50 μl of cocktail to each boiled DNA sample using a pipet-man that has not been used on PCR-amplified products. Pipette cocktail slowly onto side of sample tube. Do not immerse tip in DNA solution. Do not release plunger (i.e., aspirate until the tip is completely withdrawn from tube and then release slowly and dispose of the tip). Keep the caps closed as much as possible.
5. Add approximately 75 μl of light mineral oil slowly down side of tube, using plastic disposable Pasteur pipettes to overlay the re-action cocktail.
6. Place the tubes in a thermal cycler (do not open). For all the primer pairs listed in Table 1, the following ramping profile will produce a positive signal with 2 to 3 input molecules of target DNA following 30 cycles of amplification using the Perkin-Elmer Cetus Thermal Cycler.

File X ambient temperature to 68°C for 1 second for 1
    cycle
File Y 68 to 94°C for 1 second; hold for 15 seconds at 94°C⎤
    94 to 53°C for 1 second; hold for 30 seconds at 53° ⎬30 cycles
    53 to 68°C for 1 second; hold for 30 seconds at 68°C⎦
File Z 68°C for 10 minutes

Liquid Hybridization

1. Follow directions for PCR-DNA amplification.
2. After amplification, add 150 μl of chloroform : isoamyl alcohol (24 : 1). Shake the tubes vigorously. Let the tubes set until two layers are visible. (The amplified DNA will be the top layer.)
3. Take out 30 μl of the PCR-DNA and place it in another 0.5-ml Eppendorf tube.
4. Add 20 μl of the liquid hybridization mix. Each mix consists of:
    5 μl of a solution of 1.5 M NaCl and 25 mM Na$_2$EDTA
    250,000 cpm of ^{32}P-labeled probe
    glass-distilled H$_2$O to a volume of 20 μl
    Master mixes can be prepared.
5. Boil the probe/amplified DNA samples for 5 minutes. Place in a 55°C water bath for at least 10 minutes.
6. Add one-tenth volume (5μl) of the gel loading dye/sample.
7. Run the samples on an 8% polyacrylamide gel in TBE.
8. Load samples on gel. Run at 200 V for approximately 60 minutes.

Note: with shorter probes it may be necessary to cool the gel or run at a lower voltage to prevent the denaturation of the hybrid.

9. Autoradiography: wrap the gel with Saran Wrap. Expose to film (Kodak XAR-5) for 2 hours to overnight at room temperature.

Spot Blot Method of Hybridization

1. Take a piece of Genescreen Plus. Do not touch the Genescreen Plus with hands or powdered gloves. Using blue ball point pen only, draw circles where each sample will run.

2. Treat Genescreen Plus by soaking in water until wet. Then pretreat in 2× SSPE for 10 to 15 minutes. Air dry Genescreen Plus until the paper is damp only, not wet.

3. Dot each circle with 10 μl of PCR-amplified DNA sample. Be careful when adding the sample, as the Genescreen must be fairly dry so the sample will stay within the outlined circle. If Genescreen is saturated, the sample will run outside the circle.

4. Denature the immobilized DNA by placing the membrane, DNA side up, on a piece of Whatman 3 MM paper saturated with the denaturation solution. Be sure there are no air bubbles. Denature for 5 minutes.

5. Neutralize the DNA. Transfer the Genescreen membrane (again, DNA side up) to a piece of Whatman paper saturated with neutralization solution. Neutralize for 5 minutes.

6. Dry the Genescreen by placing it under white lamps for 15 minutes. Once the Genescreen is completely dry, proceed with the prehybridization.

Prehybridization

150 ml of 20x SSPE

0.1 g of Ficoll

0.1 g of PVP

0.1 g of bovine serum albumin

1.0 ml of 0.5 M EDTA

12.5 ml of 20% SDS

Bring up to 500 ml with ddH20. Put 300 ml into a bottle labeled "prehybridization solution." Put 200 ml into a separate beaker and add 20 g of dextran sulfate (keep in refrigerator). Stir into solution and pour into a bottle labeled "hybridization solution."

Preparation

1. Cut a heat-sealable pouch to fit the Genescreen. Put the Genescreen in the pouch, making sure that three sides of the pouch are sealed. Put prehybridization solution (amount will vary on size of blot) in the pouch. Seal the fourth side. Prehybridize at 55°C for a minimum of 60 minutes (can go overnight).

2. Hybridization: Determine the volume of solution required. Add end-labeled probe to the hybridization solution at a concentration of 10^6 cpm/ml (i.e., for a 4-ml hybridization, add 4×10^6 cpm. Cut one corner or side of the pouch and squeeze out the prehybridization solution. Add the hybridization solution with probe already added (amount will vary depending on size of blot). Seal the pouch. Hybridize at 55°C for a minimum of 3 hours (can hybridize overnight).

Washes

1. Wash at 55°C for 30 minutes with 2× SSPE, 0.1% SDS.

2. Wash a second time at 55°C for 10 minutes with 2× SSPE, 0.1% SDS. After 10 minutes, check Genescreen with a hand-held GM-counter. If Genescreen appears to be hot, place it back in the wash and continue washing until the negatives appear clean.

Acknowledgments

We greatly appreciate the input of Janice Andrews, Ginny Bryz-Gornia, and Lynn Zaumetzer toward the development of these protocols. We thank the DNA synthesis group at Cetus for the synthesis of some of the primers and probes. We thank Dr. John Sninsky and Dr. Bernard Poiesz for their help and support and Lori Raven for her excellent secretarial skills. This work was supported by U.S. Public Health Service Contract #NO1HB67021 from the NHLBI.

Literature Cited

Abbott, M., B. Poiesz, J. Sninsky, S. Kwok, B. Byrne, and G. Ehrlich. 1988. A comparison of methods for the detection and quantification of the polymerase chain reaction. *J. Infect. Dis.* **158**:1158–1169.

Bhagavati, S., G. Ehrlich, R. Kula, S. Kwok, J. Sninsky, V. Udani, and B. Poiesz. 1988. Detection of human T-cell lymphoma/leukemia virus-type I (HTLV-I) in the spinal fluid and blood of cases of chronic progressive myelopathy and a clinical, radiological and electrophysiological profile of HTLV-I associated myelopathy. *N. Engl. J. Med.* **318**: 1141–1147.

Duggan, D., G. Ehrlich, F. Davey, S. Kwok, J. Sninsky, J. Goldberg, L. Baltrucki, and B. Poiesz. 1988. HTLV-I induced lymphoma mimicking Hodgkin's disease: diagnosis by polymerase chain reaction amplification of specific HTLV-I sequences in tumor DNA. *Blood* **71**: 1027–1032.

Ehrlich, G., and B. Poiesz. 1988. Clinical and molecular parameters of HTLV-I infection. *Clinics in Lab. Med.* **8**: 65–84.

Ehrlich, G., F. Davey, J. Kirshner, J. Sninsky, S. Kwok, D. Slamon, R. Kalish, and B. Poiesz. 1989a. A polyclonal CD4+ and CD8+ lymphocytosis in a patient doubly infected with HTLV-I and HIV-1: a clinical and molecular analysis. *Am. J. Hematology* **30**: 128–139.

Ehrlich, G., J. Glaser, M. Abbott, D. Slamon, D. Keith, M. Sliwkowski, J. Brandis, E. Keitelman, Y. Teramoto, L. Papsidero, H. Simpkins, J. Sninsky, and B. Poiesz. 1989b. Detection of anti-HTLV-I Tax antibodies in HTLV-I ELISA negative individuals. *Blood* **74**: 1066–1072.

Ehrlich, G., J. Glaser, K. Lavigne, D. Quan, D. Mildron, J. Sninsky, S. Kwok, L. Papsidero, and B. Poiesz. Prevalence of human T-cell leukemia/lymphoma virus type II infection among high risk individuals: Type specific identification of HTLV's by polymerase chain reaction. *Blood,* in press.

Ehrlich, G. D. The biology and spectrum of disease associated with human retroviruses. *Sem. Neurol.,* in press.

Greenberg, S., G. Ehrlich, M. Abbott, B. Hurwitz, T. Waldmann, and B. Poiesz. 1989. Detection of sequences homologous to human retroviral DNA in multiple sclerosis by gene amplification. *Proc. Natl. Acad. Sci. USA* **86**: 2878–2882.

Kwok, S., G. Ehrlich, B. Poiesz, R. Kalish, and J. Sninsky. 1988a. Enzymatic amplification of HTLV-I viral sequences from peripheral blood mononuclear cells and infected tissues. *Blood* **72**: 1117–1123.

Kwok, S., G. Ehrlich, B. Poiesz, S. Bhagavati, and J. Sninsky. 1988b. Characterization of a HTLV-I sequence from a patient with chronic progressive myelopathy. *J. Infect. Dis.* **158**: 1193–1197.

Poiesz, B., G. Ehrlich, L. Papsidero, and J. Sninsky. 1988. Detection of human retroviruses. In *AIDS: etiology, diagnosis, treatment and prevention* (ed. V. DeVita, S. Hellman, and S. Rosenberg), p. 137–154. Lippincott, Philadelphia.

40

DETECTION OF HUMAN IMMUNODEFICIENCY VIRUS

David E. Kellogg and Shirley Kwok

Serological assays provide a rapid and sensitive procedure to screen blood and blood products for the presence of antibodies to human immunodeficiency virus type 1 (HIV-1). Samples that are repeatedly reactive are retested for HIV antibodies with supplemental tests such as Western blot, immunofluorescent assay (IFA), or radioimmunoprecipitation assay (RIPA). Although these tests are sensitive and specific for HIV, direct detection of the virus would be desirable for a number of reasons.

First, because of the latency associated with HIV, a direct assay may assist in the identification of individuals who are infected with the virus but have not seroconverted. Second, detection of HIV infection in newborns would be facilitated. Because maternal antibodies can persist up to 15 months, it is difficult to differentiate between maternal and infant antibodies. Third, direct viral detection should contribute significantly in clarifying the status of individuals with indeterminate Western blot patterns in the high- and low-risk communities. Fourth, a direct assay for the virus would provide a means to monitor both latent and actively replicating virus in patients on therapeutic drugs.

PCR Protocols: A Guide to Methods and Applications

Because of the low level of circulating free virus, direct detection of HIV in patient samples is difficult without *in vitro* propagation. Even with co-cultivation, the successful recovery of HIV varies from 10 to 75%, although some laboratories have reported close to 100% virus recovery from infected individuals (Jackson, Kwok, Sninsky, Sannerud, Rhame, Henry, Simpson, and Balfour, manuscript submitted). Virus culturing is time-consuming and expensive, and the procedures used to monitor viral growth are laborious. Often, the results from virus culturing are inconclusive. Reverse transcriptase assays, for example, are not specific for HIV. Many of the tests performed on cultured cells, such as electron microscopy, immunofluorescence, and Southern blot analysis on extracted DNA, lack sensitivity.

The heterogeneity of HIV-1 has been extensively documented (Shaw *et al.* 1984; Saag *et al.* 1988; Goodenow *et al.* 1989; Meyerhans *et al.*, in press). To ensure efficient detection of all HIV-1 variants, only highly conserved regions of the viral genomes were targeted for amplification. (Nucleic acid alignments were provided by Los Alamos National Laboratory.) Two primer pairs have been extensively used for the analysis of HIV-1-infected samples. One primer pair, designated SK38 and SK39 (Ou *et al.* 1988; Kwok *et al.* 1989) amplifies a 115-bp region of *gag* of HIV-1. The second primer pair, SK145 and SK101, amplifies a 130-bp region of *gag* that is conserved among the HIV-1 isolates and HIV-2 (isolate ROD). It should be noted that SK101 has a single-base mismatch (9 nucleotides from the 5′ end) with the consensus sequence. Although this primer when coupled with SK145 amplified efficiently, we have since replaced it with SK150 (see Table 1). The probes for the SK38−39 and SK145−101 products are SK19 and SK102, respectively. The sequences of the primers and probes are presented in Table 1.

Because of the exquisite sensitivity of PCR, it is crucial that special care be taken to avoid cross-contamination of samples or carryover of amplified products that can result in false-positive results. Because of the generally small number of HIV molecules in a clinical specimen (potentially as few as 1 : 10,000 or less in peripheral blood), the amplification protocols and detection methods described here have been optimized to provide maximum sensitivity. The demand placed on PCR to detect small numbers of molecules necessitates the avoidance of even minute quantities of contaminating target molecules.

The guidelines for avoiding false positives with PCR have been described (Kwok and Higuchi 1989; Chapter 17) but merit reiteration.

Table 1

Primers and Probes for the Amplification and Detection of HIVs

Oligonucleotide designation (orientation)	Sequence (5'-3')	Virus	Position [a]
SK38 (+)	ATAATCCACCTATCCCAGTAGGAGAAAT	HIV-1	1541-1578
SK39 (−)	TTTGGTCCTTGTCTTATGTCCAGAATGC	HIV-1	1665-1638
SK19 (+)	ATCCTGGGATTAAATAAAATAGTAAGAATGTATAGCCCTAC	HIV-1	1595-1635
SK145 (+)	AGTGGGGGGACATCAAGCAGCCATGCAAAT	HIV-1	1366-1395
		HIV-2	1150-1121
SK101 (−)	GCTATGTCAGTTCCCCTTGGTTCTC	HIV-1	1506-1482
		HIV-2	1258-1234
SK150 (−)	TGCTATGTCACTTCCCCTTGGTTCTCTC	HIV-1	1507-1480
		HIV-2	1259-1232
SK102 (+)	GAGACCATCAATGAGGAAGCTGCAGAATGGGAT	HIV-1	1403-1435
		HIV-2	1158-1190

[a]HIV-1 isolate SF2, GenBank Accession number K02007
HIV-2 isolate ROD, GenBank Accession number M15390

Perhaps most important is the use of a dedicated area that is free of PCR product for the preparation of amplification reactions. In our laboratories, biological hoods equipped with UV lights that are turned on between use serve as efficient containment areas. It is critical that amplified products are kept away from this area. Other precautions include the use of positive-displacement pipettes with disposable capillaries and pistons, dispensing of reagents into small aliquots, and the preparation of master mixes whenever possible to minimize handling. Equally important, an investigator must be cognizant of manipulations that can potentially lead to contamination. Precautions must be taken not only in the preparation of the amplification reactions but in all aspects of sample handling, from collection to DNA extraction to PCR amplification.

The protocols that follow describe the procedures used for extraction of DNA, the conditions used in the amplification of HIV, and the analysis of PCR product by oligomer hybridization (OH). Although analysis of PCR product from high-copy-number targets can be achieved by visualization on agarose gels after ethidium bromide staining, Southern blot analysis, or dot blot analysis, detection of low-copy targets is best achieved by the OH analysis. The procedure is at least 25 times more sensitive than Southern blot analysis. When PCR is coupled with OH, as few as 18 copies of HIV molecules (prior to amplification) in a background of 1 μg genomic DNA (the equivalent of 150,000 cells) can be detected by using 30 cycles of amplification (Kwok *et al.* 1989).

Protocols

Reagents

Solution A	100 mM KCl
	10 mM Tris–HCl (pH 8.3)
	2.5 mM MgC12
Solution B	10 mM Tris–HCl (pH 8.3)
	2.5 mM MgCl$_2$
	1% Tween 20
	1% NP40

Add self-digested Proteinase K to 120 μg/ml to solution B just before use.

10× *Taq* buffer   500 m*M* KCl
100 m*M* Tris–HCl (pH 8.3)
25 m*M* MgCl$_2$
0.1% gelatin

Sample Preparation

All samples, including clinical specimens, must be handled and disposed of as if they contain a transmissible infectious agent. Lab coats, masks, safety glasses, and gloves should be worn. Wash hands thoroughly when work is completed.

1. Collect 5 to 10 ml of whole blood by venipuncture in either an ACD (acid citrate dextrose) or heparin tube.
2. Pipette 5 ml of Ficoll-Hypaque into a conical centrifuge. Carefully layer an equal volume of whole blood on top of the Ficoll-Hypaque.
3. Centrifuge at room temperature for 30 minutes at 2000 rpm in a swinging bucket rotor.
4. Use a sterile transfer pipette and collect the lymphocyte/monocyte fraction (opaque band located at the gradient interface) into a 2-ml Sarstedt tube.
5. Add phosphate-buffered saline (PBS) to the sample to 2 ml. Mix and spin in a microcentrifuge at 2500 rpm for 10 minutes to pellet the cells.
6. Remove the PBS. Resuspend cells in 2 ml PBS. Mix and spin.
7. At this point, DNA can be extracted from the pelleted cells or the cells can be frozen in RPMI plus 50% fetal calf serum and 10% DMSO at −70°C or in liquid nitrogen.

DNA Extraction

This procedure describes a method for the rapid extraction of DNA from peripheral blood mononuclear cells (Higuchi 1989; Jackson *et al.* 1989). The cells are lysed in *Taq* buffer containing Tween 20 and NP40 and digested with Proteinase K. The DNA extracted with this procedure can be amplified directly, thereby obviating the need for a phenol/chloroform extraction step.

1. If cells have been stored frozen, quick thaw at 37°C. Wash once with 1 to 2 ml cold RPMI plus 10% fetal calf serum. Pellet by centrifugation at 2500 rpm for 10 minutes. Wash twice more with PBS.
2. Resuspend the cells to a concentration of approximately 6×10^6 cells/ml in solution A.
3. Add an equal volume of solution B.
4. Incubate at 60°C for 1 hour.
5. Inactivate the Proteinase K by placing the tube in a 95°C water bath for 10 minutes.
6. Cool to room temperature and refrigerate.

DNA Amplification

Each sample is initially analyzed with a primer pair that amplifies a region of HLA-DQ alpha to determine if the DNA is of sufficient quality for PCR amplification (see Chapter 32). Samples are then amplified in duplicate with each of the HIV primer pairs. The conditions for optimal amplification need to be determined for each primer pair. The conditions recommended below were optimized for SK38–39 and SK145–101. Each amplification is performed in a volume of 100 μl.

1. Add 50 μl of mineral oil to 0.5-ml microcentrifuge tubes.
2. Add 50 μl of a reaction mix that consists of:
   > 5 μl of 10× Taq buffer
   > 50 pmol upstream primer
   > 50 pmol downstream primer
   > 2 μl of 10 μM dNTPs
   > 2 units of Taq polymerase (Perkin-Elmer Cetus)
   > glass-distilled water to 50 μl.

   The amount of 10× Taq buffer used will depend on the buffer in which the DNA is resuspended. The DNAs as prepared above are already in 1× Taq buffer.
3. Add 50 μl of sample DNA to each tube (should contain about 1 to 2 μg of DNA [the equivalent of 150,000 to 300,000 cells]). To minimize aerosolization, dispense the samples under the oil. Cap each sample after the addition of DNA before proceeding to the next.

4. Spin the tubes for a few seconds in a microcentrifuge to bring the liquid to the bottom.
5. Place the samples in a DNA Thermal Cycler (Perkin-Elmer Cetus). Cycle as follows:
    a. Denature at 95°C for 30 seconds
    b. Anneal at 55°C for 30 seconds
    c. Extend at 72°C for 1 minute

    Samples are amplified for 30 cycles. At the completion of the 30 cycles, the DNA is extended for an additional 10 minutes at 72°C.
6. Store samples at 4°C until ready to assay.

Detection of Amplified Products by Oligomer Hybridization (OH)

OH is a procedure whereby a ^{32}P-end-labeled probe hybridizes in solution to one strand of the amplified product. The probe-target duplex is separated from the unhybridized probe by gel electrophoresis and is then autoradiographed.

1. Add 50 μl of mineral oil to 0.5-ml microcentrifuge tubes.
2. Add 10 μl of a probe mix consisting of:

    4 μl 60 mM NaCl
    1 μl 40 mM EDTA (pH 8.0)
    0.2 pmol ^{32}P-end-labeled probe
    glass-distilled water to 10 μl

    The oligonucleotide probes are end labeled with [^{32}P]ATP (6000 mCi/mmol) to a specific activity of 1.5 to 3 μCi/pmol by using polynucleotide kinase (Maniatis et al. 1982). The end-labeled oligonucleotides are separated from the unincorporated ATP by spin column dialysis with G50 Sephadex (Maniatis et al. 1982).
3. Add 30 μl of amplified DNA.
4. Mix and centrifuge briefly to bring down the content.
5. Denature DNA in a 95°C bath for 5 minutes.
6. Anneal the probe and target sequences by incubating the tubes at 55°C for 15 minutes.
7. Add 10 μl bromophenol blue/xylene cyanol dye mix to each tube.

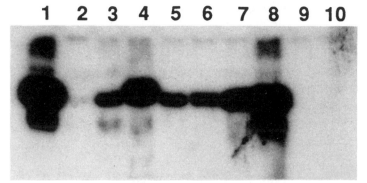

Figure 1 Representative OH analysis of samples amplified with SK38/39 and probed with SK19. Lanes 1–5 and 7–8 represent DNAs from HIV-seropositive individuals; lane 6, positive control; lane 9, seronegative individual; and lane 10, a reagent control.

8. Extract the mineral oil with 100 μl chloroform. The dye and sample will float to the top upon the addition of chloroform.

9. Load 25 μl of the sample onto a 10% polyacrylamide minigel.

10. Run in 1× TBE buffer at 100 V until the bromophenol blue dye approaches the bottom of the gel.

11. Excise the unhybridized probe from the gel by slicing just below the xylene cyanol front. Dispose of the bottom gel strip in the radioactive waste.

12. Blot the gel with tissue. Wrap the gel in plastic and expose to XAR 50 film overnight with an intensifying screen.

13. Develop the autoradiograph. The presence of a probe-target duplex is indicative of HIV infection. See Fig. 1 for a representative OH analysis.

Interpretation of Results

The HLA-DQ alpha amplifications serve as controls for these analyses. A sample can be interpreted as negative only if analysis of the DNA with the HLA primers indicates the DNA is suitable for amplification. The amount of HLA PCR product also provides a crude approximation of the number of cells being analyzed. The absence of an HLA product visible by ethidium bromide staining indicates either a presence of an inhibitor of PCR or an insufficient amount of cellular DNA. HIV-amplified products have been detected in the absence of ethidium bromide-stainable HLA amplification products, suggesting that although the amount of cellular DNA was small, the HIV molecules were at a level sufficient for detection by OH analysis. However, the absence of evidence for chromosomal DNA prevents the ability to exclude PCR carry-over as a source of HIV sequence detected.

Primer Pair		
SK38–39	SK145/150	Interpretation
+/+	+/+	Positive
–/–	–/–	Negative
+/+	+/–	
–/+	+/+	
+/+	–/–	Repeat, if pattern repeats:
–/–	+/+	positive
+/–	+/–	
–/–	–/+	Repeat, if pattern repeats:
+/–	–/–	indeterminate

Applications

Primer pairs SK38–39 and/or SK145–101 have been successfully used in the detection of HIV in greater than 96% antibody-positive individuals (Ou et al. 1988; Jackson et al., submitted; Lifson et al., in press). In addition, the identification of HIV by PCR has contributed to other areas of AIDS diagnostics and research. First, the procedure has been successfully used to test infants born to seropositive mothers (Laure et al. 1988; Rogers et al. 1989; Chadwick et al., in press). Since maternal antibodies can persist in the infants for up to 15 months, diagnosis of HIV infection in babies by serology is hampered. In several studies to date, detection of HIV by PCR in babies has correlated well with disease progression. Second, the procedure has been used to resolve the infection status of individuals with indeterminate serology (Jackson, MacDonald, Cadwell, Sullivan, Hanson, Sannerud, Stramer, Fildes, Kwok, Sninsky, Bowman, Polesky, Balfour, and Osterholm, submitted). Studies on low-risk individuals with indeterminate patterns suggest these individuals are not infected. Third, the procedure has detected HIV viral sequences in seropositive symptomatic individuals who were negative by other direct detection assays, including virus culture (Ou *et al.* 1988; Jackson *et al.* 1988). Fourth, the procedure has identified infection in a small number of high-risk individuals prior to seroconversion (Imagawa *et al.* 1989). Fifth, when PCR was used to screen seronegative sexual partners of HIV-1-infected hemophiliacs, none (0/22) was found to harbor HIV sequences, suggesting that the frequency of HIV-1 infection in antibody-negative sexual partners is probably very low. Sixth, PCR was used to confirm the first case of HIV-2 in a West African undergoing treatment in the United States (CDC 1988). In addition, the first documented case of a West African who was co-infected with HIV-1 and −2 was confirmed by PCR (Rayfield *et al.* 1988). Finally, PCR has been used to evaluate viral heterogeneity in HIV-1 isolates (Goodenow *et al.* 1989; Meyerhans *et al.*, in press).

Literature Cited

Centers for Disease Control. 1988. AIDS due to HIV-2 infection—New Jersey MMRW **37**:33−34.

Chadwick, E. G., R. Yogev, S. Kwok, J. J. Sninsky, D. E. Kellogg, and S. M. Wolinsky. Enzymatic amplification of the human immunodeficiency virus in peripheral blood mononuclear cells from pediatric patients. *J. Infect. Dis.*, in press.

Goodenow, M., T. Huet, W. Saurin, S. Kwok, J. Sninsky, and S. Wain-Hobson. HIV-1 isolates are rapidly evolving quasispecies: evidence for viral mixtures and preferred nucleotide substitutions. *JAIDS* **2**:344−352.

Higuchi, R. 1989. Simple and rapid preparation of samples for PCR. In *PCR technology: principles and applications for DNA amplification* (ed. H. Erlich), p. 31−38. Stockton Press, New York.

Imagawa, C. T., M. H. Lee, S. M. Wolinsky, K. Sano, F. Morales, S. Kwok, J. J. Sninsky, P. G. Nishanian, J. Giorgi, J. L. Fahey, J. Dudley, B. R. Visscher, and R. Detels. 1989. Long latency of human immunodeficiency virus-1 in seronegative high risk homosexual men determined by prospective virus isolation and DNA amplification studies. *N. Eng. J. Med.* **320**:1428−1462.

Jackson, J. B., K. J. Sannerud, J. S. Hopsicker, S. Y. Kwok, J. R. Edson, and H. H. Balfour. 1988. Hemophiliacs with HIV antibody are actively infected. *J. Am. Med. Assoc.* **260**:2236−2239.

Jackson, J. B., S. Y. Kwok, J. S. Hopsicker, K. J. Sannerud, J. J. Sninsky, J. R. Edson, and H. H. Balfour. 1989. Absence of HIV-1 infection in antibody-negative sexual partners of HIV-1 infected hemophiliacs. *Transfusion* **29**:265−267.

Kwok, S., D. H. Mack, J. J. Sninsky, G. D. Ehrlich, B. J. Poiesz, N. L. Dock, H. J. Alter, D. Mildvan, and M. H. Grieco. 1989. Diagnosis of human immunodeficiency virus in seropositive individuals: enzymatic amplification of HIV viral sequences in peripheral blood mononuclear cells. In *HIV detection by genetic engineering methods* (ed. P. A. Luciw and K. S. Steimer). Marcel Dekker, Inc., New York.

Kwok, S., and R. Higuchi. 1989. Avoiding false positives with PCR. *Nature (London)* **339**:237−238.

Laure, F., V. Courgnaud, C. Rouzioux, S. Blanche, F. Verber, M. Burgard, C. Jacomet, C. Griscelli, and C. Brechot. 1988. Detection of HIV-1 DNA in infants and children by means of the polymerase chain reaction. *Lancet* **2**:538–540.

Lifson, A. R., M. Stanley, J. Pane, P. M. O'Malley, J. Wilbur, A. Stanley, B. Jeffery, G. Rutherford, and P. R. Sohmer. Detection of HIV DNA using the polymerase chain reaction in a well characterized group of homosexual and bisexual men. *J. Infect. Dis.*, in press.

Maniatis, T., E. F. Fritsch, and J. Sambrook. 1982. In *Molecular cloning: a laboratory manual*. Cold Spring Harbor Laboratory, Cold Spring Harbor, New York.

Meyerhans, A., R. Cheynier, J. Alberts, M. Seth, S. Kwok, J. Sninsky, L. Morfeldt-Manson, B. Asjo, and S. Wain-Hobson. Temporal fluctuations in HIV quasispecies in vivo are not reflected by sequential HIV isolations. *Cell*, in press.

Ou, C.-Y., S. Kwok, S. W. Mitchell, D. H. Mack, J. J. Sninsky, J. W. Krebs, P. Feorino, D. Warfield, and G. Schochetman. 1988. DNA amplification for direct detection of HIV-1 in DNA of peripheral blood mononuclear cells. *Science* **238**:295–297.

Rayfield, M., K. De Cock, W. Heyward, L. Goldstein, J. Krebs, S. Kwok, S. Lee, J. McCormick, J. M. Moreau, K. Odehouri, G. Schochetman, J. Sninsky, and C.-Y. Ou. 1988. Mixed human immunodeficiency virus (HIV) infection in an individual: demonstration of both HIV type 1 and type 2 proviral sequences by using polymerase chain reaction. *J. Infect. Dis.* **158**:1170–1176.

Rogers, M. F., C.-Y. Ou, M. Rayfield, P. A. Thomas, E. E. Schoenbaum, E. Abrams, K. Krasinski, P. A. Selwyn, J. Moore, A. Kaul, K. T. Grimm, M. Bamji, G. Schochetman, and the New York City Collaborative Study of Maternal HIV Transmission and Montefiore Medical Center HIV Perinatal Transmission Study Group. 1989. Use of the polymerase chain reaction for early detection of the proviral sequences of human immunodeficiency virus in infants born to seropositive mothers. *N. Engl. J. Med.* **320**:1649–1654.

Saag, M. S., B. H. Hahn, J. Gibbons, Y. Li, E. S. Parks, W. P. Parks, and G. M. Shaw. 1988. Extensive variation of human immunodeficiency virus type-1 in vivo. *Nature (London)* **334**:440–444.

Shaw, G. M., B. H. Hahn, S. K. Arya, J. E. Groopman, R. C. Gallo, and F. Wong-Staal. 1984. Molecular characterization of human T-cell leukemia (lymphotropic) virus type III in the acquired immune deficiency syndrome. *Science* **226**:1165–1171.

41

DETECTION OF HEPATITIS B VIRUS

I. Baginski, A. Ferrie, R. Watson, and D. Mack

Hepatitus B virus (HBV) infection is diagnosed by the presence of HBsAg, anti-HBsAg, anti-HBcAg in the serum using radioimmuno-assays (RIA) or enzyme-linked immunoabsorbent assay (ELISA). The presence of the endogenous DNA polymerase activity and the find-ing of HBV DNA in the serum of patients with chronic hepatitis (CH) are indicative of active virus replication. In case of CH anti-HBs positives, the HBV DNA is generally detectable in the liver only. However, HBV sequences have been found in mononuclear blood cells. The detection of less than 0.1 pg of HBV DNA can be achieved by molecular hybridization with radiolabeled probes by slot-blot analysis. But in some cases of CH (anti-HBs positive), the HBV DNA cannot be detected in the serum due, presumably, to the low level of virus. In most of the cases of CH, PCR analysis was very useful because of its high sensitivity. The sensitivity of the PCR re-action combined with an analysis of the PCR product by Southern blot was 10-fold more than direct detection of HBV DNA in the serum by slot-blot hybridization (Mack and Sninsky, 1988). A series of studies have begun to document primer-probe combinations that have proven utility (Kaneko *et al.* 1989; Larzul *et al.* 1988; Sun *et al.* 1988; Theirs *et al.* 1988; Ulrich *et al.* 1989).

PCR Protocols: A Guide to Methods and Applications

Mack and Sninsky (1988) identified a region in the *pol* gene that is conserved on the different viruses of the Hepadnaviridae family and successfully amplified a 110 bp fragment by PCR. The oligonucleotide sequences of the specific primers, designated MD06/MD03, are complementary to the viral plus and minus strands, respectively. MD09 is the internal hybridization probe.

> MD06 (5')–CTTGGATCCTATGGGAGTGG
> MD03 (5')–CTCAAGCTTCATCATCCATATA
> MD09 (5')–GGCCTCAGTCCGTTTCTCTTGGC
> TCAGTTTACTAGTGCCATTTGTTC

The following pages describe procedures for extraction of the DNA from serum, mononuclear blood cells, and liver, the conditions for the HBV DNA amplification, and the analysis of PCR product by agarose gel electrophoresis and Southern blot hybridization. Other oligonucleotides used for the detection of the hepadnaviral genomes are tabulated (Table 1 and Table 2) and their positions indicated schematically in Figure 1.

Table 1

HBV Primers (Unique)

Primers designation (+/−)	Region	Position (adw2)
MD12/MD93	pol/sur	130-152
MD14/MD94	pol/sur	426-448
MD06/MD109	pol/sur	636-648
MD09/MD30	pol/sur probe	658-694
MD110/MD15	pol/sur	669-689
MD111/MD03	pol/sur	735-746
MD112/MD13	pol/sur	831-851
MD24/MD113	'X'	1389-1420
MD114/MD26	'X'	1606-1630
MD27/MD115	pre-core	1858-1876
MD118/119	pre-core	1873-1890
MD28	pre-core probe	1890-1920
MD116/MD25	core	1957-1975
MD75/MD76	core	2374-2397
MD79	core probe	2398-2425
MD77/MD78	core	2374-2397
MD80/MD81	pre-S1	2823-2841
MD84	pre-S1 probe	2908-2951
MD83/MD82	pre-S2	3177-3199

Table 2

Sequence of Synthetic Oligonucleotides

Primer designation	Sequence (5'–3')
MD03	CTCAAGCTTCATCATCCATATA
MD06	CTTGGATCCTATGGGAGTGG
MD09	GGCCTCAGTCCGTTTCTCTTGGCTCAGTTTACTAGTGCCATTTGTTC
MD12	GCGGGATCCGGACTGGGGACCCTGTGACGAAC
MD13	GCGAAGCTTGTTAGGGTTTAAATGTATACCC
MD14	GCGGGATCCCATCTTCTTATTGGTTCTTCTGG
MD15	GCGAAGCTTAATGGCACTAGTAAACTGAGCC
MD24	TGCCAACTGGATCCTTCGCGGGACGTCCTT
MD25	GCGAAGCTTAAGGAAAGAAGTCMGAAGG
MD26	GCGAAGCTTGTTCACGGTGGWCTCCATG
MD27	GCGGGATCCACTGTTCAAGCCTCCAAGCT
MD28	GCTTGGGTGGCTTTGGGGCATGGACATTGACCCTTATAA
MD30	CGGAGTCAGGCAAAGAGAACCCAGTCAAATGATCACGGTAAACAAG
MD75	AGAAGAAGAACTCCCTGCCCTCGG
MD76	CGGAGGCGAGGGAGTTCTTCTTCT
MD77	GAAGATCTCAATCTCGGGAATCTC
MD78	GAGATTCCCGAGATTGAGATCTTC
MD79	AGACGCAGATCTCCATCGCCGCGTCGCA
MD80	ACCATATTCTTGGGAACA
MD81	TGTTCCCAAGAATATGGT
MD82	CTCCACCTCTAAGAGACAGTCA
MD83	TGACTGTCTCTTAGAGGTGGAG
MD84	CAATCCTCTGGGATTCTTTCCCGATCATCAGTTGGACCCTGCA
MD93	GTTCGTCACAGGGTCCCCAGTCC
MD94	CCAGAAGAACCAATAAGAAGATE
MD109	CTCAAGCTTCCACTCCCATAGG
MD110	CTCGGATCCGGCTCAGTTTACTAGTGCCATTT
MD111	CTCGGATCCTATATGGATGAT
MD112	CTCGGATCCGGGTATACATTTAAACCCTAA
MD113	AAGGACGTCCCGCGAAGGATCCAGTTGGCA
MD114	CTCGGATCCATGGAGACCACCGTGAAC
MD115	CTCAAGCTTAGCTTGGAGGCTTGAACAGT
MD116	CTCGGATCCTTGTGACTTCTTTCCTT
MD118	GCGGGATCCGCTGTGCCTTGGRTGG
MD119	GCGAAGCTTCCAYCCAAGGCACAGC

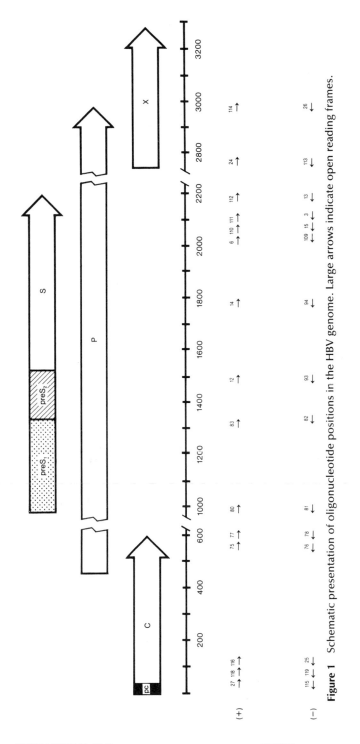

Figure 1 Schematic presentation of oligonucleotide positions in the HBV genome. Large arrows indicate open reading frames.

Protocols

Reagents

TEN buffer:	10mM Tris-HCl (pH 8)
	1 mM EDTA
	10mM NaCl
10X TAQ buffer:	500 mM KCl
	100 mM Tris-HCl (pH 8.4)
	25 mM MgC1$_2$
	0.1% (w/v) gelatin

40 mM dNTPs mix: 10 mM each of dATP, dCTP, dGTP, dTTP

DNA Extraction from the Serum

This procedure describes a method for quick preparation of DNA from the serum. The advantage of this method is the elimination of a phenol-chloroform extraction step.

1. To 20 μl of serum, add proteinase K at 250 μg/ml in 25 μl of 0.25% SDS, 5mM EDTA, 10 mM Tris-HCl (pH 8).
2. Incubate for 2 hours at 56°C.
3. Heat-inactivate the proteinase K at 95°C for 10 minutes.
4. Bring up to 450 μl in 1X TAQ buffer with 0.5% Tween 20 and 0.5% NP40.
5. Use 50 μl of above preparation for the PCR assay.
6. Store at 4°C.

DNA Extraction from Peripheral Blood Mononuclear Cells (PBMC)

Refer to Chapter 18 for rapid preparation of DNA from PBMC.

DNA Extraction from Liver Biopsy

1. Crush the biopsy specimen in liquid nitrogen.
2. In a 1.5 ml microcentrifuge tube add 0.2 ml TEN buffer, 4 μl proteinase K at 10 mg/ml, and 5 μl 20% SDS, then add the crushed biopsy.
3. Mix gently.
4. Incubate at 55°C for 4 hours or overnight at 37°C under moderate agitation.
5. Deproteinaze as follows:
   Add an equal volume of phenol-chloroform to the DNA. Cap tube and mix using a rocking motion for 10 minutes. Centrifuge for 10 minutes at 1,000 g at room temperature. Remove the lower phenol-chloroform layer and discard.
6. Repeat step 5 until the aqueous phase is clear.
7. Add 1/10 of 3 M sodium acetate and then two volumes of absolute ethanol.
8. Precipitate overnight at −20°C.

DNA Amplification

These conditions were optimized with the primers pair MD03/06. Each amplification reaction is carried out in 100 μl.

1. In a 0.5 ml microcentrifuge tube, add 50 μl of a reaction mix containing
   100 pmoles of MD03
   100 pmoles of MD06
   5 μl of 10X TAQ buffer
   2 units of TAQ polymerase
   200 μM each of dNTP
   H_2O to 50 μl
   The amount of $MgCl_2$ in the 10X TAQ buffer depends on the buffer in which the DNA is resuspended.
2. Add 50 μl of serum DNA to each tube or 1 μg of total DNA if it was extracted from PBMC or liver biopsy.

3. Cap the tube and vortex a few seconds to mix.
4. Microcentrifuge a few seconds.
5. Open the tube carefully and then add 50 μl of mineral oil.
6. Place sample in the thermal cycler and perform 30 cycles as follows:

   Denature at 95°C for 30 seconds.
   Anneal at 55°C for 30 seconds.
   Extend at 72°C for 1 minute.
7. Store sample at 4°C until ready to use.

Analysis of Amplified Product

The amplified DNA is fractionated by agarose gel electrophoresis, visualized under UV light by ethidium bromide fluorescence, and then analyze by Southern hybridization.

Agarose Gel Electrophoresis

The expected size of the amplified product using the MD03/06 primers is 110 bp; a NuSieve agarose gel will give the best resolution of a fragment of this size.

1. Prepare a 3% NuSieve agarose/ 1% Seakem agarose gel in TBE (leave the gel at least ½ hour at 4°C before using it).
2. Extract the mineral oil from the sample with 100 μl of chloroform.
3. To 15 μl of the PCR reaction mixture add 2 μl of bromophenol blue/xylene cyanol dye mix.
4. Load the sample and carry out electrophoresis in 1X TBE buffer containing 0.075 μg/ml ethidium bromide at 100 volts until the bromophenol blue dye approaches the bottom of the gel.
5. Examine the DNA fragments using a long wave UV light.

Southern Blot and Hybridization

1. Soak the gel twice for 10 minutes with 0.5 N NaOH, 1.5 M NaCl under agitation.
2. Transfer the DNA to a nylon membrane overnight in 20X SSPE buffer.

3. Crosslink the DNA to the membrane by exposure to UV.
4. Prehybridize the membrane in 3X SSPE, 5X Denhardt's solution, 0.5 SDS, 30% formamide for ½ to 1 hour at 42°C and then hybridize in the same solution containing 0.5 pmole of ^{32}P-end-labeled probe for 3–5 hours at 42°C.

   Note that these hybridization conditions are optimized for the 44-mer MD09 oligonucleotide probe end-labeled with ^{32}P-ATP (6000 Ci/mmole) to a specific activity of 1.5–3 μCi/pmole using polynucleotide kinase. The end-labeled probe is separated from the unincorporated ATP by spin column dialysis with G50 Sephadex.
5. Remove the nonspecific bound probe by washing the membrane twice in 2X SSPE, 0.1% SDS for 10 minutes at room temperature, and once in 0.2X SSPE, 0.1% SDS.
6. Dry-air the membrane, wrap it in plastic, and expose it to XAR 50 film several hours to overnight at −70°C with an intensifying screen.
7. Develop autoradiograph.

Literature Cited

Kaneko, S., R. H. Miller, S. M. Feinstone, M. Unoura, K. Kobayashi, N. Hattori, and R.H. Purcell. 1989. Detection of serum hepatitis B virus DNA in patients with chronic hepatitis using the polymerase chain reaction assay. *Proc. Natl. Acad. Sci. USA* **86**:312–316.

Larzul, D., F. Guigue, J. J. Sninsky, D. H. Mack, C. Brechot, and J.-L. Guesdon. 1988. Detection of hepatitis B virus sequences in serum by using in vitro enzymatic amplification. *Jr. Vir. Meth.* **20**:227–237.

Sun, C.-F., C. C. Pao, S.-Y. Wu, and Y.-F. Liaw. 1988. Screening for hepatitis B virus in healthy blood donors by molecular DNA hybridization analysis. *J. Clin. Microbiol.* **28**:1848–1852.

Mack, D. H., and J. J. Sninsky. 1988. A sensitive method for the identification of uncharacterized viruses related to known virus groups: Hepadnavirus model system. *Proc. Nat. Acad. Sci. USA* **85**:6977–6981.

Theirs, V., D. Kremsdorf, H. Schellekens, A. Goudeau, J. Sninsky, E. Nakajima, D. Mack, F. Driss, J. Wands, P. Tiollais, and C. Brechot. Transmission of hepatitis B from hepatitis-B-seronegative subjects. *Lancet* 1273–1276.

Ulrich, P. P., R. A. Bhat, B. Seto, D. Mack, J. Sninsky, and G. N. Vyas. 1989. Enzymatic amplification of hepatitis B virus DNA in serum compared with infectivity testing in chimpanzees. *J. Infec. Dis.* **160**:37–43.

DETECTION AND TYPING OF GENITAL HUMAN PAPILLOMAVIRUSES

Yi Ting and M. Michele Manos

The genital human papillomaviruses (HPVs) are a group of over 20 distinct virus types that are associated with a number of diseases and cancers (for review see Pfister 1987; zur Hausen and Schneider 1987). Distinct clinical manifestations are associated with different genital HPV types. For instance, types 16 and 18 are often found in cervical dysplasia and carcinoma, while types 6 and 11 are associated with benign condylomas. The detection and typing of genital HPVs in normal and diseased tissue samples are important in studies concerned with defining the role these viruses play in cancers and other abnormalities. The PCR-based methods described here provide a sensitive, accurate means of detecting a broad spectrum of genital HPVs.

The genomes of the genital HPV types are each unique, yet they share interspersed regions of DNA sequence homology, particularly within the open reading frames E1 and L1. Although over 20 genital HPV types have been identified, the complete DNA sequences of only 5 have been reported. By comparing the DNA sequences of genital HPV types 6, 11, 16, 18, and 33 (Schwarz *et al.* 1983; Dartmann *et al.* 1986; Seedorf *et al.* 1985; Cole and Danos 1987; Cole and

PCR Protocols: A Guide to Methods and Applications

Streeck 1986), we identified regions of homology 20 to 25 bp in length. From such regions we designed sets of "consensus" PCR primers that will amplify distinct regions from over 25 of the genital HPVs.

In this chapter we describe the use of an L1 consensus primer pair for the amplification of genital HPV DNA sequences, and techniques for the detection and analysis of the amplification products. This approach was first described in Manos *et al.* (1989). Tables 1 and 2 show the sequences of oligonucleotide primers and probes used in this technique. Consensus primers MY11 (primer for the positive strand) and MY09 (primer for the negative strand) are degenerate in several positions (at which one or more possible nucleotides

Table 1

L1 Consensus Primers

Positive Strand Primer	
MY11	GCMCAGGGWCATAAYAATGG
Position of first base	
HPV06	6722
HPV11	6707
HPV16	6584
HPV18	6558
HPV33	6539
Negative Strand Primer	
MY09	CGTCCMARRGGAWACTGATC
Position of first base	
HPV06	7170
HPV11	7155
HPV16	7035
HPV18	7012
HPV33	6987
Predicted Size of L1 PCR Products	
HPV06	448bp
HPV11	448bp
HPV16	451bp
HPV18	454bp
HPV33	448bp

M = A + C
R = A + G
W = A + T
Y = C + T

Table 2

Probes for L1 PCR Products

"Generic" Probe Mix			
GP1	CTGTTGTTGATACTACACGCAGTAC		
GP2	CTGTGGTAGATACCACWCGCAGTAC		
First base position			
HPV6	6771		
HPV11	6756		
HPV16	6631		
HPV18	6607		
HPV33	6588		
Type-Specific Probes			
Probe	Specificity	Sequence	Genome position
MY12	HPV6	CATCCGTAACTACATCTTCCA	6813-6833
MY13	HPV11	TCTGTGTCTAAATCTGCTACA	6800-6820
MY14	HPV16	CATACACCTCCAGCACCTAA	6926-6945
WD74	HPV18	GGATGCTGCACCGGCTGA	6905-6922
MY16	HPV33	CACACAAGTAACTAGTGACAG	6628-6648

H = A + C + T
R = A + G
W = A + T
Y = C + T

is inserted during synthesis of the oligonucleotide) to render them almost completely complimentary to each of the five sequenced HPVs (see Table 1). An additional conserved region, within the approximately 450 bp between the consensus primer regions, was used to define the consensus or "generic" probe mix. This "generic" probe mix was designed to serve as a probe for the detection of HPV L1 PCR products generated by MY11 and MY09. Oligonucleotides MY12, MY13, MY14, WD74, and MY16 serve as type-specific probes for the L1 PCR products of HPVs 6, 11, 16, 18, and 33, respectively. Each of these type-specific probes corresponds to a region that is unique to the corresponding type and that is not significantly homologous to the predicted PCR product of other types.

There are several approaches available to identify and analyze the PCR products generated from MY11 and MY09 amplification. Acrylamide or agarose gel electrophoresis of a fraction of the amplification reaction followed by ethidium bromide staining allows visualization of the PCR products from most samples that contain at least 200 copies of an HPV. Hybridization analyses with the generic and/or type-specific probes can be done using a Southern blot or dot blot format. If the amplification reaction has produced a relatively pure HPV PCR product, aliquots of the PCR can be digested with different restriction enzymes to yield the unique digestion pattern that is specific for each virus PCR product.

The L1 consensus primers MY11 and MY09 promote the amplification of an approximately 450-bp product from at least 25 distinct genital HPVs. The L1 generic probe mix described here hybridizes with many but not all of these HPVs. We are currently improving the generic probe mix to increase the spectrum of L1 PCR products with which it hybridizes. Type-specific probes for HPVs 6, 11, 16, 18, and 33 are described here. As the sequences of additional genital HPVs become available, corresponding probes will be designed. The system has been used to detect HPV in a sample containing as few as 10 copies of an HPV. The type-identification methods of the system are accurate; they agree with types determined by other methods such as *in situ* hybridization and Southern blotting.

Protocols

PCR

This procedure outlines the methods for the amplification of HPV DNA present in purified samples of DNA (e.g., HPV plasmids or DNA prepared from cell lines or clinical material), crudely prepared clinical material, or paraffin-embedded tissues. Sample preparation methods for clinical material and for paraffin-embedded tissues are thoroughly discussed in Chapters 18 and 19, respectively.

Reagents and Materials

100 μM stocks of MY11 and MY09 (see Table 1)

10× PCR buffer [500 mM KCl, 40 mM MgCl$_2$, 100 mM Tris (pH 8.5)]

10 mM dNTP stock (containing 10 mM each of dATP, dCTP, dGTP, dTTP)

5 units/μl of *Taq* polymerase

Sterile, glass-distilled H_2O

Mineral oil

Each PCR will contain:

sample DNA (crude or purified) or control DNA

500 nM MY11

500 nM MY09

200 μM dNTPs

2.5 units of *Taq* polymerase

PCR buffer

glass-distilled, sterile H_2O to bring the volume to 100 μl

HPV-positive control(s) and HPV-negative control(s) should be included in each experiment. Recombinant plasmids carrying HPVs (100 pg per reaction) or DNAs (15 ng per reaction) from cell lines containing HPVs (e.g., SiHa, HeLa) are useful positive controls. DNA from a human cell line with no HPV (such as K562) can be used as a negative control. In many of the HPV recombinant plasmids, the L1 region (between the MY11 and MY09 areas) is interrupted by plasmid sequences. In those cases, it will be necessary to excise the viral DNA from the plasmid and recircularize it by ligation to facilitate amplification.

1. Determine the total number of reactions to be run. Include at least one reaction that contains no DNA (water in place of sample).

2. Prepare a cocktail that contains all reagents for the reaction excluding sample DNA. Prepare enough cocktail for the total number of reactions plus one. A typical cocktail would contain *per reaction*:

10 μl of 10× PCR buffer

0.5 μl each of MY11 and MY09 (100-μM stocks)

2 μl of dNTP stock

0.5 μl of *Taq* polymerase (2.5 units)

75.5 μl of H_2O

90 μl total

3. Add 80 to 100 μl of mineral oil to each tube.

4. Dispense 90 μl (or other calculated amount) of reaction cocktail into each PCR tube (prelabeled with sample number).

5. Add 10 μl (or other calculated amount) of sample DNA to each labeled tube and close tube carefully.

6. Spin the tubes for 10 seconds in a microcentrifuge to collect all reaction components under the oil overlay.
7. Thermocycling conditions are as follows:

	Purified DNA or crudely prepared fresh clinical research material	Prepared paraffin-embedded tissue
Denature	95°C, 30 seconds	95°C, 1 minute
Annealing	55°C, 30 seconds	55°C, 1 minute
Extension	72°C, 1 minute	72°C, 2 minutes
Number of cycles	30	40

Extend for an additional 3 to 5 minutes at 72°C.

> NOTE: When analyzing uncharacterized samples for the presence of HPV, the use of a control PCR primer pair (for amplification of a single-copy human gene that is certain to be present in the sample) is essential. Successful amplification with a control primer pair ensures that a sample is "amplifiable" (DNA sufficiently intact, no inhibitors present) and that if an appropriate HPV were present in sufficient copy number, it would have been amplified. This is crucial for interpreting the results from samples that yield no HPV product. The PCR product of the control primer pair should be sufficiently different in size from that of the L1 HPV product. We recommend a control product size of 200 to 300 bp. The control primer pair (e.g., for actin or β-globin) may be included in the L1 amplification reaction. Control primers should be present at 50 nM concentrations each (1/10 that of the L1 primers). Alternatively, amplification with control primers may be performed in a separate reaction with aliquots of samples equal to those used in the L1 amplifications.

Analysis of PCR Products

Reagents and Materials

Hybridization Buffer: 6× SSC, 5× Denhardt's solution, 0.1% SDS, 100 μg/ml of sheared single-stranded salmon sperm DNA

^{32}P-kinased GP1 and GP2 (L1 consensus probe mix)
^{32}P-kinased oligonucleotides MY12, MY13, MY14, WD74, MY16
Blot wash solution (WASH): 2× SSC, 0.1% SDS

Gel Electrophoresis and Visualization

An aliquot representing 1/20 to 1/10 of the PCR can be analyzed by gel electrophoresis and ethidium bromide staining. The HPV amplification products are of about 450 bp in size and are resolved well on 5 to 7% polyacrylamide gels or 1.5% agarose. Polyacrylamide gels provide better resolution and greater sensitivity. Figure 1a illustrates the visualization of PCR products on acrylamide. Although the amplified products of the different genital HPV types are about the same size, some of them have aberrant mobilities on polyacrylamide gels. For instance, as seen in Fig. 1a, the amplification product of HPV33 migrates significantly slower than what would be predicted by its actual length. Some samples carrying more than one type of HPV will yield two or more distinct PCR product bands on the gel. After electrophoresis, Southern blot hybridization can be performed with the generic and/or type-specific probes. Conditions suggested for dot blot hybridization (below) can be used. Unlike dot blot hybridization, Southern blot analyses provide information about exactly what DNA in the PCR (i.e., what size DNA fragment) is hybridizing with the probe. This is particularly useful with paraffin-embedded tissue samples, where background amplification bands may preclude any interpretation of data from ethidium bromide staining. Of course, hybridization (Southern or dot blot) provides increased sensitivity as well.

Dot Blot Hybridization

Aliquots of the PCR (1/50 to 1/20 of the total) can be denatured and applied to dot blot membranes by standard procedures. Positive and negative control reactions should be included. Replicate membranes can be used for hybridizations with each of the type-specific probes of interest and with the generic probe mix. Oligonucleotide probes are kinased with [^{32}P]ATP by standard protocols to achieve a specific activity of about 10^7 cpm/pmol. An example of a dot blot analysis is shown in Fig. 1b.

1. Hybridize the membranes in hybridization buffer containing 2×10^6 to 5×10^6 cpm of probe per ml. In the case of the generic

A

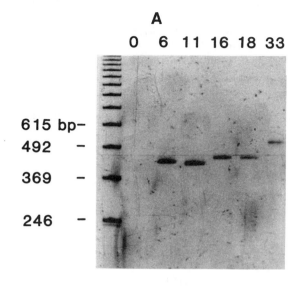

B

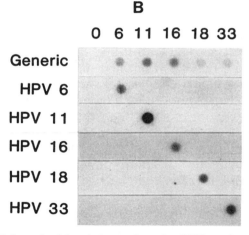

Figure 1 (A) Polyacrylamide gel electrophoresis of PCR products. Control DNAs representing the HPVs noted were amplified as described in the text, and 1/20 of each reaction was analyzed by electrophoresis on a 7% acrylamide gel. Purified genomic DNAs of K562 (0), Caski (16), and HeLa (18) or viral DNAs from recombinant plasmids (6, 11, 33) were used. Gels were stained with ethidium bromide and photographed under UV light. A negative of a photograph is shown. (B) Dot blot hybridization analyses. 1/50 of each amplification reaction was applied to each of six replicate filters. Each filter was hybridized with a ^{32}P-labeled oligonucleotide probe as described in the text and autoradiographed for 2 to 5 hours without an intensifying screen. Probes used for hybridization are noted at the left of each filter strip.

probe mix, or when using a combination of type-specific probes, use that amount of each labeled oligonucleotide. Incubate the hybridization at 55°C for 2 to 3 hours, with gentle shaking. Wash the filters in WASH (described previously) as follows.

2. Quickly rinse the membrane in 30 to 55°C WASH to remove excess probe.

3. For the generic probes: Wash in WASH at 55°C (in rotating water bath) for 10 minutes.
   For probes MY12, MY13, and MY16: Wash in WASH at 56 to 57°C for 10 minutes. Replace WASH and repeat.
   For probes MY14 and WD74: Wash in WASH at 58 to 59°C for 10 minutes. Replace WASH and repeat.

4. Autoradiography.

Restriction Enzyme Analysis of PCR Products

This analysis can only be performed on samples in which the predominant or only component of the completed amplification reaction that is larger than about 100 bp is the desired HPV product. This precludes the use of products that have much background amplification or reactions containing an internal control primer product. Digestion of the PCR product with the suggested enzymes (separately) should yield a restriction enzyme digestion pattern that is unique for each virus type (examples are shown in Fig. 2). A subtle difference between two patterns (i.e., a difference that suggests the gain or loss of one site for one enzyme) should probably be interpreted as reflecting a variant of a type rather than a distinct type.

This digestion pattern analysis can be used for typing small numbers of samples. The predicted patterns of types 6, 11, 16, 18, and 33 can be deduced from their DNA sequences in the region of the L1 PCR product. Patterns from unknown samples can be matched to the predicted patterns of the sequenced viruses or from the experimentally determined patterns of other known or yet undefined types. This is particularly useful for sorting out potential new types, so that only those that are unique to the laboratory (or to the literature) are pursued further; so the same new virus is analyzed only once in a collection of samples.

An aliquot of a PCR product that is sufficient for visualization on a gel must be used for each digestion. Some samples will not yield enough PCR product for this type of analysis. The suggested enzymes for the analysis of the L1 product include *Rsa*I, *Dde*I, *Hae*III,

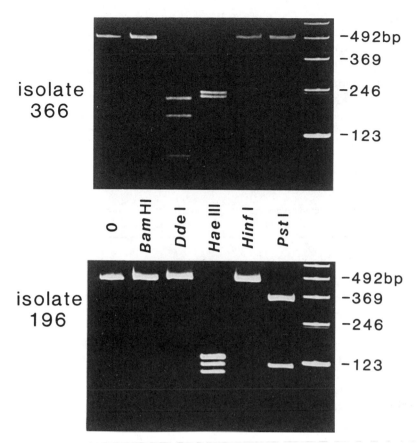

Figure 2 Example of restriction enzyme digestion patterns of two unidentified HPV types. Aliquots of amplification reactions were digested with the enzymes noted and then analyzed on a 5% acrylamide gel. A zero (0) denotes undigested.

*Hin*fI, *Pst*I, and *Sau*3a. Digestion with any of these enzymes will work in one simple buffer containing 100 mM NaCl, 10 mM MgCl$_2$, 10 mM Tris (pH 7.5), and 10 mM β-mercaptoethanol. Therefore one can make a cocktail of PCR product and buffer and then add the different restriction enzymes to aliquots of the cocktail. The PCR product does not need to be purified in any way; an aliquot of an amplification reaction can comprise up to one-half the volume of a digestion.

Table 3

Results	Conclusions
no PCR product detectable on gel or by hybridization	HPV not present, present at level below detection limit; or virus present that is not detectable by L1 primers
PCR product detected and hybridized with type-specific probe(s) and generic probe	One or more of types 6, 11, 16, 18, 33 is present, at least
PCR product detected and hybridized with generic probe, but not with any type-specific probe	Other known or novel type(s) is present
PCR product detected on gel, but no hybridization with generic probe	Other known or novel type(s) not detectable by generic probe is present

Analysis and Interpretation of Results

A guide to the analysis of results is shown in Table 3. Again we stress the importance of a control primer pair, as discussed previously.

Acknowledgments

We thank the members of the Cetus HPV Group, particularly Deann Wright and Catherine Greer, for their hard work in developing the L1 amplification scheme. We are grateful, as always, to L. Goda, D. Spasic, and C. Levenson of the Cetus DNA Synthesis Group for providing oligonucleotides.

Literature Cited

Cole, S. T., and O. Danos. 1987. Nucleotide sequence and comparative analysis of the human papillomavirus type 18 genome. Phylogeny of papillomaviruses and repeated structure of the E6 and E7 gene products. *J. Mol. Biol.* **193**:599–608.

Cole, S. T., and R. E. Streeck. 1986. Genome organization and nucleotide sequence of human papillomavirus type 33, which is associated with cervical cancer. *J. Virol.* **58**:991–995.

Dartmann, K., E. Schwarz, L. Gissmann, and H. zur Hausen. 1986. The nucleotide sequence and genome organization of human papilloma virus type 11. *Virology* **151**:124–130.

Manos, M. M., Y. Ting, D. K. Wright, A. J. Lewis, T. R. Broker, and S. M. Wolinsky. 1989. The use of polymerase chain reaction amplification for the detection of genital human papillomaviruses. *Cancer Cells* 7 : 209–214.

Pfister, H. 1987. Human papillomaviruses and genital cancer. *Adv. Cancer Res.* **48** : 113–147.

Schwarz, E., M. Dürst, C. Demankowski, O. Lattermann, R. Zech, E. Wolfsperger, S. Suhai, and H. zur Hausen. 1983. DNA sequence and genome organization of genital human papillomavirus type 6b. *EMBO J.* **2** : 2341–2348.

Seedorf, K., G. Krammer, M. Dürst, S. Suhai, and W. G. Rowekamp. 1985. Human papillomavirus type 16 DNA sequence. *Virology* **145** : 181–185.

zur Hausen, H. and A. Schneider. 1987. The role of papillomaviruses in human anogenital cancer. In *The papovaviridae: the papillomavirus* (ed. N.P. Salzman and P.M. Howley), p. 245. Plenum Publishing Corp., New York.

43

DETECTION OF HUMAN CYTOMEGALOVIRUS

Darryl Shibata

Human cytomegalovirus (CMV) is a common member of the herpes-virus group and an important pathogen in immunocompromised patients. Its genome consists of a single 240-kb linear double-stranded DNA molecule. Selective portions of the CMV genome have been sequenced and are therefore available for the generation of PCR primers and probes. The genomic diversity of CMV is unknown, although restriction fragment length polymorphisms are present. The use of a single primer pair (shown below) for CMV will amplify more than 90% of all wild-type isolates. However, the use of two primer sets against different CMV sequences may ensure the detection of virtually all CMV isolates (Shibata *et al.* 1988).

Protocols

Primers and Probes (Shibata *et al.* 1988 [other CMV PCR primer sets are given in Demmler *et al.* (1988)]

CMV immediate early gene (Stenberg *et al.* 1984) (159-bp PCR product)

IE1 CCACCCGTGGTGCCAGCTCC upstream primer

IE2 CCCGCTCCTCCTGAGCACCC downstream primer

IE3 CTGGTGTCACCCCCAGAGTCCCCTGTACCCGCGACTATCC hybridization probe

CMV late antigen gp64 (Pande *et al.* 1984; Rüger *et al.* 1987) (139-bp PCR product)

LA1 CCGCAACCTGGTGCCCATGG upstream primer

LA2 CGTTTGGGTTGCGCAGCGGG downstream primer

LA3 TTCTTCTGGGACGCCAACGACATCTACCGCATCTTCGCCG hybridization probe

Reagents

Taq DNA Polymerase	5 units/μl (Perkin-Elmer Cetus)
10× Reaction Buffer	100 mM Tris–HCl (pH 8.4)
	500 mM KCl
	25 mM MgCl₂
	2 mg/ml of gelatin
10× dNTPs	2.0 mM of each of four dNTPs
Primers	100 pmol/μl
Reaction Mix (for 100-μl reaction)	
10× Reaction Buffer 10	μl
10× dNTPs	10 μl
Taq polymerase	0.2—0.4 μl (1 to 2 units)
Primers	0.5 μl each (50 pmol/reaction)

Sample and distilled water to 100 μl, overlay with mineral oil (1 to 2 drops).

(Two primer pairs are usually used in one reaction, one for a CMV sequence and one for a genomic sequence.)

Thermal Profile

The tube (including *Taq* enzyme) is first heated at 94°C for 7 minutes before cycling at 95°C for 25 seconds, 42°C for 15 seconds, and 72°C for 60 seconds. Fifty cycles are usually performed.

Hybridization and Washing Conditions

The prehybridization solution is 3× SSPE, 5× Denhardt's, 0.5% SDS, and 25% formamide. Prehybridization is at 42°C for 30 to 60 minutes.

Labeled probe is added (10 cpm/μg, 2 ng/ml) and hybridized for 30 to 60 minutes. The filter is washed with 0.2× SSPE with 0.1% SDS for 5 minutes, three times at room temperature, once at 60°C for 10 minutes, and once more for 5 minutes at room temperature.

Specimens for CMV Detection

Blood-DNA may be isolated by conventional methods or by the quick extraction method described in Chapter 18. Tissue-DNA may be isolated by conventional methods or from fixed tissue by the method of Manos (Chapter 19). Alternatively, in the clinical laboratory, tissue is often frozen and stored in OCT at −20 to −70°C. This tissue may be quickly processed for the PCR by the following method.

1. Cut one 5- to 10-μm slice of OCT frozen tissue on a cryostat and place in a 1.5-ml Eppendorf tube.
2. Add 10% buffered formalin for 10 minutes.
3. Centrifuge in a microfuge for 1 to 2 minutes, decant, and wash twice with ethanol.
4. Desiccate (usually 10 to 60 minutes).
5. Add an extraction solution (100 mM Tris−HCl, 4 mM EDTA (pH 8.0), 400 μg/ml Proteinase K) to cover the tissue pellet (usually 50 to 100 μl). Disrupt the tissue pellet with the pipette tip. (Single slices of formalin-fixed, paraffin-embedded tissue can also be used after they are deparaffinized and desiccated.)
6. Incubate at 37°C overnight.
7. Boil for 7 minutes to inactivate the Proteinase K. Residual tissue fragments may still be present.
8. Centrifuge and use the supernatant (1 to 10 μl) as the PCR substrate. The extraction may be monitored by visualizing the genomic DNA on a 0.6% agarose, ethidium bromide-stained gel. The DNA is usually degraded in size.

Urine-DNA can be purified from urine by conventional techniques, or used directly (Demmler *et al.* 1988).

Comments

Reaction conditions can be varied for optimal sensitivity. Clinical samples may vary in amplification efficiency, and amplification for a genomic sequence such as globin should be included to prevent false negatives (i.e., CMV present, but like a genomic sequence, not amplified because of poor PCR efficiency).

CMV infections are characterized by latency, and the majority of adults are infected. Therefore, interpretation of a positive CMV PCR may be difficult. However, CMV is more readily detected from the blood or tissues of immunocompromised patients. Refinements of the PCR may allow the detection of latent CMV in normal patients. Conversely, CMV can be detected by the PCR from virtually all tissues in some immunocompromised patients having evidence of CMV viremia (Shibata *et al.*, in press).

Literature Cited

Demmler, G. J., G. J. Buffone, C. M. Schimbor, and R. A. May. 1988. Detection of cytomegalovirus in urine from newborns by using polymerase chain reaction DNA amplification. *J. Infect. Dis.* **158**: 1177–1184.

Pande, H., S. W. Baak, A. D. Riggs, B. R. Clark, J. E. Shively, and J. A. Zaia. 1984. Cloning and physical mapping of a gene fragment coding for a 64-kilodalton major late antigen of human cytomegalovirus. *Proc. Natl. Acad. Sci. USA* **81**: 4965–4969.

Rüger, B., S. Klages, B. Walla, J. Albrecht, B. Fleckenstein, P. Tomlinson, and B. Barrell. 1987. Primary structure and transcription of the genes coding for the two virion phosphoproteins pp65 and pp71 or human cytomegalovirus. *J. Virol.* **61**: 446–453.

Shibata, D. and E. C. Klatt. Analysis of HIV and CMV infection by polymerase chain reaction in the acquired immunodeficiency syndrome: an autopsy study. *Arch. Pathol. Lab. Med.*, in press.

Shibata, D., W. J. Martin, M. D. Appleman, D. M. Causey, J. M. Leedom, and N. Arnheim. 1988. Detection of cytomegalovirus DNA in peripheral blood of patients infected with human immunodeficiency virus. *J. Infect. Dis.* **158**: 1185–1192.

Stenberg, R. M., D. R. Thomsen, and M. F. Stinski. 1984. Structural analysis of the major immediate early gene of human cytomegalovirus. *J. Virol.* **49**: 190–199.

PCR AMPLIFICATION OF ENTEROVIRUSES

Harley A. Rotbart

The human enteroviruses (EVs) comprise a genus of more than 60 serotypes of the family *Picornaviridae*. Included among these viruses are the polioviruses, coxsackieviruses, echoviruses, and the "newer" numbered EVs, e.g., hepatitis A (enterovirus 72). With the advent of vaccines, the polioviruses are less clinically significant in developed countries but continue to cause paralysis in 4 of every 1000 children born in underdeveloped countries. In the United States, the nonpolio EVs account for 10 to 30 million infections annually, the vast majority in children, making these viruses the most common and among the most important viral infections in pediatrics. The spectrum of disease due to these agents ranges from benign febrile illness to meningitis, myocarditis, and overwhelming neonatal sepsis.

The traditional diagnostic approach to the EVs has been with tissue culture. This method may take more than 1 week for the detection of cytopathic effect due to the EVs and is only useful for the 65 to 75% of serotypes that grow in the usual cell lines. Suckling mice inoculation will recover many of the other serotypes but is a cumbersome and time-consuming technique that is not widely available. Immunoassays and serologic techniques have been hampered

PCR Protocols: A Guide to Methods and Applications

by the antigenic diversity among the serotypes. Our own approach of nucleic acid hybridization using probes derived from highly conserved sequences has been very promising in that a single probe or a combination of two probes has recognized multiple diverse EV serotypes (Rotbart *et al.* 1984; Rotbart *et al.* 1985; Rotbart *et al.* 1988a; Rotbart *et al.* 1988b). Clinical application, however, has been limited by inadequate sensitivity due to the extremely low titer of EVs present in certain body fluids, particularly cerebrospinal fluid (Wilfert and Zeller 1985; Rotbart 1989). The need for increased sensitivity due to insufficient target nucleic acid, coupled with the knowledge of genomically conserved regions among many of the EVs, led us to the development of a PCR protocol for the diagnosis of many (perhaps all) EVs with a single primer pair and an oligomeric probe.

Protocols

Specimen Preparation

Clinical specimens should be rapidly transported to the laboratory on ice. Store aliquots of virus-containing samples at $-70°C$ until ready for use. Although EVs are relatively resistant to freeze-thawing, small aliquots that minimize the number of freeze-thaw episodes are desirable. Body fluids and tissues are heavily contaminated with RNases, which poses a risk in attempting to purify EV RNA (Rotbart *et al.* 1987). To every 100 μl of specimen, add 40 units of RNasin (Promega, Madison WI) before beginning the extraction.

RNA Extraction

Viral RNA is extracted by the addition of sodium dodecyl sulfate (SDS) to a final concentration of 0.5%, followed by 1 volume of phenol : chloroform (1 : 1 mixture). After gentle mixing by hand, centrifuge at 15,000 $\times$ g for 5 minutes in a tabletop Eppendorf centrifuge. The aqueous phase is then removed to a tube, and the remaining organic phase is back extracted with the addition of an equal volume of a solution consisting of 10 mM Tris–HCl (pH 7.5), 100 mM NaCl, 1 mM EDTA, and 0.5% SDS. This mixture is centrifuged as above, and the aqueous phase is combined with the earlier aqueous phase. Ammonium acetate is added to a final concentration of 2 M, followed by the addition of 2.5 volumes of cold 100% ethanol. This preparation is placed at −20°C for 2 or more hours, followed by centrifugation for 30 minutes as above. The supernatant is discarded.

Primers and Probe

Figure 1 illustrates the location and specific sequences of the primer pair and probe that we use in this protocol. These genomic regions were chosen because of their 100% sequence conservation among the six serotypes of EVs that have been fully sequenced to date (poliovirus types 1, 2, and 3, and coxsackieviruses B1, B3, and B4) (Toyoda *et al.* 1984; Iizuka *et al.* 1987; Lindberg *et al.* 1987; Jenkins *et al.* 1987). Primer 1 and the probe are antisense to genomic RNA; primer 2 is sense.

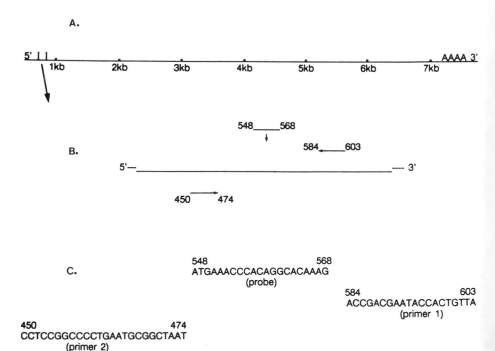

Figure 1   (A) Schematic representation of 7.5-kb single-stranded genomic RNA of the enteroviruses. The first 740 nucleotides from the 5′ end are noncoding. (B) Enlargement of the 5′ noncoding region to illustrate the orientation of the two primers (primer 1 equals 584–603; primer 2 equals 450–474) and probe (548–568). (C) Base sequences of primers and probe.

Reverse Transcription

To the pellet from the extraction above add 40 units RNasin, 2 μl of 5× reverse transcription buffer [250 mM Tris–HCl (pH 8.3), 15 mM MgCl$_2$, 350 mM KCl, 50 mM DTT], 1 μl each of 10 mM ATP, CTP, GTP, and TTP, and 2 μl of diethylpyrocarbonate-treated H$_2$O. Transfer this mixture to a siliconized tube. One μl of the downstream primer (primer 1) (10 pmol/μl) and 1 μl avian myeloblastosis virus reverse transcriptase (Life Sciences, Inc., St Petersburg, Florida) (5 units/μl) are added last. The mixture is overlayed with 100 μl of mineral oil and incubated 90 minutes at 37°C.

Polymerase Chain Reaction

The following reagents are added directly to each of the reverse transcription mixtures: 10 μl of double-distilled sterile H$_2$O, 4 μl of 10× PCR buffer [560 mM KCl, 100 mM Tris–HCl (pH 8.3), 15 mM MgCl$_2$], 0.1% gelatin (w/v), 6.5 μl of diluted mix of deoxynucleotides (125 μl each of 10 mM dATP, dCTP, dGTP, and TTP diluted in 500 μl of double-distilled sterile H$_2$O), 4 μl of primer 1 (10 pmol/μl), 4 μl of primer 2 (10 pmol/μl), and 0.5 μl of *Taq* polymerase (AmpliTaq, Perkin-Elmer Cetus). A lambda phage template and primer pair provided in the GeneAmp kit (Perkin-Elmer Cetus) provide the positive PCR control, which is run per manufacturer's directions. A 5-minute denaturation step at 95°C is followed by 25 cycles of annealing (50°C, 2 minutes), primer extension (72°C, 2 minutes), and denaturation (95°C, 2 minutes). The final step is a 9-minute primer extension at 72°C. To enhance the sensitivity of testing in certain samples, a second set of 25 PCR cycles may be performed by removing 1 μl of the first PCR mixture, diluting with 10 μl of H$_2$O, and adding a new complement of PCR reagents as previously described.

Analysis of Results

Minigel electrophoresis is performed with 5 μl of the PCR product. We use 3% NuSieve agarose (FMC Corp., Rockland, Maine) and 0.5% Ultrapure electrophoresis grade agarose (Bethesda Research Laboratories, Inc., Gaithersburg, Maryland) for our minigels. We find this combination of agarose to give optimum resolution of amplification product in the size range required. A reference sizing "ladder" of known fragment lengths of control DNA is run with each minigel. A band of 154 base pairs in length is considered positive for enteroviral RNA (109-base intervening sequence plus the incorporated 25- and 20-base primer molecules). Confirmatory hybridization using the oligomeric probe (Fig. 1) end labeled with ^{32}P is performed using conditions previously reported (Rotbart *et al.* 1988b).

Example

Four serotypes of EVs (poliovirus type 1, coxsackieviruses A16 and B1, and echovirus 11), two with fully determined sequence (PV1, CB1) and two of unknown sequence (CA9, E11), were added at a titer of 10^3 TCID$_{50}$ to phosphate-buffered saline. Gel analysis performed according to the previously described procedure revealed bands of 154 base pairs in length for all four serotypes (Fig. 2). Hybridization with the probe described confirmed these findings (Fig. 2).

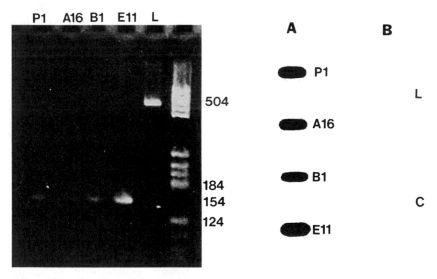

Figure 2 Results of PCR of four serotypes of EVs. Left panel: Agarose gel electrophoresis of 10 μl of PCR products. A band of 154 bp is consistent with amplification of EVs. Right panel: Confirmatory slot blot hybridization using ^{32}P-labeled probe and 10 μl of PCR product. (A) EV samples; (B) λ phage [L] and no DNA [C] controls. Abbreviations used: P1, poliovirus type 1; A16, coxsackievirus A16; B1, coxsackievirus B1; E11, echovirus 11. Numbers along right-hand margin indicate band sizes in base pairs.

Literature Cited

Iizuka, N., S. Kuge, and A. Nomoto. 1987. Complete nucleotide sequence of the genome of coxsackievirus B1. *Virology* **156**:64–73.

Jenkins, O., J. D. Booth, P. D. Minor, and J. W. Almond. 1987. The complete nucleotide sequence of coxsackievirus B4 and its comparison to other members of the picornaviridae. *J. Gen. Virol.* **68**:1835–1838.

Lindberg, A. M., P. O. K. Stalhandske, and U. Pettersson. 1987. Genome of coxsackievirus B3. *Virology* **156**:50–63.

Rotbart, H. A., M. J. Levin, and L. P. Villarreal. 1984. Use of subgenomic poliovirus DNA hybridization probes to detect the major subgroups of enteroviruses. *J. Clin. Microbiol.* **20**:1105–1108.

Rotbart, H. A., M. J. Levin, L. P. Villarreal, S. M. Tracy, B. L. Semler, and E. Wimmer. 1985. Factors affecting the detection of enteroviruses in cerebrospinal fluid with coxsackievirus B3 and poliovirus 1 cDNA probes. *J. Clin. Microbiol.* **22**:220–224.

Rotbart, H. A., M. J. Levin, N. L. Murphy, and M. J. Abzug. 1987. RNA target loss during solid phase hybridization of body fluids—a quantitative study. *Mol. Cell. Probes* **1**:347–358.

Rotbart, H. A., M. J. Abzug, and M. J. Levin. 1988a. Development and application of RNA probes for the study of picornaviruses. *Mol. Cell. Probes* **2**:65–73.

Rotbart, H. A., P. S. Eastman, J. L. Ruth, K. K. Hirata, and M. J. Levin. 1988b. Nonisotopic oligomeric probes for the human enteroviruses. *J. Clin. Microbiol.* **26**:2669–2671.

Rotbart, H. A. 1989. Human enterovirus infections—molecular approaches to diagnosis and pathogenesis. In *Molecular aspects of Picornavirus infection and detection* (ed. B. Semler and E. Ehrenfeld), p. 243–264. American Society of Microbiology, Washington, D.C.

Toyoda, H., M. Kohara, Y. Kataoka, T. Suganuma, T. Omata, N. Imura, and A. Nomoto. 1984. Complete nucleotide sequences of all three poliovirus serotype genomes: implication for genetic relationship, gene function and antigenic determinants. *J. Mol. Biol.* **174**:561–585.

Wilfert, C. M., and J. Zeller. 1985. Enterovirus diagnosis. In *Medical Virology IV: Proceedings of the 1984 International Symposium on Medical Virology* (ed. L. de la Maza and E. M. Peterson), p. 85–107. Lawrence Erlbaum Associates, London, England.

45

NOVEL VIRUSES

David Mack, Oh-Sik Kwon, and Fred Faloona

Despite intense efforts over the past decade to understand the growth
of viruses in cells and illness due to viral replication, viruses continue to be a major cause of disease. It is probable that only a fraction of the viruses responsible for disease have been identified and
that many viruses cause more than one type of clinical infection.
The association of disease with viruses that replicate through DNA
and RNA intermediates (retro- and hepadnaviruses, respectively)
and the wide distribution of these viruses among vertebrate species
suggest that these viruses represent a particularly important class
of pathogens.

Approaches to identify novel viral agents generally rely on the
presence of substantial numbers of viral particles in the infected organism or the *in vitro* propagation of the virus. Hence, the detection
of viruses with restricted cell specificities has proven particularly difficult. Serologic cross-reaction and nucleic acid cross-hybridization
have been used to detect viruses that are related to characterized viruses. These approaches, though, have had limited success because
of the often small number of infected cells and the minimal complementarity between target and probe sequences. Although striking
sequence similarities have been noted between several viruses (Chiu

PCR Protocols: A Guide to Methods and Applications

et al. 1984; Kamer and Angos 1984; Mack and Sninsky 1988; Sonigo *et al.* 1985; Toh *et al.* 1983), the homology frequently extends over only a small number of nucleotides. The complexity of the mammalian genome presents additional problems, since the use of DNA hybridization at low stringencies often results in unacceptable backgrounds.

The difficulties encountered by the current methods of identifying and characterizing novel viruses can be circumvented with the use of PCR. Short stretches of conserved amino acids can serve to define degenerate oligonucleotide primer sequences (Mack and Sninsky 1988). Enzymatic amplification of low-copy viral sequences flanked by these primers allows the exponential accumulation of a novel sequence defined at its ends by the primer sequences. The overall complexity of the DNA is decreased by greatly reducing the ratio of nontarget to target sequences, hence allowing for low-stringency probe hybridizations. In addition, PCR primers with 5' extensions that incorporate convenient restriction endonuclease sites allow for cloning and characterizing novel sequences (as well as increasing amplification efficiency).

Regional amino acid sequence similarities among viruses are often evident only after comparative computer-assisted analyses. The conserved regions typically encode portions of proteins that are functionally important in the viral life cycle. The *pol* gene (reverse transcriptase) products that possess at least three enzymatic activities (polymerase activity utilizing an RNA template, RNase H activity, and a DNA endonuclease activity) are required for viral replication. Numerous investigators have noted regions of sequence similarity in this gene within and between many retroviral subfamilies (see Literature Cited).

The key to designing primers based on conserved but degenerate sequence regions for the detection of novel viruses is to minimize the degeneracy of the oligonucleotide primers. In theory, an extremely degenerate primer might increase the chance of amplifying a divergent (uncharacterized) sequence. However, one must consider the total number of primer species. A primer with too many permutations (>400?) may have too low a concentration of the particular sequence(s) to allow effective amplification to take place. For example, a degenerate 20-mer oligo with five positions containing all four nucleotides would have 1024 different sequence permutations. A typical 100-μl PCR usually contains 100 pmoles of each primer. If one were to use this concentration of a degenerate primer, the reaction would effectively contain about 100 fmoles of each species of primer. Of course, some of the individual degenerate primer species will be very similar, so many different primer permutations might anneal to the same target.

One generally uses higher concentrations of degenerate primer mixes than for homogeneous primers. One can expect in most cases a bell-shaped curve representing the efficiency of amplification of desired target versus the concentration of degenerate primer. Hence, there is a different and optimal primer concentration to use for each particular degenerate primer (more does not necessarily mean better).

When designing your degenerate primer, try to keep the degeneracy to a minimum at the extreme 3' end. Since *Taq* polymerase extends in a 5'-to-3' direction, amplifications are more tolerant of mismatches toward the 5' end of the primer. A primer with a breathing 3' end could preclude the enzyme from extending it on the template.

Typically, a decrease in primer length is more desirable than the introduction of extreme degeneracy. A 20-mer annealing sequence with less than 300 permutations and a 5' extension is optimal. The 5' extension should contain a six-base endonuclease site with three extra bases (e.g., CTC) at the extreme 5' end in order to assist duplex formation at the sites that the restriction enzymes will cut. This 5' linker not only allows for potential ease of cloning of the amplified product but also functionally transforms the 20-mer primer into a longer (thus, more stable) 29-mer primer after one cycle of amplification with polymerase. Bear in mind that the presence of upstream and downstream primers with asymmetric endonuclease sites can simplify eventual cloning of the amplified products.

If the number of sequence permutations is too great to achieve a 20-mer annealing sequence, begin to shorten your primer length, remembering to minimize the degeneracy at the 3' end if possible. Use an 11-base length for the annealing sequence and the 9-base 5'-extension length as the smallest primer for this approach (20-mer). The use of data bases with cumulative statistics for preferred codon usage for amino acids from various viral and human proteins may be of help.

We have identified a conserved tetrapeptide, Tyr-Met/Val-Asp-Asp, found in the reverse transcriptase gene of all exogenous viruses that replicate through an RNA or DNA intermediate, that efficiently serves as a degenerate primer sequence for PCR (Mack and Sninsky 1988). The fact that polypeptides with mutations at the tyrosine or aspartic acid moieties in this sequence are incapable of catalyzing reverse transcription supports a critical role for this region in replication (Larder *et al.* 1987). An octapeptide conserved region located 25 amino acids upstream from the tetrapeptide sequence also serves as an efficient degenerate primer sequence (Mack and Sninsky 1988). This region, although conserved between viral subfamilies, displays particular conservation within a viral group. Thus, this region can be used to identify and associate new viruses within a particular subfamily (i.e., defined by members that cause similar clinical disease). The sequences flanked by these primer domains are highly variant among and between retroviral subfamilies. This region can then serve as a "fingerprint" for a specific virus and will enable one to easily identify novel viruses by sequence comparisons.

Amplification of Novel Viral Sequences Using Degenerate Oligonucleotide Primers

Setting Up the Reaction

1. Prepare microgram amounts of purified genomic DNA potentially harboring proviral forms of putative virus from suspect infected cells. (As few as 30 copies of viral genome equivalents per 150,000 cells have been detected by this approach.)
2. In a 0.5-ml Eppendorf tube prepare a 100-μl reaction mixture by adding 1 μg of DNA to 50 mM KCl, 10 mM Tris (pH 8.4), 2.5 mM MgCl$_2$, containing 50 μM each dNTP and 2.5 units of *Taq* polymerase. Complete the amplification mixture by adding the upstream and downstream degenerate primers at a minimum concentration of 20 nM for each possible sequence permutation. Overlay sample(s) with ~100 μl of mineral oil to prevent condensation.

Performing the Amplification

1. Heat reactions from annealing temperature (see next step) to 95°C over a 1.0-minute period to denature the DNA.
2. Cool reactions over 2 minutes to 37°C (annealing temperature

for first two cycles) and then cool over 2 minutes to 55°C (annealing temperature for remaining cycles).

It is advisable to use a low annealing temperature in early cycles with your degenerate primers. Higher-stringency annealing allows for reduced backgrounds and better product yields with most homogeneous primers, but since the target sequence is not known *a priori*, a low-stringency annealing temperature is a prudent step. This will allow for annealing of primers to semi-complementary target sequences in early cycles, which after one successful round of amplification results in a template 100% complementary to the primer sequence. In latter cycles, then, a high-stringency annealing temperature can be applied without fear of precluding successful priming.

3. After annealing of primer to template, heat reactions to 70°C and incubate at that temperature for 0.5 minutes/500 bp of predicted template length. This is the optimal temperature for *Taq* polymerase activity, i.e., for the extension of annealed primers at their 3' ends.

4. Perform a total of 32 cycles of amplification: 2 cycles at 37°C annealing followed by 30 cycles of 55°C annealing.

More cycles or fewer cycles can be used depending on the amounts of specific ethidium bromide-stainable target fragment and background.

Visualizing the Amplified Product

1. Remove ¹⁄₁₅ of the PCR amplification and visualize it by ethidium bromide/gel electrophoresis. Depending on the expected size of the amplified product, one can use DNA grade ultra-pure agarose (>1 kb), polyacrylamide (25 bp to 2 kb), or high-density agarose (NuSieve) gels.

Use an oligonucleotide probe based on conserved sequences within the amplified target to analyze a Southern blot of the amplified reaction. Since the sequences of interest are selectively amplified, and background sequences are relatively reduced, low-stringency hybridization conditions can be successfully used. This allows high signal-to-background detection with probes possessing at most 65% homology with target sequences. End-labeled probes of 30 to 40 bases in length are optimal for this approach.

Alternatively, one can end label the upstream and downstream degenerate primers and hybridize each to Southern blots separately. Amplified product of predicted molecular weight and hybridizing with both labeled primers should define your target.

2. Use 0.5 pmol of ^{32}P-end-labeled probe (2×10^6 cpm/pmol) per ml of hybridization solution ($3\times$ SSPE, $5\times$ Denhardt's solution, 30% formamide, 0.5% SDS).

3. Incubate blots at 42°C for 3 to 5 hours with agitation.

4. Remove excess probe by two washes with $2\times$ SSPE/0.1% SDS for 10 minutes at room temperature. From this point the blot can be subjected to a number of washes of increasing stringencies and autoradiographed after each treatment until the desired signal-to-background level is achieved. For example, wash with $2\times$ SSPE/0.1% SDS for 10 minutes at 42°C, air dry, and autoradiograph. Follow by a wash with $0.2\times$ SSPE/0.1% SDS for 10 minutes at room temperature, air dry, and autoradiograph. Follow by a wash with $0.2\times$ SSPE/0.1% SDS for 10 minutes at 42°C, . . . $0.1\times$ SSPE/0.1% SDS for 10 minutes at 42°C, air dry, and autoradiograph.

Cloning of the Amplified Product

1. If gel analysis of the amplified reaction displays a single ethidium bromide-stainable fragment, take an aliquot of the reaction representing $\sim$1 μg of product and digest it with 10 units of the restriction enzyme(s) defined by the sites added to the 5' end of the primers.

2. Check the restricted product by gel electrophoresis.

   Since the sequences flanked by the primers are not known, they may contain the same endonuclease sites present in the primers; thus, a fragment of altered mobility will be observed. In this situation one must resort to cloning the amplified sequence by blunt-end ligation (see Chapter 11).

   If multiple products are visible from the undigested PCR, or no well-defined product is observable, one can selectively purify the fragment of the expected molecular weight after electrophoretic separation. The purified DNA can either be cloned directly or serve as the target DNA for additional amplification reactions. In the latter case, one can often obtain clonable quantities of desired fragment when efficiencies are too poor from genomic target DNA alone (because of nonspecific priming of background sequences).

DEAE Cellulose (in Membrane Form) Provides a Simple and Rapid Method for Recovery of PCR Fragments

1. After separating and visualizing the entire PCR products on an agarose gel, cut a horizontal slit below the fragment to be recovered.
2. Wet the DEAE membrane (Schleicher & Schuell NA-45) in running buffer and insert it into the slit. One may want to make an additional slit above the desired fragment and insert the DEAE membrane. This will prevent any higher-molecular-weight DNA from contaminating the preparation.
3. Resume electrophoresis and run the DNA fragment into the membrane.
4. Check for the presence of the fragment on the membrane with a UV transilluminator. Trim the paper and place the membrane with bound DNA in a 1.5-ml Eppendorf tube containing 400 to 500 μl elution buffer [1 M NaCl, 1 mM EDTA (pH 8.0), 10 mM Tris (pH 8.0)].
5. Incubate at 68°C for 30 minutes.
6. Remove supernatant from paper and ethanol-precipitate the DNA.
7. Resuspend in TE buffer and continue as per the section on Visualizing the Amplified Product, or use as target DNA for reamplification.

Literature Cited

Chiu, I.-M., R. Callahan, S. R. Tronick, J. Schlom, and S. A. Aaronson. 1984. Major *pol* gene progenitors in the evolution of oncoviruses. *Science* **223**:364–370.

Kamer, G., and P. Argos. 1984. Primary structural comparison of RNA-dependent polymerases from plant, animal and bacterial viruses. *Nucleic Acids Res.* **12**:7269–7282.

Larder, B. A., D. J. M. Purifoy, K. L. Powell, and G. Darby. 1987. Site-specific mutagenesis of AIDS virus reverse transcriptase. *Nature (London)* **327**:716–717.

Mack, D. H., and J. J. Sninsky. 1988. A sensitive method for the identification of uncharacterized viruses related to known virus groups: hepadnavirus model system. *Proc. Natl. Acad. Sci. USA* **85**:6977–6981.

Sonigo, P., M. Alizon, K. Staskus, D. Klatzmann, S. Cole, O. Danos, E. Retzel, P. Tiollais, A. Haase, and S. Wain-Hobson. 1985. Nucleotide sequence of the visna lentivirus: relationship to the AIDS virus. *Cell* **42**:369–382.

Toh, H., H. Hayashida, and T. Miyata. 1983. Sequence homology between retroviral reverse transcriptase and putative polymerases of hepatitis B virus and cauliflower mosaic virus. *Nature (London)* **305**:827–829.

46

ANALYSIS OF *ras* GENE POINT MUTATIONS BY PCR AND OLIGONUCLEOTIDE HYBRIDIZATION

John Lyons

ras genes have been implicated in the oncogenesis of many tumors over the past few years and appear to be activated by single point mutations. These point mutations can occur in all three *ras* genes (N-*ras*, Harvey *ras* and Kirsten *ras*) at codons 12, 13, and 61, with corresponding amino acid substitutions in the *ras* proteins (p21 proteins). These substitutions were originally detected by DNA transfection experiments followed by molecular cloning and sequencing analysis. (For reviews see Barbacid 1987; Bos 1988).

The simplest method of detecting point mutations in the *ras* genes is a two-step procedure involving PCR. In the first step, by using synthetic oligonucleotides that flank the sequences of choice (i.e., the DNA regions around codons 12, 13, or 61 in the *ras* genes), the target is amplified *in vitro* many thousandfold. In the second step, an aliquot of the *in vitro*–amplified DNA is applied to filters as a dot blot and screened by radioactively labeled synthetic oligomers whose sequences correspond to all possible amino acid substitutions that can be generated by a single base change at codons 12, 13, and 61 of each *ras* gene. When hybridizing, only fully matched hybrids between the filter-bound PCR product and the probe used will correspond to a positive dot on an X-ray film.

The *in vitro* amplification of *ras*-specific sequences in genomic DNA by PCR has vastly improved and simplified the analysis of the frequency of *ras* point mutations in many different tumors. Since neither the quality nor the quantity of DNA required for amplification seems to be of great importance (even cell lysates can be used), the method has evolved to be a powerful means of detecting point mutations in neoplasia, as well as in premalignant states.

Principle of the Method

Two steps are involved in the detection of point mutations. The first step is the amplification of sequences around the codons 12, 13, and 61 of the three *ras* genes (Table 1; see Janssen *et al.* 1987). Since codon 61 is separated from codons 12 and 13 by large intron sequences in each *ras* gene, it is necessary to amplify DNA around these regions by using separate pairs of primers. These reactions may therefore be performed by one of these methods: (1) amplifying each region in a separate tube, (2) amplifying the region around

Table 1

List of Amplimers for Human *ras* Gene Amplification

Amplimer	Sequence	Strand
N−12 + 13	CTTGCTGGTGTGAAATGACT	S
N−12 + 13	ACAAAGTGGTTCTGGATTAG	A
N61	GTTATAGATGGTGAAACCTG	S
N61	AAGCCTTCGCCTGTCCTCAT	A
Ki−12 + 13	TTTTTATTATAAGGCCTGCT	S
Ki−12 + 13	GTCCACAAAATGATTCTGAA	A
Ki−61	ACCTGTCTCTTGGATATTCT	S
Ki−61	TGATTTACTATTATTTATGG	A
Ha−12 + 13	GAGACCCTGTAGGAGGACCC	S
Ha−12 + 13	CGTCCACAAAATGGTTCTGG	A
Ha−61	CCGGAAGCAGGTGGTCATTG	S
Ha−61	ACACACACAGGAAGCCCTCC	A

codon 12 in all three *ras* genes in one tube and the region around codon 61 of all three genes in another tube, or (3) amplifying both regions around 12 and 61 of each *ras* gene individually. One can therefore choose which form of PCR to perform depending on the nature of the investigation. Usually, 30 pmols of each primer gives good results when used on 0.1 to 1 μg of genomic DNA; after 30 cycles the amplified fragments appear as discrete bands on ethidium bromide-stained agarose gels. The number of cycles used does depend on the quality of the tumor sample under investigation. Thus, deparaffinized formalin-fixed tissue sections may require up to 50 cycles, probably as a result of spurious amounts of formic acid in formalin that causes depurination of the DNA. Subsequent sizing of the PCR product on a 2.5% agarose gel confirms the amplification.

Protocol

PCR (100-μl reactions)

Buffer: GenAmp *Taq* buffer (10× is 500 m*M* KCl, 100 m*M* Tris (pH 8.3), 15 m*M* MgCl$_2$, and 0.1% gelatin)

10 m*M* dNTPs (Pharmacia 100 m*M* stock solutions), final concentration in the reaction equals 100 μM of each dNTP

30 pmols each oligonucleotide primer

0.5 to 1 unit of *Taq* polymerase (Cetus)

0.1 ng to 1 μg of total genomic DNA

(30 to 50 cycles)

In the second step in the detection of point mutations, the PCR product is applied to filters and then hybridized with specific oligonucleotides that are capable of detecting any possible mutation in codons 12, 13, and 61.

Filter Preparation

1. Nylon filters (Pall-Biodyne-B; 0.4 μm) are soaked in distilled water for 5 minutes and then for a further 5 minutes in 10× SSC.

The filters are subsequently allowed to dry either at room temperature or in an incubator at 50°C for 30 minutes.

2. The PCR products (50 or 100 μl) are denatured at 95°C for 3 minutes and are placed immediately on ice.

3. Two μl of the denatured product is then applied using a pipette to the filter. The filter should have six duplicates for hybridization with codons 12 or 13, and seven duplicates for hybridization with codon 61.

4. After applying 2 μl of the product to each filter the filters are allowed to dry at room temperature. They are then placed under a UV light and illuminated at 254 nm for 3 minutes.

Prehybridization

The filters may then be prehybridized in a 3 M TMACl solution (1× TMACl hybridization mixture is 3 M tetramethylammonium chloride, 50 mM Tris–HCl (pH 8), 2 mM EDTA, 0.1% SDS, 5× Denhardt's, and 100 μg/ml of salmon sperm DNA) at 56°C for 30 minutes. The TMACl stock solution is made up by dissolving the TMACl in H_2O to a concentration of 4 to 4.5 M. The exact concentration may be calculated by measuring the refractive index of this solution at 20°C using the following formula:

$$C = \frac{\text{refractive index} - 1.331}{0.018}$$

Hybridization

The filters can now be incubated with radioactively labeled oligonucleotides. Briefly, 1 pmol of a 20-mer (6.6 ng) is kinased in 10 μl by using approximately 1 unit of T4 polynucleotide kinase over 30 to 45 minutes (1× kinase buffer equals 5 mM Tris–HCl [pH 8], 1 mM $MgCl_2$, 0.1 mM DTT, 50 μg/ml of bovine serum albumin) and 1 μl of [γ-^{32}P]dATP (3000 Ci/mmol). The addition of 90 μl of 10 mM Tris–HCl, 0.1 mM EDTA, and 0.1% SDS stops the reaction. Separation of the labeled oligonucleotide from the unincorporated nucleotide triphosphate is achieved by a Sephadex-G 50 column. One-sixth of the eluted product is added to a polyethylene bag containing the filter immersed in 5 ml of 3 M TMACl hybridization mixture. Hybridization is performed as before at 56°C for 2 to 3 hours in a shaking water bath.

Washing Procedures

After hybridization the filters are washed twice in a 2× SSPE, 0.1% SDS solution at room temperature [1× SSPE equals 10 mM sodium phosphate (pH 7), 180 mM NaCl, and 1 mM EDTA]. Subsequently, the filters are washed in a 5× SSPE, 0.1% SDS solution for 20 minutes at 56°C. A stringent wash is then performed for 10 minutes. The temperatures for the stringent washes for each oligonucleotide set are as follows:

N-12 + 13	63°C	N-61	59°C
Ki-12 + 13	64°C	Ki-61	59°C
Ha-12 + 13	71°C	Ha-61	64°C

Note: These temperatures have been obtained empirically by washing until point mutations can be discerned on the filters. The temperatures apply only to oligonucleotides that are 20 bases long.

Another possibility is to wash the filters in a 3 M TMACl solution (3 M TMACl wash solution equals 50 mM Tris–HCl (pH 8), 0.2% SDS, and 3 M tetramethylammonium chloride solution). One should start washing all filters at 58°C and then at 1°C increments until only wild-type and mutant signals may be detected. Under both conditions (i.e., SSPE and TMACl) the chosen temperature allows only fully complementary hybrids to stay formed, resulting in a positive dot on a filter.

Autoradiography for up to 12 hours gives wild-type signals (for most human tumors when only one allele is involved in activation) that also serve as an internal control for positive amplification, and on the duplicate filters possible mutant signals. This method has an overall detection limit of approximately 10%; i.e., samples in which mutant alleles represent less than 10% total DNA appear to be wild-type.

Although *in vitro* amplification using PCR and subsequent direct sequencing of the PCR product is an alternative to the use of oligonucleotides, it remains to be seen how many actual tumor samples contain point mutations, if any, outside of the already well-documented 12, 13, and 61 codons in the *ras* genes. In addition, this method requires separate amplification of the target sequence and the gene, precluding the screening of large numbers of samples simultaneously.

Literature Cited

Barbacid, M. 1987. *ras* genes. *Ann. Rev. Biochem.* **56**:799–827.

Bos, J. L. 1988. The *ras* gene family and human carcinogenesis. *Mut. Res.* **195**: 255–271.

Janssen, J. W. G., J. Lyons, A. C. M. Steenvoorden, H. Seliger, and C. R. Bartram. 1987. Concurrent mutations in two different ras genes in acute myelocytic leukemias. *Nucleic Acids Res.* **15**:5669–5680.

47

B-CELL LYMPHOMA: t(14;18) CHROMOSOME REARRANGEMENT

Marco Crescenzi

The 14;18 translocation is found in the majority of follicular and in some nonfollicular lymphomas. It is thus present, even if not always cytogenetically detectable, in a large part of the non-Hodgkin's lymphomas (Yunis 1983; Levine *et al.* 1985). In more than half of the cases, the breakpoint on chromosome 18 falls within a 150-bp region (major breakpoint region, MBR). The chromosome 18 fragment always joins chromosome 14 at one of the six J segments of the immunoglobulin heavy chain locus (Bakhshi *et al.* 1987). This allows, in most cases, the PCR to amplify a small, translocation-specific fragment. This fragment, encompassing the chromosomal juncture, is unique to the malignant cells, which are the only translocation-bearing cells. This gives one the possibility of detecting very specifically and with a high sensitivity the presence of minimal numbers of malignant cells among normal tissues. This has an obvious importance for the study, diagnosis, and treatment of this group of diseases.

The following is a description of a method (Crescenzi *et al.* 1988) that can be readily applied in a laboratory equipped with standard molecular biology reagents and instruments. This method achieves the optimal goal of detecting even a single translocated DNA mole-

PCR Protocols: A Guide to Methods and Applications

significantly modified, and this must be taken into consideration, especially if probes are not used to visualize results. Shorter primers (for example, 15-mers composed of the 5′ ends of the primers shown above) yield a much lower background but are somewhat less efficient in amplification (Crescenzi *et al.* 1988).

PCR Conditions

1× PCR buffer
    10 mM Tris–HCl (pH 8.3)
    50 mM KCl
    10 mM MgCl$_2$
    0.1% gelatin (DNase free)
    1.5 mM each dNTP
    1 μM each primer

Other buffers (for example, with lower Mg concentrations) might work even better, but we have optimized the reaction by using this one. It can be conveniently prepared 10× and stored refrigerated or frozen.

Each PCR tube should contain 5 units of *Taq* polymerase (2.5, 7.5, or 10 units yield worse results).

Maximal amplification is achieved in 45 cycles when either 1 or 10 μg of DNA is used. Cycles are as follows:

Denaturation	94°C	2 minutes	(5 minutes before the first cycle)
Annealing	61°C	3 minutes	
Synthesis	72°C	3 minutes	

The above steps include the time necessary for the content of tubes to equilibrate their temperature with the heating block.

> NOTE: While not directly tested with these primers, recent results indicate that lower dNTP (0.05 to 0.2 mM each) and lower MgCl$_2$ (about 1.5 mM) concentrations increase the yield of the desired PCR product and permit lower concentration of *Taq* DNA polymerase (1 to 2 units) in the PCR. In addition, reducing the primer concentration (about 0.2 μM each) will conserve primer and generally increase the specificity of the PCR without reducing the yield of the desired PCR product (see Chapter 1).

Detection of Amplification Products

PCR products can be analyzed in different ways. A 6% polyacryl-amide or a 4% agarose gel (both nondenaturing) are effective in sep-arating the expected PCR products (using the primers described above) of about 80 to 250 bp. Because of their high resolution, poly-acrylamide gels are preferable when bands are to be identified by ethidium bromide staining only. Agarose gels make transfer to paper easier when hybridization of products to a probe is required.

Ethidium bromide staining is often sufficient, particularly when the size of the expected band is known, on the basis of previous re-sults obtained from samples from the same patient (bands differ in size from patient to patient, depending on the specific rearrange-ment). However, hybridization with an internal, labeled primer allows unambiguous identification of the specific band over back-ground amplification products.

For transfer of gel-separated products, the use of nylon mem-branes is advisable because nitrocellulose paper does not bind small fragments efficiently. In our hands, GeneScreen*Plus* (New England Nuclear Corp.) used according to the manufacturer's instructions has worked efficiently, with no apparent loss of even 100-bp fragments.

Cloned genomic DNA probes (Crescenzi *et al.* 1988) or oligonu-cleotide probes can be employed. An internal hybridization probe shown in Fig. 1 detects any product found so far. Nevertheless, it could conceivably miss rearrangements in which the breakpoint falls within the probe sequence itself. It should be possible to devise an oligonucleotide probe in the J region, immediately upstream of the second primer, in which a considerable homology among the six J segments still exists (Fig. 2). With the appropriate hybridization conditions, such a probe should recognize every possible amplified fragment. However, we have not tried this probe.

Given the abundance of the PCR products, a radioactive probe is not required. The use of biotinylated probes should accelerate the process of detection and make it safer and less cumbersome.

In particular cases, direct sequencing of PCR products could pro-vide a "fingerprint" of the rearrangement in a given patient, con-clusively prove the identity of an ambiguous band, or rule out the possibility of a contaminated sample (see below).

```
                      OPEN READING FRAMES
┌─────────────────────────────────────────────────────────┐
  GCTGAATACTTCCAGCACTGGGGCCAGGGCACCCTGGTCACCGTCTCCTCAGGT         J1
        ••  •       •••••••••  ••  •••••••••••••••••••••••

  CTACTGGTACTTCGATCTCTGGGGCCGTGGCACCCTGGTCACTGTCTCCTCAGGT        J2
        ••••    •••••••••••  •    ••••••••••••  •••••••••••

    ATGCTTTTGATGTCTGGGGCCAAGGGACAATGGTCACCGTCTCTTCAGGT           J3
        •  ••   ••••••••••••••••    ••••••••••••••  •••••••

    ACTACTTTGACTACTGGGGCCAAGGAACCCTGGTCACCGTCTCCTCAGGT           J4
        •  ••••   •••••••••••••   ••••••••••••••••••••••••

   ACACTGGTTCGACTCCTGGGGCCAAGGAACCCTGGTCACCGTCTCCTCAGGT          J5
        ••••••   •••••••••••••   •••••••••••••••••••••••••

AT(TAC) GGTATGGACGTCTGGGGGCAAGGGACCACGGTCACCGTCTCCTCAGGT         J6
    5      •  •••  ••••••••  ••••••••••   •••••••••••••••••••
```

```
CONSENSUS        TCGACTTCTGGGGCCAAGGGACCCTGGTCACCGTCTCCTCAGGT
SEQUENCE

                      3'-  TGGGACCAGTGGCAGAGGAGTCCA  -5'
                      ─────────────────────────────────
                              SECOND PRIMER
```

Figure 2   J segments and consensus sequence: identities to the consensus sequence are shown by dots under the sequence. The second primer is shown with its orientation.

Standardization of the Procedure

To ensure that the procedure achieves its maximal sensitivity, it is important to run a dilution experiment. This can be done by progressively diluting cells bearing the translocation with normal cells. DNA prepared from these mixtures can then be used for the standard curve. Alternatively, DNA containing the translocation can be diluted with normal DNA.

Ideally, a cell line bearing the 14;18 translocation should be used for standardization purposes. If this is not available, DNA from neoplastic tissues (e.g., lymph nodes) is a reasonable alternative, provided that the particular translocation is amplifiable (it must fall in the MBR) and the tissue is mainly composed of neoplastic cells. The latter can be verified on a Southern blot, in which the band derived from the translocated chromosome 14 (or 18) should show roughly the same intensity as that of the germline band (Bakhshi *et al.* 1987).

Contamination

As with any other extreme amplification by PCR, the danger is present that trace amounts of contaminating DNA are amplified in samples that would otherwise be negative. Sources of contamination are countless (see Chapter 17). Microtiter pipets, glassware (avoid it), and phenol are the most likely ones in our experience.

Because of the likelihood of contamination problems, *at least* one negative control should *always* be run with each set of tests. A contamination problem may not originally be present in a given laboratory but may develop later as more and more samples are amplified, and translocation-containing DNA inevitably tends to taint equipment and reagents.

The method described here should be useful for laboratories involved in basic and/or clinical investigations of human lymphomas. It is certainly possible to improve it. For example, other primers could be designed to include other translocations not occurring within the MBR. The more we learn about the molecular mechanisms involved in this translocation, the more we will be able to refine and improve this protocol.

Literature Cited

Bakhshi, A., J. J. Wright, W. Graninger, M. Seto, J. Owens, J. Cossman, J. P. Jensen, P. Goldman, and S. J. Korsmeyer. 1987. Mechanism of the t(14;18) chromosomal translocation: structural analysis of both derivative 14 and 18 reciprocal partners. *Proc. Natl. Acad. Sci. USA* **84**:2396–2400.

Crescenzi, M., M. Seto, G. P. Herzig, P. D. Weiss, R. C. Griffith, and S. J. Korsmeyer. 1988. Thermostable DNA polymerase chain amplification of t(14;18) chromosome breakpoints and detection of minimal residual disease. *Proc. Natl. Acad. Sci. USA* **85**:4869–4873.

Levine, E. G., D. C. Arthur, G. Frizzera, B. A. Peterson, D. D. Hurd, and C. D. Bloomfield. 1985. There are differences in cytogenetic abnormalities among histologic subtypes of the non-Hodgkin's lymphomas. *Blood* **66**:1414–1422.

Ravetch, J. V., U. Siebenlist, S. J. Korsmeyer, T. Waldmann, and P. Leder. Structure of the human immunoglobulin μ locus: characterization of embryonic and rearranged J and D genes. 1981. *Cell* **27**:583–591.

Yunis, J. J. 1983. The chromosomal basis of human neoplasia. *Science* **221**:227–236.

48

DETECTING BACTERIAL PATHOGENS IN ENVIRONMENTAL WATER SAMPLES BY USING PCR AND GENE PROBES

Ronald M. Atlas and Asim K. Bej

Various bodies of water that are contaminated by human fecal matter are involved in the transmission of infectious bacteria (for example, *Salmonella* [gastroenteritis and typhoid fever], *Shigella* [dysentery], and *Vibrio cholerae* [cholera]). Environmental water supplies can also serve as reservoirs for bacterial pathogens; for example, air conditioning cooling towers have been involved in the transmission of *Legionella pneumophila* (Legionnaires' disease). Disinfection treatment is performed when levels of detectable pathogens indicate a high probability of an outbreak of disease. Public health measures require bacteriological surveillance of environmental water supplies; e.g., monitoring for levels of enteric pathogens that can be transmitted via the fecal–oral route (Hoadley and Dutka 1977). The traditional methods used to detect specific pathogens or the bacterial indicators of fecal contamination are direct microscopy and viable plate count (Greenberg 1985). These detection methods may be quantitative or may be used to simply establish the presence or absence of pathogenic bacteria at a given threshold level.

Pathogens can be detected in water samples by the direct application of gene probes, but the sensitivity achievable is about 10^4 cells/ml (Jain *et al.* 1988). This is several orders of magnitude less sensitive than what is required for environmental monitoring purposes. However, amplification of a target gene sequence by PCR coupled with gene probes can achieve the necessary sensitivity at the same time the specificity required for monitoring bacterial pathogens in environmental waters is retained. The detection of pathogenic bacteria in water samples by using PCR and probes involves the recovery of bacterial cells, release of DNA from all recovered cells, PCR amplification of DNA sequences associated with the target bacteria, and specific probe detection of the amplified target DNA. PCR-gene probe methods have been developed for the detection of "total coliform bacteria" [based upon a region of *Escherichia coli lacZ* (Atlas *et al.* 1989) (Fig. 1A)], "fecal coliform bacteria" [based upon a coding region of *E. coli lamB* (Atlas *et al.* 1989) (Fig. 1B)], enteropathogenic *E. coli* [based upon the heat-labile toxin gene (Olive 1989)], *Legionella* [based upon a region coding for 5S rRNA (Mahbubani *et al.* 1989) (Fig. 1C)], *L. pneumophila* [based upon the macrophage-infecting pro-

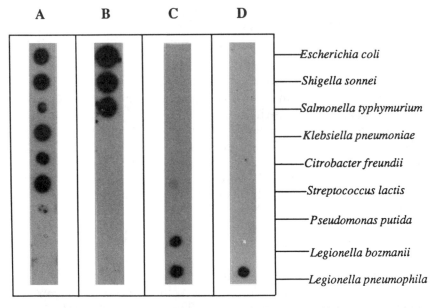

A   B   C   D

—Escherichia coli

—Shigella sonnei

—Salmonella typhymurium

—Klebsiella pneumoniae

—Citrobacter freundii

—Streptococcus lactis

—Pseudomonas putida

—Legionella bozmanii

—Legionella pneumophila

Figure 1 Examples of specificity of PCR-gene probe detection of bacterial pathogens showing dot blot analysis of the PCR-amplified samples of coliforms and *Legionella* using radiolabeled probes. One tenth (10 μl) from each PCR-amplified sample was denatured and spotted onto the membrane. (A) Total coliforms were amplified and detected by using a probe for the *lacZ* gene of *E. coli*; (B) fecal coliforms were detected by using a probe for the *lamB* gene of *E. coli*; (C) total *Legionella* was detected by using a probe for a DNA sequence coding for 5S rRNA; and (D) *L. pneumophila* was detected by using a probe specific for the *mip* gene.

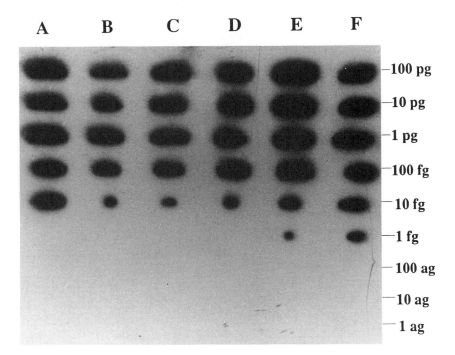

Figure 2 Examples of sensitivity of PCR-gene probes for detecting bacterial pathogens showing signal bands of slot blots for replicate samples of various amounts of total genomic DNA of *L. pneumophila* (A-C) and *E. coli* (D-F). The *E. coli* genomic DNA was amplified by using primers and probes of the *lamB* genomic gene, and *L. pneumophila* was amplified by using primers and probe specific for the *mip* gene.

tein (*mip*) gene (Mahbubani *et al.* 1989) (Fig. 1D)], and *L. pneumophila* [based upon a *L. pneumophila*-associated DNA sequence of unknown function (Starnbach *et al.* 1989)]. The sensitivity of detection that has been achieved is 1 to 10 fg of genomic DNA (Fig. 2) and 1 to 10 cells of target organism per 1- to 100-ml water sample (Fig. 3).

Protocols

Cell and DNA Recovery

Cells can be recovered from a 1- to 100-ml water sample by centrifugation at 10,000 × *g* for 15 minutes, and the cells can be trans-

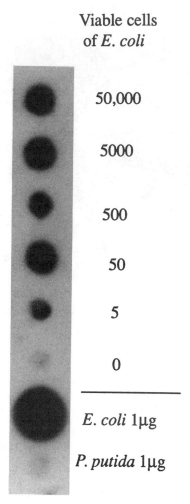

**Viable cells
of *E. coli***

50,000

5000

500

50

5

0

_____

E. coli 1μg

P. putida 1μg

Figure 3   Example of sensitivity of PCR-gene probes for detecting bacterial pathogens in water showing dot blot analysis following PCR amplification of serial dilutions of an *E. coli* culture by using primers for *lacZ* amplification. Viable cells were determined by viable plate count. Controls included pure *E. coli* and *Pseudomonas putida* DNAs.

ferred to a 0.6-ml Eppendorf tube and resedimented at 10,000 × *g* for 5 minutes. Cells also can be recovered by filtration with a 13-mm diameter, 0.4-μm pore size teflon filter, and the filter can be placed directly into a 0.6-ml Eppendorf tube.

The direct lysis procedure of Li *et al.* (1988) can be used to release DNA from the bacterial cells. Lysis buffer (50 μl for cells recovered by centrifugation or 150 μl for cells recovered by filtration) containing 1× PCR buffer, 0.05 mg/ml of Proteinase K, 20 mM dithiothreitol (DTT), and 1.8 μM SDS is used to lyse the bacterial cells; the samples are vortexed for 15 seconds and incubated at 37°C for 1 to 1.5 hours, after which they are heated to 80°C for 5 minutes to inactivate the Proteinase K.

PCR Amplification

PCR mixture (50 μl) is added directly to the lysis buffer, which should contain 1 fg to 1 μg of template DNA. The PCR mixture contains 1× PCR amplification buffer [10× buffer contains 50 mM KCl, 100 mM Tris–HCl (pH 8.13), 15 mM MgCl$_2$, and 0.1% gelatin], 200 μM each of the dNTPs, 0.2 to 1 μM of each of the primers, 2.5 units of Taq polymerase, and double-distilled water containing 0.1% diethylpyrocarbonate (DEPC). For coliform bacterial detection, native Taq polymerase is used, since preparations of AmpliTaq may contain up to 100 molecules of $E.$ $coli$ DNA/2.5 units; for $Legionella$ detection, either native Taq polymerase or AmpliTaq is used.

In some cases the DNA is labeled during the PCR amplification by including 66 μM biotin-11-dUTP (Sigma Chemical Co., St. Louis, Missouri) (1 : 2 ratio biotin-dUTP : dTTP) or by including 5 μCi radiolabeled [α-^{32}P]dCTP (specific activity 250 μCi/ml) and 10 μM unlabeled dCTP.

For PCR DNA amplification, template DNA is initially denatured at 94°C for 2 minutes. Then a total of 30 PCR cycles are run using either a two-step PCR procedure (denaturation at 94°C for 1 minute, followed by primer annealing and extension at 65°C for 1 minute) or a three-step PCR procedure (denaturation at 94°C for 1 minute, primer annealing at 50°C for 1 minute, extension at 72°C for 1 minute).

Detection of Amplified DNA

PCR-amplified DNAs are detected by using 5'-end-radiolabeled 50-mer oligonucleotide probes. The probes are labeled by using T4 polynucleotide kinase (PNK), PNK buffer [50 mM Tris–HCl (pH 7.5), 10 mM MgCl$_2$, 5 mM DTT, 1 mM KCl], 20 pmol oligonucleotide, and

25 μCi [γ-^{32}P]ATP (specific activity >3000 Ci/mmol). For Southern blots, the DNAs are separated by using 1% horizontal agarose gels run in TAE buffer [0.04 M Tris-acetate and 0.001 M Na$_2$EDTA (pH 8.0), transferred onto nylon membranes (ICN Biomedicals, Costa Mesa, California, or Bio-Rad Laboratories, Richmond, California) by using 0.4 M NaOH solution, and fixed onto the membranes either by baking for 1 hour at 80°C or by UV irradiation. For dot and slot blots, the amplified DNA is denatured by adding a solution containing 0.1 volume 3 M NaOH and 0.1 M Na$_2$EDTA, incubating at 60°C for 15 minutes, and neutralizing with 1 volume cold 2 M ammonium acetate; the samples are then spotted onto Zeta probe nylon membranes (Bio-Rad) by using a BioRad dot or slot blot manifold at a 4 to 5 lb/in^2 vacuum pressure.

The amplified DNA immobilized on the nylon membrane is prehybridized with a hybridization solution containing (for Biotrans membranes) either 5× SSPE [1× SSPE is 10 mM sodium phosphate (pH 7.0), 0.18 M NaCl, 1 mM Na$_2$EDTA], 0.5% SDS, 5% Denhardt's solution, and 100 μg/ml of phenol-extracted, denatured salmon sperm DNA (Sigma). For Zeta probe membranes, 0.5 M NaH$_2$PO$_4$ (pH 7.2), 1 mM Na$_2$EDTA, and 7% SDS solution are used. Prehybridization is at 55°C for 15 minutes. The prehybridization buffer is removed and the membrane is hybridized with fresh hybridization solution containing 200 to 500 ng of denatured radiolabeled probe and incubated at 55°C for 3 to 16 hours. The blots are washed twice in 2× SSPE, 0.5% SDS at 25°C for 10 minutes each and once in 0.1× SSPE, 0.1% SDS at 55°C for 5 minutes with gentle agitation. To detect ^{32}P-labeled DNAs, the blots are covered with Saran Wrap and exposed to X-ray film at −70°C for 1 to 48 hours.

Alternately, DNA that is biotin- or ^{32}P-labeled during the PCR is detected by using unlabeled trapping probe (reverse hybridization). In this method, a poly(dT) tail (200 to 400 bases) is added to the probe enzymatically by using dTTP and deoxyribonucleotidyl terminal transferase or chemically during oligonucleotide synthesis. The poly(dT)-tailed probe is then affixed to a Zeta probe or Hybond-N (Amersham Corp., Arlington Heights, Illinois) membrane by exposure to UV light. Hybridization is done as described above, and ^{32}P-radiolabeled hybridized DNA is detected by autoradiography. Biotin-labeled hybridized DNA is detected by using either streptavidin-peroxidase development (See-Quence, Cetus Corp., Emeryville, California) with Zeta Probe membranes or streptavidin alkaline phosphatase development (BlueGene, Bethesda Research Labora-

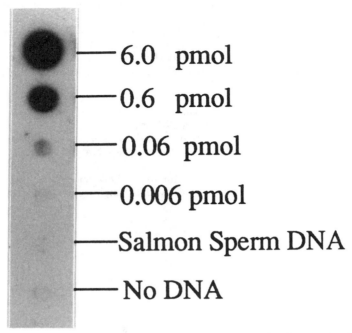

Figure 4   An example of detection of bacterial indicators of fecal contamination by using PCR-"reverse hybridization" (capture probe). Various amounts (0.006–6.0 pmol) of a 50-mer probe with a 400-T-tail specific for the detection of the *lamB* gene of *E. coli* were affixed onto a nylon membrane and subjected to hybridization with a PCR-amplified sample from 1 pg of genomic DNA of *E coli* with a total of 5 μg of nonspecific DNA from different microorganisms. The DNA was labeled with [α-^{32}P]dCTP during PCR amplification.

tories, Inc., Gaithersburg, Maryland) with Hybond-N membranes (Fig. 4).

Literature Cited

Atlas, R., A. Bej, R. Steffan, J. DiCesare, and L. Haff. 1989. Detection of coliforms in water by polymerase chain reaction (PCR) and gene probe methods. Poster presented ASM Annual Meeting.

Greenberg, H. (ed.). 1985. Standard methods for the examination of water and wastewater. American Public Health Association, Washington, D.C.

Hoadley, A. W., and B. J. Dutka (ed.). 1977. Bacterial indicators/health hazards associated with water. ASTM Publication 635. American Society for Testing and Materials, Philadelphia.

Jain, R. K., R. S. Burlane, and G. S. Sayler. 1988. Methods for detecting recombinant DNA in the environment. *Crit. Rev. Biotechnol.* **8**:33–84.

Li, H., U. B. Gyllensten, X. Cui, R. K. Saiki, H. A. Erlich, and N. Arnheim. 1988. Amplification and analysis of sequences in single human sperm and diploid cells. *Nature (London)* **335**:414–417.

Mahbubani, M., A. Bej, L. Haff, J. DiCesare, R. Miller, and R. M. Atlas. 1989. Detection of *Legionella* by using polymerase chain reaction. Poster presented ASM Conference on Biotechnology, Orlando, Florida.

Olive, D. M. 1989. Detection of enterotoxigenic *Escherichia coli* after polymerase chain reaction amplification with a thermostable DNA polymerase. *J. Clin. Microbiol.* **27**:261–265.

Starnbach, M. N., S. Falkow, and L. S. Tompkins. 1989. Species specific detection of *Legionella pneumophila* in water by DNA amplification and hybridization. *J. Clin. Microbiol.* **27**:1257–1261.

49

PCR IN THE DIAGNOSIS OF RETINOBLASTOMA

Sang-Ho Park

Retinoblastoma is an intraocular malignancy that occurs in children at a frequency of about 1 in 20,000 live births (for review, see Hansen and Cavenee 1988; Sparkes 1985). It can occur sporadically or can be inherited as a dominant trait with high penetrance. In the sporadic (nonhereditary) form, it is generally unilateral (occurring in only one eye), whereas in the hereditary form it is usually bilateral and multifocal. Statistical data compiled by Knudson suggest that retinoblastoma is caused by two mutational events (Knudson 1971). Knudson postulated that in the dominantly inherited form, one mutation was inherited through the germ cells, whereas in the nonhereditary form, both mutations have to occur in the somatic cells. The simplest model consistent with Knudson's data would be if there were a single gene whose inactivation caused retinoblastoma. Since the human genome is diploid, this would require mutations in both alleles. The sporadic form of retinoblastoma would require two somatic mutational events, while in the hereditary form, a predisposing mutation would be inherited through the germ line, thus requiring only a single mutation in the normal allele.

A gene has been cloned that exhibits all the properties of being the Retinoblastoma (*Rb*) gene (Friend *et al.* 1986; Lee *et al.* 1987). Data

PCR Protocols: A Guide to Methods and Applications

indicate that retinoblastoma occurs when the *Rb* gene is mutated in such a way that no functional *Rb* gene product is made (Cavenee *et al.* 1983; Murphree and Benedict 1984; Dryja *et al.* 1986). The *Rb* gene is also suspected to play a role in the development of other cancers in addition to retinoblastoma. This is based on two lines of evidence. The first is that patients diagnosed with hereditary retinoblastoma contract certain secondary tumors at a very high frequency. Osteosarcomas occur most frequently, but a number of other soft-tissue sarcomas occur as well (Friend *et al.* 1987). The second line of evidence is that the *Rb* gene has been found to be mutated in a number of different types of cancers, including small-cell lung carcinoma, breast cancer, and bladder carcinoma (Harbour *et al.* 1988; T'Ang *et al.* 1988; Horowitz *et al.* 1989).

Treatment of retinoblastoma varies depending on which form of retinoblastoma is diagnosed. In hereditary cases of retinoblastoma, because of the possibility of secondary tumors arising, the patient should be monitored for other cancers. In the case of nonhereditary retinoblastoma, subsequent monitoring of the patient is not necessary. In the hereditary form of retinoblastoma, there is a 50% chance that the child of an affected parent will inherit the predisposing gene, with a 90% chance of developing retinoblastoma. This makes it necessary to frequently examine the eyes of children that have a parent who had hereditary retinoblastoma. Ocular examinations are conducted under general anesthesia usually once every 3 months for the first few years of their lives. Even in the sporadic cases of retinoblastoma (about 94% of all cases of retinoblastoma), about 25 to 35% of these are germinal mutations, representing new carriers of a predisposing gene (Yanoff and Fine 1982). Thus it is necessary to have a means of determining which form of retinoblastoma a patient has and to be able to determine if a child of such a parent has inherited the mutated *Rb* gene. Also, recognition of a patient with hereditary retinoblastoma can direct the correct observation of other siblings who might be at risk for retinoblastoma.

The development of the PCR technology now makes it relatively easy to screen patients or possible carriers for aberrations in the *Rb* gene. By making PCR primers throughout the *Rb* gene, one can amplify and analyze tumor or tissue (i.e., blood) samples. The amplification is preceded by a reverse transcriptase reaction in which cDNA complementary to the RNA is made; this is amplified in the subsequent PCR. The PCR products can then be analyzed by using denaturing gradient gels (Lerman *et al.* 1984; Smith *et al.* 1986) or by doing RNA : DNA hybrid RNase protection assays (Myers *et al.* 1985). When looking for specific mutations, one can sequence or

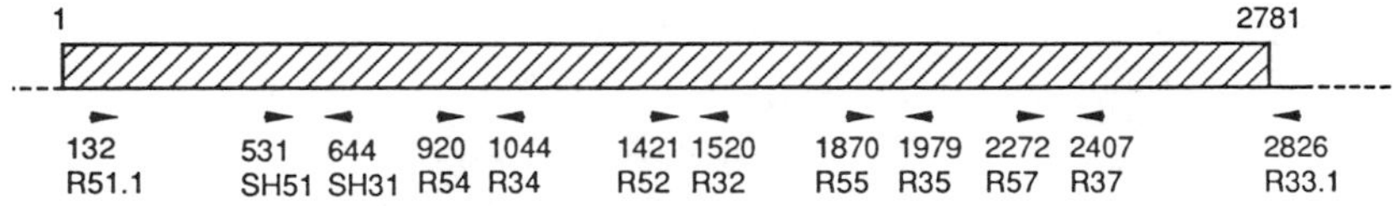

Figure 1   PCR primers for the *Rb* gene. The numbering starts at the first AUG of the coding region. (Not drawn to scale.)

probe the PCR products with oligonucleotides specific for those mutations. The PCR primers that I have synthesized were designed to generate 500- to 600-bp fragments with about a 100-bp overlap (Fig. 1). All were 20-mers, ranging from 8 to 12 GC per oligonucleotide. The 5' end of the coding region of the *Rb* gene could not be readily amplified because it is a very G+C-rich region (position 3 to 139; the numbering is relative to the first AUG). Several primers were used with little success, so an alternate primer just 3' of the G + C-rich region has been used. This means that the PCR products from the primers currently being used and their subsequent analysis are going to be blind to mutations in the 150 bp at the 5' end of the gene.

By using PCR, we have been able to identify several mutations in the *Rb* gene, one of which is explained in Horowitz *et al.* (1989). In this chapter we show a cell line in which there is an in-frame deletion of 105 bps in the *Rb* gene. (Fig. 2) This mutation was initially discovered as a cell line that gave an altered *Rb* gene product as identified by immunoprecipitation using an anti-*Rb* antibody. The RNA from this cell line was then amplified as explained in this chapter. Sequence analysis of the PCR product showed precisely where this deletion had occurred and how many base pairs it involved.

Protocols

Enzymes and Buffers

   Taq DNA polymerase—Perkin-Elmer Cetus
   Reverse transcriptase (RT)—Bethesda Research Laboratories (BRL), cloned M-MuLV reverse transcriptase
   RNasin—Promega Biotec
   $10\times$ *Taq* polymerase buffer
      500 mM KCl
      200 mM Tris–HCl (pH 8.0)

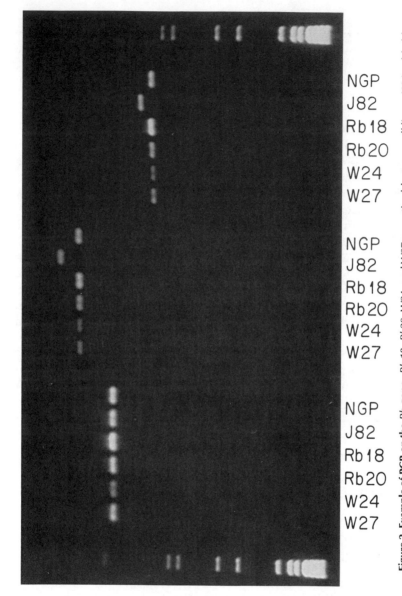

NGP
J82
Rb18
Rb20
W24
W27

NGP
J82
Rb18
Rb20
W24
W27

NGP
J82
Rb18
Rb20
W24
W27

Figure 2. Example of PCR on the *Rb* **gene.** *Rb* 18, *Rb20*, *W24*, and *W27* are retinoblastoma cell lines; J82 is a bladder carcinoma; NGP is a control cell line for the *Rb* gene. The primers used were R55-R33 for the first six, R55-R37 for the second, and R57-R33 for the last six. One-tenth of each reaction was loaded on each lane. The size marker is λ-BstE II.

25 mM MgCl$_2$
200 μg/μl bovine serum albumin
5× RT buffer
    250 mM Tris–HCl (pH 8.3)
    375 mM KCl
    50 mM DTT
    15 mM MgCl$_2$
    (The 5× RT buffer provided by BRL was used)
dNTPs and oligo(dT)$_{16}$—Pharmacia

cDNA Reaction

1. Mix the following:

5× reverse transcriptase buffer	4 μl
H$_2$O	10.7 μl
10 mM dNTP	2 μl
RNasin 40 units/μl	0.3 μl
primer	1 μl
M-MuLV reverse transcriptase 200 units/μl	1 μl
RNA 1 to 5 μg/μl	1 μl
	20 μl

    primer—0.5 μg/μl oligo(dT)$_{16}$ or 10 pM/μl downstream oligo
2. Incubate at 37°C for 1 hour when priming with oligo(dT) or 30 minutes when priming with the downstream oligo.
3. Heat to 75°C for 5 minutes.
4. Chill and spin down.

Polymerase Chain Reaction

1. To the 20-μl cDNA reaction, add the following:

10× *Taq* Buffer		8 μl
H$_2$O		61.8 μl
upstream oligo	10 pM/μl	5 μl
downstream oligo	10 pM/μl	5 μl
Taq polymerase	5 units/μl	0.2 μl
		100 μl

(when using the downstream oligo to prime the reverse transcriptase, add 4 μl of the downstream oligo and 62.8 μl H$_2$O)

2. Cover with 3 to 4 drops of mineral oil (enough to cover the reaction mixture).
3. PCR (explained in section on PCR Temperature Profile)
4. Add chloroform (twice the volume of the mineral oil).
5. Vortex. Spin.
6. Take 10 μl of the aqueous phase and run it out on an agarose
   gel to check the reaction.

PCR Temperature Profile

Forty cycles of amplification are performed using the Thermo-Cycle
program of the DNA Thermal Cycler (Perkin-Elmer Cetus). A cycle is
1 minute at 94°C, 30 seconds at 55°C, and 2 minutes at 72°C. A 30-
second time ramp is inserted between 55 and 72°C and 72 and 94°C,
and a 1-minute time ramp is inserted between 94 and 55°C. The
samples are initially denatured for 1 minute at 94°C before the cycling and subjected to a 10-minute final extension at 72°C at the end
of the cycling.

   A routine reaction (for six PCRs) would go as follows:

1. Pellet RNA and resuspend in DEPC-treated H$_2$0.
2. Measure the OD$_{260}$.
3. Taking enough for 6 μg of RNA, bring the total volume up to
   64.2 μl with H$_2$O.
4. Add the buffer, RNasin, dNTPs, primer, and enzyme; let the reaction go for 1 hour.
5. In the meantime, set up and label the tubes for the PCR.
6. Add the primers to each tube.
7. In a large microfuge tube combine 48 μl PCR buffer, 370.8 μl
   H$_2$O, and 1.2 μl *Taq* polymerase. Aliquot 70 μl of this to each of
   the PCR tubes.
8. Heat, cool, and spin down the tube with the reverse transcrip-

tase reaction. Aliquot 20 μl into each of the PCR tubes, add mineral oil, and spin down.

9. PCR.

Discussion

In most of the reactions, 1 μg of total cellular RNA was used. The RNA was prepared using the guanidinium isothiocyanate method (Davis *et al.* 1986). Although the amount of RNA per reaction has not been optimized, occasional instances where less than 0.5 μg were used gave visibly fainter bands, whereas too much RNA seemed to give more background. Oligo(dT)$_{16}$ was used routinely to prime the reverse transcriptase reaction. The amount used was 0.5 μg per reaction. Using the downstream oligonucleotide to prime the reverse transcriptase reaction generally resulted in a higher background compared to that of the oligo(dT)-primed reactions, at least for the primers that have been used in this lab. The reason for this is still not clear. Since multiple PCRs were run on each RNA sample, being limited to using oligo(dT) to prime the reverse transcriptase reaction was not a problem. It was possible to scale up the reverse transcriptase reaction sixfold for the six different primer pairs, priming with oligo(dT) in a single reaction instead of having to do six separate reverse transcriptase reactions. The concentration of primers used in the PCRs was generally 50 pM per reaction. While this could be cut down to 25 pM with no significant loss in the specificity or intensity of the PCR product when run out on a gel, increasing the concentration usually resulted in a higher background and decrease in final specific product.

An example of PCR on the *Rb* gene is given in Fig. 2. PCR was conducted on the RNA from four different retinoblastoma cell lines, one bladder carcinoma cell line, and one neuroblastoma cell line used as a wild-type control for the *Rb* gene. Total cellular RNA was prepared from each of the cell lines, of which 1 μg was used for each reaction. The primers used were R55-R33 for the first set of six, R55-R37 for the second set, and R57-R33 for the last set. As is evident from Fig. 2 alone, the bladder carcinoma cell line J82 contains a small deletion of about 100 bp. This was sequenced and shown to be the result of a splicing error due a point mutation in the splice acceptor site (for more information, refer to Horowitz *et al*. 1989). Some of the other cell lines that do not show any aberrations in size are now being analyzed more carefully for more minute mutations.

Literature Cited

Cavenee, W. K., T. P. Dryja, R. A. Phillips, W. F. Benedict, R. Godbout, B. L. Gallie, A. L. Murphree, L. C. Strong, and R. L. White. 1983. Expression of recessive alleles by chromosomal mechanisms in retinoblastoma. *Nature (London)* **305**: 779–783.

Davis, L. G., M. D. Dibner, and J. F. Battey. 1986. *Basic methods in molecular biology*. Elsevier Science Publishing Co., Inc., New York.

Dryja, T. P., M. M. Rapaport, J. M. Joyce, and R. A. Petersen. 1986. Molecular detection of deletions involving band q14 of chromosome 13 in retinoblastomas. *Proc. Natl. Acad. Sci. USA* **83**: 7391–7394.

Friend, S. H., R. Bernards, S. Rogelj, R. A. Weinberg, J. M. Rapaport, D. M. Albert, and T. P. Dryja. 1986. A human DNA segment with properties of the gene that predisposes to retinoblastoma and osteosarcoma. *Nature (London)* **323**: 643–646.

Friend, S. H., J. M. Horowitz, M. R. Gerber, X.-F. Wang, E. Bogenmann, F. P. Li, and R. A. Weinberg. 1987. Deletions of a DNA sequence in retinoblastoma and mesenchymal tumors: organization of the sequence and its encoded protein. *Proc. Natl. Acad. Sci. USA* **84**: 9059–9063.

Hansen, M. F., and W. F. Cavenee. 1988. Retinoblastoma and the progression of tumor genetics. *Trends in Genetics* **4**(5): 125–128.

Harbour, J. W., S.-L. Lai, J. Whang-Peng, A. F. Gazdar, J. D. Minna, and F. J. Kaye. 1988. Abnormalities in structure and expression of the human retinoblastoma gene in SCLC. *Science* **241**: 353–357.

Horowitz, J. M., D. W. Yandell, S.-H. Park, S. Canning, P. Whyte, K. Buchkovich, E. Harlow, R. A. Weinberg, and T. P. Dryja. 1989. Point mutational inactivation of the retinoblastoma antioncogene. *Science* **243**: 937–940.

Knudson, A. G. 1971. Mutation and cancer: statistical study of retinoblastoma. *Proc. Natl. Acad. Sci. USA* **68**(4): 820–823.

Lee, W.-H., R. Bookstein, F. Hong, L.-J. Young, J.-Y. Shew, and E. Y.-H. P. Lee. 1987. Human retinoblastoma susceptibility gene: cloning, identification, and sequence. *Science* **235**: 1394–1399.

Lerman, L. S., S. G. Fischer, I. Hurley, K. Silverstein, and N. Lumelsky. 1984. Sequence-determined DNA separations. *Ann. Rev. Biophys. Bioeng.* **13**: 399–423.

Murphree, A L., and W. F. Benedict. 1984. Retinoblastoma: clues to human oncogenesis. *Science* **223**: 1028–1033.

Myers, R. .M., Z. Larin, and T. Maniatis. 1985. Detection of single base substitutions by ribonuclease cleavage at mismatches in RNA:DNA duplexes. *Science* **230**: 1242–1246.

Smith, F. I., J. D. Parvin, and P. Palese. 1986. Detection of single base substitutions in influenza virus RNA molecules by denaturing gradient gel electrophoresis of RNA-RNA or DNA-RNA heteroduplexes. *Virology* **150**: 55–64.

Sparkes, R. S. 1985. The genetics of retinoblastoma. *Biochimica et Biophysica.* **780**:95.

T'Ang, A., J. M. Varley, S. Chakraborty, A. L. Murphree, and Y.-K. T. Fung. 1988. Structural rearrangement of the retinoblastoma gene in human breast carcinoma. *Science* **242**:263–266.

Wiggs, J., M. Nordenskjöld, D. Yandell, J. M. Rapaport, G. Valerie, M. Janson, B. Werelius, R. A. Petersen, A. Craft, K. Riedel, R. Liberfarb, D. Walton, W. Wilson, and T. P. Dryja. 1988. Prediction of the risk of hereditary retinoblastoma, using DNA polymorphisms within the retinoblastoma gene. *N. Eng. J. Med.* **318**:151–157.

Yanoff, M. and B. S. Fine. 1982. Retinoblastoma and pseudoglioma. *Biochemical Foundation Ophthalmology* **3**:1–12.

DETERMINATION OF FAMILIAL RELATIONSHIPS

Cristián Orrego and Mary Claire King

The examination of sequence polymorphisms in DNA is now the most powerful tool to genetically distinguish among individuals or to confirm familial relationships. Until recently only blood group antigens, HLA antigens, and the electrophoretic variants of a large set of polymorphic proteins had been used for such purposes. Protein markers require assay procedures that are tailored to each protein and demand a diversity of skills and reagents that can be quite daunting. On the other hand, the detection of DNA polymorphisms is susceptible to analytical procedures, each of which simultaneously reveals an extraordinarily high number of alleles. For example, restriction fragments that differ in length because of the existence of tandem repeats that vary in number at different loci dispersed throughout the genome (Jeffreys 1987; Nakamura et al. 1987) can generate, with only one probe, fingerprint patterns that are individual-specific (Jeffreys et al. 1985a; Fowler et al. 1988). In addition to making it possible to scan large numbers of sites in the genome for genes linked to those responsible for inherited diseases (Jeffreys 1987), DNA fingerprinting has expanded the scope of determination of familial relationships (Jeffreys et al. 1985b) and the genetic analysis of forensic samples (Gill et al. 1985).

However, DNA recovered from environmentally compromised

samples is often of short length (Pääbo 1989; Pääbo *et al*. 1989), which precludes the elucidation of genetic variation by the restriction fragment length polymorphism approach. As an alternative, Higuchi *et al*. (1988) used PCR to amplify discrete DNA regions covering the multiallelic HLA gene *DQα* and analyzed the resulting fragments by hybridization with allele-specific oligonucleotide probes.

The rapidly evolving noncoding region containing the D loop in the mitochondrial (mt) genome (Aquadro and Greenberg 1983; Cann *et al*. 1987) is another DNA sequence that appears suitable for ascertaining genetic individuality (Higuchi *et al*. 1988). The maternal inheritance of the animal mitochondrial genome (Giles *et al*. 1980) makes D loop sequencing via PCR an attractive tool for testing whether individuals are maternally related. However, no significant population data have been available to assess the potential of the D loop region in this respect. This chapter presents the first such assessment, using sequences obtained from a 347-bp section at one end of the D loop from a panel of unrelated individuals of Caucasian origin. This short span of the D loop was chosen, as it reveals more than half of the variation found in the entire noncoding region (Greenberg *et al*. 1983; Vigilant *et al*., in press). The distribution of pairwise differences presented here indicates that the probability that two unrelated individuals are genetically identical for this section of mtDNA is 0.27%. A feature of this approach that distin-

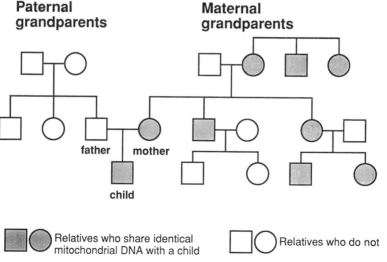

Figure 1 Family tree displaying maternal inheritance of the mitochondrial chromosome through three generations.

guishes it from all others emerges when there is only one surviving relative and he or she is removed by more than one generation; disputed identity can still be resolved provided that kinship has preserved the maternal lineage (Fig. 1).

Strategy for the Amplification and Direct Sequencing of the D Loop

The desired fragment from the mitochondrial noncoding region is obtained in two stages to secure optimal specificity of the amplification. The first stage amplifies the entire noncoding region by using

Table 1

Amplification and Sequencing Primers for the Human Mitochondrial D Loop

Primer[a] (gene location; otherwise in D loop)	Sequence (5' to 3')	Bp	Designer
L 15926 (tRNA[thr])	TCAAAGCTTACACCAGTCTTGTAAACC	27	T.D.K.[b]
H 00580 (tRNA[phe])	TTGAGGAGGTAAGCTACATA	20	M.S.[b]
L 15997 (tRNA[pro])	CACCATTAGCACCCAAAGCT	20	C.O.
H 16401	TGATTTCACGGAGGATGGTG	20	R.G.H.[b]
L 16099	AACCGCTATGTATTTCGTAC	20	R.G.H.[b]
L 16209	CCATGCTTACAAGCAAGT	18	T.D.K.[b]
H 16255	CCTAGTGGGTGAGGGGTGGC	20	R.G.H.[b]

[a] Primers are identified by a letter designating the strand of mtDNA and a number corresponding to the reference sequence coordinate (Anderson *et al.* 1981) of the base at the 3' end of the primer. L is the noncoding strand of mtDNA presented in the reference sequence. Primers L 15926 and L 15997 have their 3' ends in the sequence coding for the anticodon stem of threonine tRNA and proline tRNA, respectively (Sprinzl and Gauss 1984). Primer H 00580 has its 3' end in the sequence coding for the aminoacyl stem of phenylalanine tRNA (Sprinzl and Gauss 1984). Primers positioned within the D loop were directed to sequences displaying minimal variation from a comparison of seven human individuals (Aquadro and Greenberg 1983).

[b] Initials of investigators mentioned in the acknowledgments. Otherwise those of authors.

primers that are complementary to the relatively well-conserved flanking tRNA^{thr} and tRNA^{phe} genes (Table 1 and Fig. 2A). This amplification product is then further purified by electrophoresis on an agarose gel. Primers internal to sequences in this fragment are used in a second stage of amplification to obtain the first 443-bp of the noncoding region (Fig. 2B, Table 1) in which most of the variation is found (Greenberg *et al.* 1983; Vigilant *et al.*, in press). In this case the PCR is performed with unequal concentrations of primers (Table 2) to generate single strands (Gyllensten and Erlich 1988) suit-

A. First-Stage Amplification

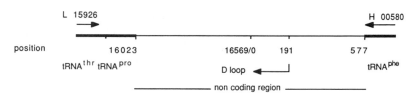

B. Second-Stage Amplification

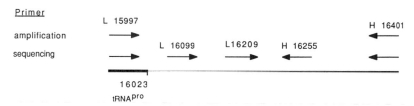

Figure 2 Amplification and sequencing primers for the D loop of human mtDNA. The tip of the arrows corresponds to the 3′ end of the primer.

Table 2

Protocol for Primer Use

Amplification stage	Amplification primers	Relative primer amounts	Strand amplified	Product size (bp)	Sequencing primers
First	L 15926 H 00580	1 : 1	L : H	1269	
Second	L 15997 H 16401	50 : 1	L	443	H 16401 H 16255
	H 16401 L 15997	50 : 1	H	443	L 15997 L 16099 L 16209

able for sequencing by the dideoxy chain-termination technique (Sanger *et al.* 1977).

Protocols

DNA Extraction Procedure

Blood samples in Acid Citrate Dextrose tubes (ACD, Solution A, Becton Dickinson Vacutainer Systems) were subjected to differential centrifugation to enrich for white blood cells within three days after collection. This fraction was frozen at −20°C prior to total DNA extraction by the sodium N-lauroylsarcosine-Proteinase K, phenol, and RNase method (Maniatis *et al.* 1982).

Reagents for PCR

10× Amplification Buffer (store frozen)
  670 mM Tris–HCl (pH 8.8) at 25°C
  20 mM MgCl$_2$
  166 mM ammonium sulfate
  100 mM β-mercaptoethanol
20× dNTP Stock Mixture (store frozen)
  5 mM each of dATP, dGTP, dTTP, dCTP in water (diluted from neutralized stocks purchased from Pharmacia)
Amplification Primers
  Oligonucleotides received in ammonium hydroxide are dried under vacuum centrifugation and resuspended in sterile glass-distilled water. Primer concentration is determined by spectrophotometry at 260 nm using extinction coefficients calculated on the basis of the nucleotide composition of each primer. Working stocks of primers at 10 and 0.2 μM concentrations are prepared in water.

PCR Procedure

First-Stage Amplification

1. Amplification is performed with 0.25 to 1 μg of total DNA as template. The entire noncoding region is amplified with primers L 15926 and H 00580 at 1 μM final concentration (Table 1 and Fig. 2A) in 25-μl reactions with 1.25 units of the *Thermus aquaticus* DNA polymerase (Saiki *et al*. 1988). Cycling under the control of a Perkin-Elmer Cetus Thermal Cycler is performed as follows: denaturation at 94°C for 45 seconds; annealing at 50°C for 1 minute; and extension at 72°C for 5.5 minutes, for a total of 25 cycles.

2. One-fifth of the reaction is subjected to electrophoresis in a 2- to 3-mm thick gel of 3% NuSieve agarose in Tris-borate-EDTA buffer (Maniatis *et al*. 1982). Electricity is applied at 10 V/cm of gel for 4 hours to obtain optimal separation of the expected product (1333 bp) from other unintended amplified sequences. The desired fragment, stained with ethidium bromide and visualized by fluorescence under ultraviolet light, is pierced twice from the gel with a sterile Pasteur pipette, and the gel is delivered into a microfuge tube. The material is then melted at 65°C for 10 minutes, and a fourfold dilution is obtained in glass-distilled water (approximate concentration: 2 ng DNA/μl). Care should be taken not to expose the amplified products to UV irradiation for longer than 15 seconds.

Second-Stage Amplification

1. The fragment (1 to 2 ng) obtained from the previous procedure is used to generate an excess of single-stranded template by unbalanced priming of the PCR (Gyllensten and Erlich 1988) with primers L 15997 and H 16401 (Fig. 2B, Table 1). A ratio of 50 : 1 for the two primers (final concentrations 1 and 0.02 μM, respectively) is used in two separate reactions to independently generate an excess of each strand with 5 units of *Taq* DNA polymerase in 100 μl. Cycling parameters: 94°C, 45 seconds; 50°C, 1 minute; 72°C, 3 minutes, for 31 cycles.

2. The reaction mixture is subjected to four cycles of centrifugation dialysis with 2 ml of water on a Centricon 30 device (Ami-

con/W. R. Grace) to remove salts and unincorporated nucleotides. The final volume is 40 to 50 μl.

DNA Sequencing

A portion (7 λ) of the fraction obtained from the previous procedure is used for sequencing with the Sequenase Kit (U.S. Biochemical protocol). Five sequencing primers (10 pmol) spaced about every 120 bp and directed to both the H and L strands are used (Fig. 2B, Table 1) to generate sequences uncomplicated by tight spacing in the ladder and to resolve sequence ambiguities that may be found in one or the other strand. Electrophoresis on a 6% polyacrylamide sequencing gel was conducted to obtain sequence data starting as close to each of the sequencing primers as possible.

Population Genetics of the Human D Loop

The first mtDNA sequence published (Anderson *et al.* 1981) is generally used as the reference sequence for human populations. In addition to the D loop from this Caucasian individual (ascertained by phylogenetic analysis; Cann *et al.* 1987 and A. di Rienzo, personal communication), sequences from 13 other unrelated Caucasians can now be compared from sequence positions 16042 to 16388 (Fig. 3). These 14 individuals yield 91 pairs of sequences for comparison. No pair was identical for the 347 nucleotide sequence. The average number of nucleotide differences between such pairs was 5.90, with standard deviation 2.43. The distribution of the number of nucleotide differences per pair of individuals is shown in Fig. 4. The Poisson distribution (also shown in Fig. 4) with mean = 5.90 provides a good fit to these data: $\chi^2 = 12.81$ with 12 degrees of freedom, $P > 0.40$ (the P value is only approximate because the 91 comparisons are not strictly independent). On the basis of the Poisson distribution, the probability that a pair of unrelated Caucasian individuals will be identical by chance over this genetic region is

$$P(O) = e^{-5.9} = 0.0027$$

Base#	16069	16092	16111	16114	16124	16126	16129	16145	16160	16163	16172	16176	16184	16187	16189	16192
SEQ. 1	C	T	C	C	T	T	G	G	A	A	T	C	C	C	T	C
2	T	.	.	.	.	C	.	.	G	.	.	.	.	.	.	.
3	.	.	T	.	.	.	.	.	.	.	.	.	.	.	.	.
4	.	.	.	.	.	.	A	.	.	.	G	.	.	.	.	.
5	.	.	.	.	.	.	.	.	.	.	.	.	.	.	.	.
6	.	.	T	.	.	.	.	.	.	.	.	.	.	.	C	.
7	.	.	.	A	.	.	A	.	.	.	.	.	.	.	C	T
8	.	C	.	.	.	.	.	.	.	.	.	.	.	.	.	.
9	.	.	.	.	.	C	.	.	G	.	.	.	T	.	-	.
10	.	.	.	.	.	.	.	.	.	.	.	T	.	.	.	.
11	.	.	.	.	.	.	.	.	.	.	.	.	.	.	.	T
12	o	o	o	o	C	.	.	.	.	.	.	.	.	.	.	.
13	o	o	o	o	o	o	o	.	.	.	C	.	.	.	.	.
14	o	o	o	o	o	o	o	.	.	.	.	.	.	.	.	.

BASE#	16222	16223	16224	16256	16270	16278	16290	16292	16293	16294	16298	16304	16311	16319	16320	16362
SEQ. 1	C	C	T	C	C	C	C	C	A	C	T	T	T	G	C	T
2	T	.	.	.	.	.	.	.	.	.	.	.	.	.	.	.
3	.	T	.	.	.	.	T	.	.	.	.	.	C	A	.	C
4	.	T	.	.	.	.	.	.	.	.	.	.	.	.	.	.
5	.	.	.	.	.	.	.	.	.	.	C	.	.	.	.	.
6	.	T	.	.	.	.	T	.	.	.	.	.	.	A	T	C
7	.	.	.	T	T	.	.	.	.	T	.	.	.	.	.	.
8	.	.	.	.	.	.	.	.	G	.	.	.	C	.	.	.
9	.	.	.	.	.	.	.	.	.	T	.	.	.	.	.	.
10	.	T	.	.	.	.	.	T	.	.	.	.	.	.	.	.
11	.	.	.	.	.	.	.	.	.	.	.	.	C	.	.	.
12	.	-	.	.	.	T	.	.	.	.	.	.	.	.	.	.
13	.	.	.	.	.	.	.	.	.	.	.	C	.	.	.	.
14	.	.	C	.	.	.	.	.	.	.	.	.	C	.	.	.

Figure 3 Variable sites in the D loop of Caucasians. Base # is the position coordinate of the noncoding region in the reference sequence (Anderson *et al.* 1981) presented here as sequence 1. Dots indicate identity with such sequence, dashes represent deletions, and empty circles represent regions not sequenced. Sequences 2–6 are from Argentineans. Sequences 7–9 are from three grandmothers from Family 981, Utah pedigree K-1329 (National Institute of General Medical Sciences Human Genetic Mutant Cell Repository: 1988–1989 Catalog of Cell Lines, p. 592–593). Sequence 10 is from a member of the third generation from Family 981. Sequence 11 is an anonymous sequence from the DNA collections at the Centre d'Étude du Polymorphisme Humain. Sequences 12, 13, and 14 are from KB, pBHK2, and pLKK3, respectively (Greenberg *et al.* 1983). Sequence information for sequences 1–11 is available from position 16042 to 16388 and for sequences 12–14 from at least the earliest variable position for which sequence information is presented in the figure up to position 16388.

or one in 370. This is a preliminary estimate that requires additional sequences from Caucasians and other groups to more fully assess the randomness or independence of distribution of D loop sequences in human populations suggested by the goodness of fit of the data presently available to a Poisson curve. Amplification and sequencing of the rest of the noncoding region should also be of help in this regard.

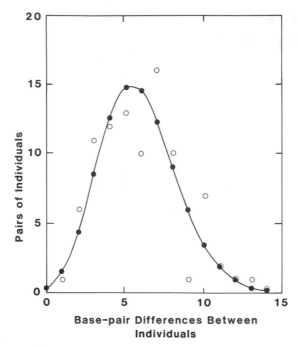

Figure 4   mtDNA differences among pairs of Caucasians (Figure 3). Observed (○) and expected (·) number of base pair differences. The expected curve is based on a Poisson distribution with mean = 5.90.

Determination of Maternal Relationships via mtDNA D Loop Sequencing

To test the power of mtDNA sequencing to identify maternal relationships, we sequenced the 347-bp segment from the maternal grandmother and a child from each of three Caucasian families. The six individuals were sequenced without family identification. The grandmother–grandchild pairs were unambiguously assigned (Fig. 3: sequences 7, 8, and 9).

As with any other genetic identification technique, mtDNA sequencing is most informative in large populations without inbreeding. If a population has only a few maternal lineages, apparently unrelated individuals will be more likely to share mtDNA sequences. Also, it is quite possible that not all mtDNA types are equally represented in human populations and that we have not yet found from such a small sample size as is presented in this survey a type from the D loop that is more frequent in the population and thus shared among otherwise unrelated individuals. This preliminary study indicates that mtDNA amplification and sequencing of the D loop appears uniquely suitable for determining familial relationships when only one relative is available, provided that the maternal lineage has not been broken.

As breeding habits in most mammals range from polygynous to promiscuous (Greenwood 1980), amplification and direct sequencing of an extremely variable region such as the D loop from the mitochondrial genome should also prove useful in the elucidation of complex social organizations. Knowledge of the mother of an individual will enhance the power of DNA fingerprinting (Hill 1987) for the determination of paternity in field populations (Lynch 1988).

Acknowledgments

We thank T. D. Kocher, M. Stoneking, R. G. Higuchi, E. M. Prager, A. C. Wilson, and J. L. Patton for advice; B. A. Malcolm for primer synthesis, J. Crandall for sample processing, J. Hall for facilitating the DNA matching study, M. Lee for calculations; the Laboratorio de Immunogenética, Hospital Durand, and Abuelas de Plaza de Mayo in Buenos Aires, Argentina, for samples. This work received support from National Science Foundation grant BSR8415867 and National Institutes of Health grant GM21509 to A. C. Wilson and National Institutes of Health grants CA27632 and GM28428 to M. C. King and a grant from the J. Roderick MacArthur Foundation to the American Association for the Advancement of Science and thence to C. Orrego.

Literature Cited

Anderson, S., A. T. Bankier, B. G. Barrell, M. H. L. de Bruijn, A. R. Coulson, J. Drouin, I. C. Eperon, D. P. Nierlich, B. A. Roe, F. Sanger, P. H. Schreier, A. J. H. Smith, R. Staden, and I. G. Young. 1981. Sequence and organization of the human mitochondrial genome. *Nature (London)* **290**:457–465.

Aquadro, C. F., and B. D. Greenberg. 1983. Human mitochondrial DNA variation and evolution: analysis of nucleotide sequences from seven individuals. *Genetics* **103**:287–312.

Cann, R. L., M. Stoneking, and A. C. Wilson. 1987. Mitochondrial DNA and human evolution. *Nature (London)* **325**:31–36.

Fowler, S. J., P. Gill, D. J. Werrett, and D. R. Higgs. 1988. Individual specific DNA fingerprints from a hypervariable region probe: alpha-globin 3'HVR. *Hum. Genetics* **79**:142–146.

Giles, R. E., H. Blanc, H. M. Cann, and D. C. Wallace. 1980. Maternal inheritance of human mitochondrial DNA. *Proc. Natl. Acad. Sci. USA* **77**:6715–6719.

Gill, P., A. J. Jeffreys, and D. J. Werrett. 1985. Forensic application of DNA "fingerprints." *Nature (London)* **318**:577–579.

Greenberg, B. D., J. F. Newbold, and A. Sugino. 1983. Intraspecific nucleotide sequence variability surrounding the origin of replication in human mitochondrial DNA. *Gene* **21**:33–49.

Greenwood, P. J. 1980. Mating systems, philopatry and dispersal in birds and mammals. *Animal Behav.* **28**:1140–1162.

Gyllensten, U. B., and H. A. Erlich. 1988. Generation of single-stranded DNA by the polymerase chain reaction and its application to direct sequencing of the *HLA-DQα* locus. *Proc. Natl. Acad. Sci. USA* **85**:7652–7656.

Higuchi, R., C. H. von Beroldingen, G. F. Sensabaugh, and H. A. Erlich. 1988. DNA typing from single hairs. *Nature (London)* **332**:543–546.

Hill, W. G. 1987. DNA fingerprinting applied to animal and bird populations. *Nature (London)* **327**:98–99.

Jeffreys, A. J., V. Wilson, and S. L. Thein. 1985a. Individual-specific "fingerprints" of human DNA. *Nature (London)* **316**:76–79.

Jeffreys, A. J., J. F. Y. Brookfield, and R. Semeonoff. 1985b. Positive identification of an immigration test-case using human DNA fingerprints. *Nature (London)* **317**:818–819.

Jeffreys, A. J. 1987. Highly variable minisatellites and DNA fingerprints. *Biochem. Soc. Trans.* **15**:309–317.

Lynch, M. 1988. Estimation of relatedness by DNA fingerprinting. *Mol. Biol. Evol.* **5**:584–599.

Maniatis, T., E. F. Fritsch, and J. Sambrook. 1982. *Molecular cloning: a laboratory manual.* Cold Spring Harbor Laboratory, Cold Spring Harbor, New York.

Nakamura, Y., M. Leppert, P. O'Connell, R. Wolff, T. Holm, M. Culver, C. Martin, E. Fujimoto, M. Hoff, E. Kumlin, and R. White. 1987. Variable number of tandem repeat (VNTR) markers for human gene mapping. *Science* **235**:1616–1622.

Pääbo, S. 1989. Ancient DNA: extraction, characterization, molecular cloning, and enzymatic amplification. *Proc. Natl. Acad. Sci. USA* **86**:1939–1943.

Pääbo, S., R. G. Higuchi, and A. C. Wilson. 1989. Ancient DNA and the polymerase chain reaction. The emerging field of molecular archeology. *J. Biol. Chem.* **264**:9709–9712.

Saiki, R. K., D. H. Gelfand, S. Stoffel, S. J. Scharf, R. Higuchi, G. T. Horn, K. B. Mullis, and H. A. Erlich. 1988. Primer-directed enzymatic amplification of DNA with a thermostable DNA polymerase. *Science* **239**:487–491.

Sanger, F., S. Nicklen, and A. R. Coulson. 1977. DNA sequencing with chain-terminating inhibitors. *Proc. Natl. Acad. Sci. USA* **74**:5463–5467.

Sprinzl, M., and D. H. Gauss. 1984. Compilation of tRNA sequences. *Nucleic Acids Res.* **12** (supplement) r1-r58.

Vigilant, L., R. Pennington, H. Harpending, T. D. Kocher, and A. C. Wilson. Mitochondrial DNA sequences in single hairs from a Southern African population. *Proc. Natl. Acad. Sci. USA,* in press.

Part Five

INSTRUMENTATION AND SUPPLIES

51

PCR IN A TEACUP

A Simple and Inexpensive Method for Thermocycling PCRs

Robert Watson

PCRs require repeated thermocycling between at least two temperatures, a high temperature adequate for sample-template and subsequent product denaturation, and a low temperature appropriate for primer annealing and polymerase extension. A third, intermediate temperature near the optima for polymerase extension is often used. While thermocycling can be easily accomplished by alternating the samples between two or more water baths, the tedium involved may provoke rebellion and can lead to the premature termination of thermocycling. A variety of commercial devices have been developed and are being marketed for thermocycling; most are designed for use with 0.5-ml centrifuge tubes. The use of nonstandard containers requires an alternative to the commercially available instruments. Also, colleagues and collaborators unable to afford or obtain automated thermocyclers that use local voltages and frequencies have asked for advice on implementing a semiautomated method of thermocycling. Finally, the challenge of reducing the required hardware and using methods as simple and inexpensive as possible has lead to the concept presented here.

The required hardware includes a timer/controller, a heater, a solenoid water valve, and a reaction chamber. A pressure regulator and

PCR Protocols: A Guide to Methods and Applications
429

flow meter with an adjusting valve may be added to improve the performance. A means for monitoring the temperature is also described. The timer activates both the heater that heats a small volume of water (about 150 ml) in the reaction chamber and a solenoid water valve that controls cooling tap water. Samples are supported by a piece of foam rubber or plastic and are floated on the surface of the water in the reaction chamber. This approach does not actively regulate the temperature but rather achieves reproducible temperature profiles by precise timing and consistent heating and cooling rates. Reproducible cooling requires that the water pressure and flow rate be kept constant. Room temperature changes will affect both heating and cooling and can be controlled in part by insulation of the reaction vessel.

The timer/controller, a Lindberg Industries ChronTrol CD4, is an inexpensive, microprocessor-controlled timer that can be programmed to turn several electrical outlets on or off and to loop or repeat its instructions indefinitely. It is available for use with 115 V, 60 Hz or 240 V, 50 Hz line current. The heater is a 200-W submersible heater intended for use with teacups and coffee cups and is also available for both voltages and frequencies. Unfortunately, the heaters come in two forms that cannot be differentiated in appearance, those which do not self destruct for many cycles and those which do. Fortunately, the wide availability and low price of the two types minimize this problem. The solenoid valve is a washing machine water valve that is widely available, inexpensive, and very reliable. The reaction chamber is an open design of modest size to minimize the volume of water contained and the time to heat it. It has an entry port for water from the solenoid valve and a large overflow port near its top, room for a magnetic stirring bar in the bottom, and a means to secure the heater. Local machine shops can fabricate such a chamber out of plastic, or local glassblowers may construct or modify a glass container. Alternatively, a 150-ml-tall form beaker or a ceramic coffee cup or teacup may be used. The chamber should be placed on a magnetic stirrer and, if necessary, a "catch basin" to collect the spillover and direct it to a drain. A ring stand and clamps can help support the plumbing and heater. Most hardware and plumbing supply stores have small and inexpensive pressure regulators that will reduce the variable domestic water pressure to a constant. A flow meter and adjusting valve placed after the pressure regulator will allow for fine tuning of the flow rate. Reducing the water pressure to 10 lb/in^2 and adjusting the flow rate to 8 gallons/hour (about

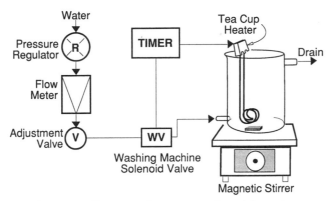

Figure 1   A diagrammatic representation of the system.

0.5 liters/minute) has worked well. Figure 1 is a diagrammatic representation of the system.

The thermal performance of the system can be measured by using temperature probes made of Type K thermocouple wire immersed in a similar volume and container to that used for PCRs. The ends of the wire are stripped of their insulation and twisted tightly together. The probe is more stable if the junction is welded or sealed with a drop of solder. The weak voltage generated by the thermocouple is amplified with a battery-powered amplifier, and the signal is converted into 1 mV/°C. This voltage is read using a digital volt meter and can be recorded on a strip chart recorder. It is important both to use small-diameter wire with a small thermal mass to minimize the effect of the probe on the temperature being measured and to make sure that the tip of the probe is kept at the bottom of the tube and immersed in liquid.

The timer/controller is programmed with on and off times, initially determined manually, for the two electrical outlets being used. They are entered into individual statements or "program lines," each of which controls one outlet. An outlet may be controlled by more than one program line. Each time an outlet is turned on or off, a "vary time" is added on to the original, programmed time. The value of the vary time is the length of a complete thermocycle, and its addition causes the complete set of instructions to be repeated indefinitely. A typical program is shown in Table 1, assuming that the heater is connected to outlet number 1, the solenoid valve is connected to outlet number 2, the thermocycle (and vary time) is for

Table 1

Sample ChronTrol Program

Program line no.	Outlet no.	On time	Vary time	Off time	Vary time	Note
1	1	1:00	+650	2:45	+650	a
2	1	3:00	+650	3:05	+650	b
3	1	3:20	+650	3:25	+650	
4	1	3:40	+650	3:45	+650	
5	1	4:00	+650	6:16	+650	c
6	1	6:22	+650	6:26	+650	d
7	1	6:32	+650	6:36	+650	
8	1	6:42	+650	6:46	+650	
9	2	6:52	+650	7:12	+650	e
10	1	7:30	+650	7:32	+650	f

All of the times are given as min:sec or, with the vary times, as plus 6 min and 50 sec. This usage is consistent with that used in programming the timer.

[a]The heater is turned on for 1 min and 45 sec to raise the temperature from 60 to 75°C.

[b]Instructions 2 through 4 turn the heater on for 5 sec and off for 15 sec to hold the temperature at 75°C.

[c]Instruction 5 turns the heater on for 2 min and 16 sec to raise the temperature from 75 to 96°C.

[d]Instructions 6 through 8 turn the heater on for 4 sec and off for 6 sec to hold the temperature at 96°C.

[e]Instruction 9 turns the solenoid value on for 20 sec to bring the temperature down to 55°C.

[f]Instruction 10 turns the heater on for 2 sec and off for 18 sec to hold the temperature at 55°C.

6 minutes and 50 seconds, and thermocycling starts with the reaction chamber at the annealing temperature.

The apparatus was used with the program listed in Table 1 to thermocycle a PCR designed to amplify a 500-base-pair region of the bacteriophage lambda *CI*857 *ind*1 *Sam*7 between positions 7131 and 7630 at the end of the E gene and through the F gene. The two primers, PCR01 and PCR02, are those supplied in the GeneAmp kit from Perkin-Elmer Cetus. The reaction contained 1× *Taq* buffer, 200 μM each dNTP, 1 μM each primer, 10^6 copies of lambda template DNA, 10 units of *Taq* pol 1 in a reaction volume of 400 μl contained in a 1.5-ml microcentrifuge tube overlayed with 200 μl of light mineral oil. Aliquots were withdrawn during thermocycling and subsequently analyzed by electrophoresis on a 3% NuSieve 1% agarose gel in Tris-Borate buffer, as shown in Fig. 2, with a portion of the thermoprofile shown in Fig. 3.

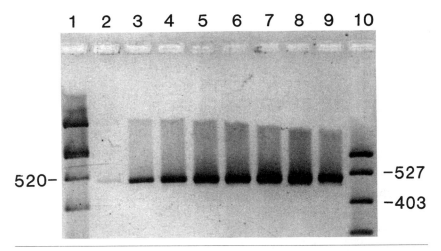

Figure 2 Reverse print of the fluorescence of an ethidium bromide-stained 3% NuSieve 1% agarose gel in Tris borate buffer. Lane 1 is pBR322/*AluI*, and lane 10 is pBR322 *MspI*. Lanes 2–8 are 3-μl aliquots of the reaction taken after 15, 18, 20, 22, 24, 26, 28, and 30 cycles.

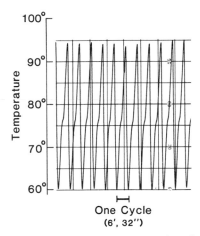

Figure 3 A portion of the thermal profile measured as described during the thermocycling of the reaction.

Supplies

Timer/Controller:
ChronTrol CD4
Lindberg Enterprises
9707 Candida St.
San Diego, California 92126
(619) 566-5656

Solenoid water valve:
Type 525 washing machine valve
Appliance Parts & Equipment Distributors, APED
2901 San Pablo Ave
Berkeley, California 94710

Heater:
Most hardware and small-appliance stores

Flow meter and valve:
VFA-42-BV
Dwyer Instruments, Inc.
P.O.Box 373
Michigan City, Indiana 46360

Type K temperature adapter/amplifier (battery powered):
#L608B101
Jensen Tools
7815 South 46th St.
Phoenix, Arizona 85044-5399

Type K thermocouple wire
#30-K-Tex/Tex
Duro-Sense Corp.
20801 Higgins Court
Torrance, California 90501
(213) 515-1234

Acknowledgments

I wish to thank David Gelfand and John Sninsky for their support and encouragement and Michael Innis for advice, comments, and occasional provocation.

A LOW-COST AIR-DRIVEN CYCLING OVEN

Peter Denton and H. Reisner

PCR is a process that begs for automation. This statement will not be gainsaid by any who have transferred tubes between heat blocks for 30 reaction cycles. Having done so several times, we decided to design an automated system that would meet our particular needs at a reasonable price.

The biochemical reactions that constitute PCR require at least two, or more generally, three incubation ("soak") temperatures of about 55, 70, and 90°C with rapid "ramping" between them. Because optimal temperatures (particularly that of the extension reaction) vary with the nature of the sequence amplified, the soak temperature must be variable within a reasonable range and be maintained accurately (Saiki *et al.* 1986; Saiki *et al.* 1988). Optimal ramp rates remain unclear, but extreme rapidity is desirable if only to minimize the total reaction times. Minimal overshoot of the high soak (denaturation) temperature is essential to prevent heat denaturation of the *Taq* polymerase.

Such basic design considerations apply to most implementations of PCR and can be accommodated in many ways. Our particular application required a variety of reaction vessels (microcentrifuge tubes, microtiter plates, and custom-designed cuvettes). We also desired great flexibility in the programming of operating parameters. Ease of programmability was not considered essential in a developmental design, but economic realities of academic life being what they are, low component cost was a major consideration.

The most obvious approach to an automated PCR device would appear to be a single heat block design with cooling provided by circulation of refrigerant or tap water. Such designs work well at prices ranging from about $4000 to $9000. Several major constraints are intrinsic to single heat block designs. (1) The reaction vessel must be in intimate contact with the walls of the heat-transfer block. This requires an accurately machined block for each type of reaction vessel. Such blocks will constitute a significant percentage of total system cost and require a sophisticated machining ability. (2) Even accurately machined blocks may require a thermally conductive oil to improve heat transfer, as reaction-vessel geometry is not rigidly controlled. (3) Since the entire reaction vessel is not contained within the block, internal temperature gradients can result in condensation on the cooler end of the vessel (commonly the cap of microcentrifuge tubes). For this reason, an oil overlay is generally used to cover the aqueous reaction components. The use of oil is aesthetically unpleasant and can lead to contamination of downstream reactions.

Although we are conditioned by the common use of solid heat blocks and liquid baths as temperature-control devices in the laboratory, air can also serve as an effective heat-transfer medium. Dedicated gas chromatography ovens are adaptable to PCR applications but are inconvenient and uneconomical (Hoffman and Hundt 1988). A custom-built convection oven constructed with easily available components has proved to be an inexpensive device for PCRs in our laboratory. The lack of flowing liquids leads to a potentially safer and simpler design, particularly in a "homebrew" device.

System Overview

The cycling oven system consists of casing, heat source, cool-air source, controller, and auxiliary electrical components. Relatively little thought has been given to optimizing the geometry of the system. A commercial "convection" oven designed for home use provided a convenient case to house the components. The double walled, thin sheet metal oven casing prevents excessive heating of the outer wall but does not prevent rapid ramping.

For a variety of reasons we elected not to use any of the internal electromechanical components of the oven. The internal resistance elements acted as a radiative heat source while the small motor-driven fan did not support a sufficient airflow to allow for rapid convective heating and cooling. A commercial cycling convection oven design we have examined (Biothern Inc., Arlington, Virginia) uses internal-resistance elements as a heat source with the addition of external blowers to provide airflow.

In our design, turbulent heated air is conveniently provided by the integrated resistance element/blower assembly from a hot-air gun. A high-capacity muffin fan provides ambient external air for cooling. Cold air could be provided by running the unit in a cold room or by more sophisticated refrigeration units. Temperature sensor design proved to be extremely critical. The sensor chosen was loaded so as to attempt to duplicate the thermal mass of the reaction vessel. The RTD sensor controller combination used allows for a $\pm 1°C$ accuracy of soak temperature.

Many approaches to controller design are possible based on the abilities and resources of the builder. We chose an integrated proportionating temperature controller that provides sophisticated programmability of soak, ramp, and cycle parameters. The intricacies of the initial programming of the controller are offset by its extreme flexibility, reliable design, and compact size. Microcomputer enthusiasts could easily adapt commercially available sensor controller boards and software or single-board computer/controllers for use in the oven, albeit at a somewhat higher initial cost.

Details of Construction

The following instructions apply specifically to the Toastmaster convection oven (model number 7071). If that device is not locally available, these directions should still be helpful as a general guide for construction and component placement.

Disassembly

1. Lift off the door from its hinges. Store in a safe place. Remove the adjustment knobs and discard. Remove sheet metal screws that secure top/side casing. Save screws and casing. With rubber gloves remove insulation and discard carefully.

2. All the internal components must now be removed and discarded. This includes the entire convection fan assembly, switches, wire, and the oven heating elements. The latter will come out easily if the wire is snipped as close as possible to the ceramic spacer. One might want to retain the power cord and its grommet.

3. The oven core must be separated from the remaining casing. Remove the galvanized back from the core and the base. Remove the six sheet metal screws that secure the core to the chrome front piece. It is not necessary to separate the front and the bottom.

Alterations of Oven Core and Body

1. *Ventilation Port.* With a chassis punch cut a 2-inch-diameter hole in the oven core back. The hole center should be 3 inches from the top and 3 inches from the left side facing into the oven. If a chassis punch is not available, make an array of smaller holes with the same aggregate cross section.

   Inspect the galvanized back sheet and notice the two holes that connect to the raised brackets on the oven core. Draw two lines on the back parallel to the sides and 0.5 inches from those holes toward the center. Extend the line from the bottom up to the first set of perforations. Cut along the lines and the first set of perforations. This removes a large segment of the back but leaves the holes to reattach the remainder to the oven core.

2. *Fan Port.* The fan will be located on the outside of the right-hand-side wall of the oven core (refer to Fig. 1). You will notice that there is already an opening (2 by 3 inches). Center the fan over this opening snug up against a lip of metal on the top. With the fan clamped in place, drill four holes for machine screws to secure the fan. Remove the fan and with sheet metal shears cut away a square hole that does not encroach on the bolt holes. Dress the opening with a file and install the fan.

3. *Heater Port.* Position a hole as shown in Fig. 1. It is important that it be done accurately. A 2.25-inch chassis punch aids a great deal. Otherwise, one might seek the services of a local machine shop. Carefully remove the metal horn of the heater from the fan and heating element. Place the horn through the port and fasten it to the oven wall with four sheet metal screws through the tabs on the horn. When restoring the fan assembly of the heater, position the wires so that they emerge from the top. On

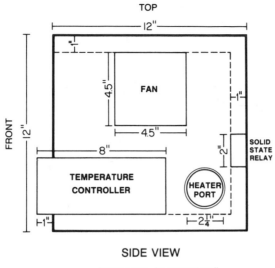

TOP

SIDE VIEW

Dashed line = inner oven wall
Heavy solid line = outer oven wall

Figure 1 Component placement guide. View is of right face of oven with side and top removed.

the bottom of the side there is a second hole (2 by 3 inches). This should be covered over with an appropriate-sized piece of sheet metal.

4. *Controller Port.* If you look at the oven core from the outside on the face that will house the heater and fan, you will see a latch device held in a metal lip. Cut away that lip from 0.25 inches below the latch to the bottom. On the bottom of the chrome panel from which you removed the control switches, there is a small rectangular opening. Along its edge scribe a line exactly 100 mm long, measuring from the right-hand edge of the chromed sheet. Scribe a second line parallel and 95 mm above the first. Connect the two lines with a third on their left-hand ends. Cut away the material defined by the scribed lines. Reassemble the oven core, front, bottom, and cutaway back.

Wiring and Installation of Electrical Components

1. *Component Installation.* Install the solid-state relay (SSR) on inside of back galvanized wall a few inches above the heater. Using the supplied bracket, install the controller.

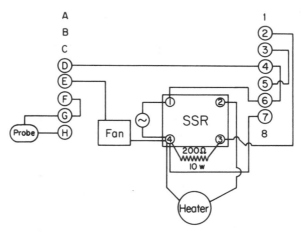

Figure 2   Wiring diagram. Vertical columns of letter and numbers are terminals of controller.

2. *Wiring.* Wiring should be done with good-quality 14-gauge multistrand wire. Proper terminals should be used and a color-coding scheme should be devised (Fig. 2).

3. *Thermal Probe.* Attach 18 inches of small-gauge single-strand wire to the RTD probe. It is best to solder the connections. Check for 100 ohms resistance after connecting. Fill a microfuge tube with silicone rubber by adding a bit at a time and spinning a few seconds on a microfuge. Repeat until full. Put the probe in the bottom of the tube. Allow time for the silicone rubber to set.

Setting Parameters on the Temperature Controller

1. *Calibration.* This section will require reference to the "Series CN-2010 Operator's Manual." Enter PAR/TUNE/LAST simultaneously (paragraph 6.1 of manual), reconfigure the part number by entering AUTO/LAST simultaneously and queuing through the number with PARAM CHECK. The new number is DO 34 21 11 01. Continue toggling through with PARAM CHECK and set C and M:S (paragraphs 6.5 and 6.6). Set LO SPAN at 20 and HI SPAN at 100. Enter RETURN.

2. *Tuning.* Enter tune loop by depressing TUNE/LAST/YES sequentially. The example cited is a typical configuration suitable for a 30-cycle PCR with a denaturation temperature of 90°C for 10 seconds (SP0-TIME 1-SP1), an annealing temperature of 55°C

for 2 minutes (SP2-TIME 3-SP3), and an extension temperature of 70°C for 3 minutes (SP4-TIME 5-SP5). The Operator's Manual can be consulted for the significance of the programmed commands. The controller's prompt is underlined and the suggested reply is set off in brackets. *Alarm 1 and 2* [100]; *Tune outputs?* [YES] (See paragraph 4.2); *Cycle time 1* [15 seconds]; *PR band* [2%]; *Reset 1* [.5 R/M]; *Rate 1* [.33 M]; *Aux SP* [0 DC]; *Cycle time 2* [7 seconds]; *PR band* [3%]; *Reset 2* [.8 R/M]; *Rate 2* [.21 M]; *Ramp and soak* [YES] (See par. 4.4.1 for explanation of the following program loop set off in ; *SP0* [90°C]; *TIME 1* [10 seconds]; *SP1* [90°C]; *TIME 2* [10 seconds]; *SP2* [55°C]; *TIME 3* [2 minutes]; *SP3* [55°C]; *TIME 4* [10 seconds]; *SP4* [70°C]; *TIME 5* [3 minutes]; *SP5* [70°C]; *TIME 6* [00 seconds]; *Cycles* [30]; *Assured soak?* [YES]; *Soak tolerance* [1.0]; [RETURN].

Sources for Specialized Materials

From Omega Engineering (P.O. Box 2669, Stamford, Connecticut 06906): (1) Programmable Temperature controller with Dual Output and Pt RTD, 100 ohm input, 0.1 degree resolution (Model CN-2012[P2]), $755; (2) Solid State Relay (Model SSR 240 A25), $37; and (3) Thin Film Detector (Model TFD), $50.

From Newark Electronics (282 Moody St., Waltham, Massachusetts 02154): (1) Pamotor AC fan (Model 2500S Stock #57F1019), $51.35.

From Black and Decker Service and Parts Center (check the Yellow Pages): (1) Heater and Horn Assembly for Black and Decker Heat Gun Model 9751. The unit consists of nine components that can be ordered and assembled. WARNING: do not substitute other units in this design.

Literature Cited

Hoffman, L. M., and H. Hundt. 1988. Use of a gas chromatograph oven for DNA amplification by the polymerase chain reaction. *BioTechniques* **6**:932–936.

Saiki, R. K., T. L. Bugawan, G. T. Horn, K. B. Mullis, and H. A. Erlich. 1986. Analysis of enzymatically amplified β-globin and HLA-DQα DNA with allele-specific oligonucleotide probes. *Nature (London)* **324**:163–166.

Saiki, R. K., D. H. Gelfand, S. Stoffel, S. Scharf, R. Higuchi, G. T. Horn, K. B. Mullis, and H. A. Erlich. 1988. Primer-directed amplification of DNA with a thermostable DNA polymerase. *Science* **239**:487–491.

53

MODIFICATION OF A HISTOKINETTE FOR USE AS AN AUTOMATED PCR MACHINE

N. C. P. Cross, N. S. Foulkes, D. Chappel, J. McDonnell, and L. Luzzatto

PCR has become established as a powerful technique in molecular biology. Our main interest is to use it as a method for the rapid diagnosis of human genetic disease. The procedure may be performed by manually transferring the reactions between water baths, but this is extremely laborious, especially when the reaction conditions for a new pair of amplimers need to be optimized by trial and error. Several commercial machines are now available for automation of the PCR, but their cost may be beyond the resources of many small laboratories. We have previously reported the low-cost adaptation of a Histokinette for use as an automatic thermocycling apparatus (Foulkes *et al.* 1988). Here we describe in more detail how these modifications were made and our experiences in performing PCR with this machine.

Modification of a Histokinette

A Histokinette (Elliots Ltd., Liverpool, United Kingdom) is a tissue embedding instrument that has been employed in histopathology

PCR Protocols: A Guide to Methods and Applications

laboratories for over three decades. Early models are now often being discarded in favor of more modern equipment.

The Histokinette that we adapted (Fig. 1) consists of a lower unit that contains an electric motor, gearbox, and associated electrical components. One side of this has a panel on which heater outlet sockets and timers are mounted. Located centrally above this unit is a rigid disk to which a circular metal column is attached.

Equally spaced around this are twelve cylindrical containers, three

A

B

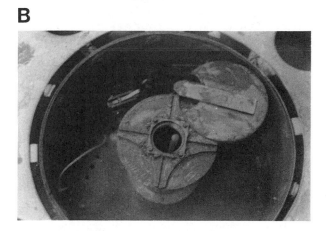

Figure 1   (A) Front view of the modified Histokinette showing the three heated water baths, mechanical arm, and control panel. (B) Inside view showing the new Geneva mechanism.

of which are heated. These rest on the disk and are attached to the column by spring clips. A hardboard disk is located on the top of this assembly, and this rotates. An arm, hinged at the outer edge of this disk, extends across the center to the radial position of the containers and carries the sample holder.

A push rod activated by a motor-driven crank in the lower unit extends centrally up through the whole assembly, contacting the hinged arm to raise and lower the samples into the containers. On the underside of the upper disk a "Geneva" mechanism, driven through a gearbox from the motor, indexes the rotating disk to each container position when actuated by the timer.

Procedure

1. Eight of the nine unheated containers were removed, and the remaining three heated baths were upgraded by changing the heating elements from 100 to 500 W. The insulation material between the inner and outer casing was changed to fiberglass, and a platinum resistance temperature probe was mounted in each container. The containers were also fitted with a safety heat fuse set at 110°C.

2. As only three heated stages were required and the "between stage" time ideally was to be kept to a minimum, the indexing mechanism had to be remade. A four-stage system was chosen to provide a nonheated position for loading and unloading the samples. A new Geneva mechanism, which required precision machining, was constructed. This has the advantage that it gives positive locking and is not affected by any motor overrun. The original motor and gearbox were retained. In this way the "between stage" time was reduced to 15 seconds.

3. An electromechanical timer was installed to control the period of incubation in the water baths. This was triggered by the action of a second timer that controlled the "between stage" time.

4. It proved necessary to fit a specially formed tray to direct condensation and any water that might drip from the sample holder back into the containers. A circular tray was made to fit between the top of the containers and the upper disk. It was fixed by brackets to the circular column. The tray had an outer up-

turned rim and holes corresponding to the container positions, with formed edges to direct any water.

5. A two-tier circular steel rack that can accommodate microfuge tubes (36 by 0.5 ml) was constructed and resuspended from the mechanical arm. During incubation the tubes are completely submerged; this reduces condensation and obviates the need for mineral oil. The microfuge tubes must have tightly fitting lids; we find that those produced by Sarstedt and Scotlab work well.

Notes

After about a month of operation, the hardboard lid that supports the mechanical arm had warped to the point at which its rotation was severely impaired. This problem was solved by fashioning a new lid out of PVC and raising it by 3 mm.

With only a single timer to control the period of incubation, our device is at present limited in that the period at each temperature has to be the same. In many cases this does not produce any apparent difference in the efficiency or specificity of amplification. However, it would be easy to add two extra timers that would allow the times of the three incubations to be preset individually. A revolution counter or a timed on/off switch may also be used to shut down the machine after the required number of cycles.

Examples of Use

We have used the modified Histokinette for over a year in our laboratory. Examples of experiments in which PCR has been successfully performed using the adapted Histokinette are listed as follows:

1. Amplification of specific exons from genomic DNA
   a. Sequencing the coding regions of human G6PD (Foulkes *et al.* 1988; Vulliamy *et al.* 1988; Poggi *et al.*, in preparation), aldolase B (Cross *et al.* 1988; de Franchis *et al.* 1988), and β-globin genes to determine the nature of disease-associated mutations.

    b. Probing with allele-specific oligonucleotides to examine the distribution and frequency of such mutations.

2. Amplification of gene fragments from reverse-transcribed mRNA
    a. Rapid detection of G6PD variants.
    b. Cloning of human acid phosphatase type V using mixed oligonucleotides derived from protein sequence (Lord *et al.*, in preparation).

3. Site-directed mutagenesis
    a. To examine the catalytic properties of the mutant proteins by expression in *Escherichia coli*.

During this time the machine has required no maintenance, apart from lubrication every 3 months. In a comparison between reactions performed on the adapted Histokinette and ones performed on commercially available machines, the results were found to be indistinguishable. The only disadvantage our machine suffers from is that the denaturing water bath needs to be filled up two or three times during a run; therefore, amplifications cannot be performed overnight.

Acknowledgments

We are grateful to Jaspal Kaeda, Tom Vulliamy, Phil Mason, and members of the Medical Physics workshop for valuable contributions. This work was supported by the Medical Research Council of Great Britain.

Literature Cited

Cross, N. C. P., D. R. Tolan, and T. M. Cox. 1988. Catalytic deficiency of human aldolase B in hereditary fructose intolerance caused by a common missense mutation. *Cell* **53**:881–885.

de Franchis, R., N. C. P. Cross, N. S. Foulkes, and T. M. Cox. 1988. A potent inhibitor of *Taq* polymerase copurifies with human genomic DNA. *Nucleic Acids Res.* **16**:10355.

Foulkes, N. S., P. P. Pandolfi de Rinaldis, J. Macdonnell, N. C. P. Cross, and L. Luzzatto. 1988. Polymerase chain reaction automated at low cost. *Nucleic Acids Res.* **16**:5687–5688.

Vulliamy, T. J., M. D'Urso, G. Battistuzzi, M. Estrada, N. S. Foulkes, G. Martini, V. Calabro, V. Poggi, R. Giordano, M. Town, L. Luzzatto, and M. G. Persico. 1988. Diverse point mutations in the human glucose-6-phosphate dehydrogenase gene cause enzyme deficiency and mild or severe hemolytic anemia. *Proc. Natl. Acad. Sci. USA* **85**:5171–5175.

54

ORGANIZING A LABORATORY FOR PCR WORK

Cristián Orrego

The current integration of molecular genetics into almost every field of the biomedical and natural sciences has been intensified by the simplicity and versatility of PCR (White *et al.* 1989). Scientists without a formal background in molecular biology can now quickly learn generic DNA isolation methods applicable to a great variety of biological samples and produce via PCR copious amounts of a specific gene sequence, even from extremely small quantities of complex DNA templates. Refinements in DNA sequencing of amplified products performed with the aid of commercially available kits have made the technique accessible to most laboratories. Recent protocols designed to use nucleotides labeled with the ^{35}S isotope, a low-radiation emitter, have diminished shielding requirements and concerns about radiation exposure (for qualification see Smith *et al.* 1989) that were common with ^{32}P-labeled nucleotides used in earlier sequencing procedures. Finally, the advent of DNA sequencing at high temperature and low ionic strength with the thermoresistant *Thermus aquaticus* (*Taq*) DNA polymerase promises to reduce or eliminate a common problem with DNA sequencing. Until now, template secondary structures have interfered with the sequencing reaction, resulting in termination bands across all four sequencing

PCR Protocols: A Guide to Methods and Applications
Copyright © 1990 by Academic Press, Inc. All rights of reproduction in any form reserved.

lanes or, at best, "ghost" bands that complicate sequence interpretation. Thermal destabilization of such structures eliminates these sequence abnormalities (Innis *et al.* 1988).

The following guidelines are intended to aid researchers planning to include PCR in their work. They highlight some features of the equipment that automate the performance of PCR, steps that can be taken to prevent cross-contamination with amplified DNA sequences, and the material costs of PCR and direct sequencing once the equipment is in place.

A costly item in setting up the PCR is the instrumentation for automated, timed exposures of the reaction tubes to different temperatures for a preset number of cycles. Though manual transfer of reaction tubes through a series of constant-temperature water baths or dry-heat aluminum blocks is an alternative, variations inherent to manual PCR appear to compromise reproducibility of amplifications (Kazazian and Dowling 1988), not to mention the tedium of such a labor-intensive approach.

The cost of commercially available microprocessor-controlled "DNA amplifiers" or "temperature cyclers" varies at present, between U.S. $3000 and $8000. These instruments differ in such features as the design of the cooling system (internal refrigeration unit as opposed to cooling from an external water line), control of ramping time between temperature steps, memory capacity for program storage, sequential linking of different programs, and capacity of the heating block for 0.5-ml conical bottom microcentrifuge tubes.

Low-cost devices adapted from equipment and materials generally available in biomedical laboratories have been described (see Chapters 51–53). Another device consists of a chamber in which samples are exposed to flowing hot or cold water under control of a microcomputer or, in its place, a simple timing device connected to washing machine valves that modulate water flow (Rollo *et al.* 1988; Torgersen *et al.* 1989). Another arrangement calls for performing the amplification in microcapillary tubes with thermal control accomplished with what otherwise is a gas chromatograph (Hoffman and Hundt 1988).

Programmable thermocyclers suffer from a certain lack of thermal uniformity across the block at short step times (a water chamber would, of course, not be subject to this effect). Comparison of temperature profiles in mock reaction fluid at multiple sites in the blocks of two commonly used designs suggests that such temperature gradients occur in all block instruments when programmed for brief step times. However, concerns about temperature heterogeneity diminish when one realizes that successful amplification can be obtained with step times a fraction of those programmed into the instrument and sometimes at temperatures not even reaching the intended values (Fig. 1). Brief temperature pulses for template denaturation, primer annealing, and extension are sufficient for most PCR experiments (Saiki *et al.* 1988a).

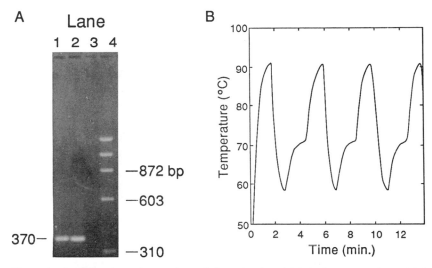

Figure 1 Amplification of a section of the mitochondrial (mt) genome from total *Akodon subfuscus* (a South American rodent) DNA and from purified mt human DNA as determined by agarose gel electrophoresis (A) and a representative temperature profile in the amplification fluid (B). (A) Expected amplification products (370 bp) obtained following loading of 10% of the reaction on a 4% NuSieve agarose gel subjected to electrophoresis and staining by ethidium bromide: *Akodon* (lane 1); human (lane 2); and negative control lacking template (lane 3). Lane 4 contains a molecular weight standard. Amplification was achieved in a 50-μl volume under mineral oil through 27 cycles of PCR with *Taq* DNA polymerase and "universal" amplification primers directed to the *cyt* b gene of the mt genome (Kocher *et al.* 1989). Initial template concentrations were 50 ng of *Akodon* DNA and 12 pg human mt DNA. A Perkin-Elmer Cetus DNA Thermal Cycler was programmed as follows: 93°C for 30 seconds (denaturation), 60°C for 30 seconds (annealing), and 72°C for 1 minute (extension). Reaction tubes were placed in the center of the block. (B) Temperature profiles were obtained with a linear recorder linked to a thin-tip thermocouple inserted into 50 μl of a mock reaction under mineral oil in a 0.5-ml Eppendorf-tip tube placed in the center of the dry block. The thermocouple was calibrated in the block. Agreement within 1°C was obtained between the probe and the temperature readings of the thermocycler at various representative temperatures for PCR.

Preventing Cross-Contamination with Amplified DNA Sequences

Successful amplification via PCR yields at least 1 pmol of a desired product or about 6×10^9 molecules/μl. As this technique permits amplification starting from one template molecule (Saiki *et al.* 1988b; Li *et al.* 1988), steps must be taken to prevent unintended transfer of amplified sequences to items used in the preparation of DNA extracts and PCRs. Cross-contamination becomes evident when amplification occurs in negative controls that did not receive template and should be suspected when more than one ladder is obtained upon sequencing an amplified fragment expected to reveal only one genotype. Precautions should be intensified when performing PCR and sequencing from samples with a small number of initial template molecules. A single sequence might be obtained, but because of infiltration by an overwhelming number of DNA copies from a previous amplification, it will not correspond to the template of interest.

A likely mode of cross-contamination with amplified sequences is via aerosols normally generated during sample processing, as in mixing by vortex action, opening of microcentrifuge tubes, pipetting, ejecting pipette tips from micropipettors, and centrifuging under vacuum. Aerosol particles can be as large as 20 μm in diameter (*Encyclopaedia Britannica* 1986) or about 4×10^{-6} μl in volume. Therefore, one microdroplet of this size produced from a typical PCR will carry about 24,000 copies of an amplified product [transport by airborne particulates is well known in microbiology; in one experiment, about a dozen bacteria were recovered by air-sampling within 2 feet of a fluid decanting operation (National Institutes of Health Biohazards Safety Guide 1974)].

Laboratory Space Allocation

Segregation of laboratory space for all procedures up to and including the setting up of amplification reactions from those following amplification (Table 1) is helpful in preventing undesirable contamination by PCR products. Enhanced containment of amplified sequences is provided by allocating a hood for the handling of PCR products.

Table 1

Laboratory Space Allocation for PCR

Area A Pre-PCR work	Area B Post-PCR work
• Weighing chemicals • Buffer preparation • DNA extraction[a] • Preparation of stocks for PCR • Preparation of PCR	• Electrophoresis of PCR products • Processing of PCR products • Restriction analysis and sequencing of PCR products • Loading of amplified products for second-stage amplification

[a] Segregation of Area A from the area used for growth and extraction of bacterial plasmids having inserts with a target sequence is recommended (Kwok 1989; Chapter 17). The same applies to processing restriction fragments representing such sequences.

Micropipettes

The shaft of a conventional air-displacement micropipette is a likely vehicle for carrying over amplified sequences. This part of the pipette is exposed to aerosols produced during suction of samples into and out of the disposable tip. Therefore, a set of pipettors used exclusively for handling amplified DNA is recommended. In this regard, an extra precaution consists of assigning a set of positive-displacement micropipettors for preparing and pipetting stocks of the individual components of the PCR (buffer, deoxynucleotides, primers, enzyme [Table 2]). Such pipettors employ a disposable piston that moves within a precisely molded capillary, making direct contact with the sample and thus eliminating the air-to-liquid interface between the sample and the shaft of the device. The initial investment for three sets of micropipettors in a nucleic acid laboratory performing PCR (Table 2) is about $1500.

Additional Potential Sources of DNA Carry-Over

Other potential sources of DNA resulting in the intrusion of sequences that will confound amplification experiments are plasmid preparations with inserts containing a sequence amplified often via PCR (Table 1), purified restriction fragments containing a target sequence (Table 1), centrifuge chambers, and instruments (e.g.,

Table 2

Micropipette Sets for the PCR Laboratory

Set A Air displacement	Set B Positive displacement	Set C Air displacement
• DNA extraction[a]	• Preparation of stocks for PCR • Preparation of PCR reactions • Pipetting amplification primers used for DNA sequencing • Preparation of stocks (Proteinase K, RNase) for DNA extraction • Dilution of template DNA	• Analysis of amplified products by minigels, etc. • Sequencing of amplified DNA

[a] Some micropipettes have fluorocarbon shafts resistant to most organic solvents and corrosive materials as well as to autoclaving at 120°C. Therefore, the shaft of the pipettor can be removed for cleaning with a 10% dilution of household bleach (sodium hypochlorite) or by autoclaving. This is recommended when extraction of DNA from forensic or ancient samples is to be undertaken.

forceps, scissors, and microtome blades) used to prepare and analyze tissues.

A simple way of avoiding cross-contamination is to use reagents prepared in small aliquots so that stocks can be discarded when evidence of contamination arises. As far as possible, stocks should be prepared and stored in disposable plastic ware. Extractions for DNA destined to be amplified should also be performed in disposable tubes. The budget for disposable gloves should also be generous, as shed skin cells might very well contribute "background" template activity, especially when amplifying sequences present in multiple copies per cell, e.g., mitochondrial DNA.

Estimated Cost of PCR and Direct Sequencing

The estimated cost (Table 3) is based on sequencing both strands of an amplified 400-bp fragment, using a single sequencing gel (two loadings). The list includes the materials for amplification in two stages: an amplified double-strand product is fractionated by electrophoresis on agarose gels and then used in a second PCR to generate an excess of single-stranded template by unbalanced priming (Gyllensten and Erlich 1988). The single-strand product is purified

Table 3

Estimated Cost of PCR and Direct Sequencing

Material	Cost
Taq polymerase (0.63 units per 25 μl reaction) and other PCR components (except primers) for generation of double-strand product	$ 0.53
Agarose minigel (for product fractionation)	0.75
Polaroid film (two exposures)	1.90
Taq polymerase and other PCR components (except primers) for unbalanced, single-strand amplification (two 50-μl reactions)	1.72
Oligonucleotide primers (125 pmol total for first- and second-stage amplifications	0.04
Microconcentrators for centrifugation dialysis of single-strand templates prior to sequencing	3.20
Sequencing of both strands ($1.85/strand)	3.70
Polyacrylamide sequencing gel	1.20
^{35}S-labeled nucleotide for sequencing	0.50
X-ray film (two exposures)	4.00
	$17.54

by centrifugation dialysis before being subjected to sequencing with the modified form of T7 DNA polymerase supplied in commercially available kits. The list does not include the cost of supplies for DNA extraction, which is negligible for 20- to 25-mg quantities of tissue processed in a microcentrifuge tube.

Savings (approximately $5) can be obtained if components such as buffer and deoxynucleotides are prepared in-house (instead of purchased in kits) and if single-strand amplified products are processed by alcohol precipitation instead of centrifugation dialysis (Chapter 10).

Once optimal conditions for double- and single-strand amplification have been found for a given set of primers, 4 to 5 normal working days are required to run the course from DNA extraction to obtaining sequence from amplified products.

Acknowledgments

I thank Monica Frelow for the experiment displayed in Fig. 1A and W. Rainey for thoughtful review of this article.

Literature Cited

Encyclopaedia Britannica. 1986. Fifteenth Edition. p. 121–122.

Gyllensten, U. B., and H. A. Erlich. 1988. Generation of single-stranded DNA by the polymerase chain reaction and its application to direct sequencing of the HLA DQA locus. *Proc. Natl. Acad. Sci. USA* **85**:7652–7656.

Hoffman, L. M., and H. Hundt. 1988. Use of a gas chromatograph oven for DNA amplification by the polymerase chain reaction. *BioTechniques* **6**:932–936.

Innis, M. A., K. B. Myambo, D. H. Gelfand, and M. A. D. Brow. 1988. DNA sequencing with *Thermus aquaticus* DNA polymerase and direct sequencing of polymerase chain reaction-amplified DNA. *Proc. Natl. Acad. Sci. USA* **85**:9436–9440.

Kazazian, H. H., Jr., and C. E. Dowling. 1988. Laboratory implications of automated polymerase chain reaction. *Biotechnology Laboratory* **6**:23, 26, 28.

Kocher, T. D., W. K. Thomas, A. Meyer, S. V. Edwards, S. Pääbo, F. X. Villablanca, and A. C. Wilson. 1989. Dynamics of mitochondrial DNA evolution in animals: Amplification and sequencing with conserved primers. *Proc. Natl. Acad. Sci. USA* **86**:6196–6200.

Kwok, S. Perkin-Elmer Cetus Polymerase Chain Reaction Workshop. San Francisco, CA. January 31, 1989.

Li, H., U. B. Gyllensten, X. Cui, R. K. Saiki, H. A. Erlich, and N. Arnheim. 1988. Amplification and analysis of DNA sequences in single human sperm and diploid cells. *Nature (London)* **335**:414–417.

National Institutes of Health Biohazards Safety Guide. 1974. U.S. Department of Health, Education, and Welfare. Section IV-20. GPO Stock #1740–00383.

Rollo, F., A. Amici, and R. Salvi. 1988. A simple and low cost DNA amplifier. *Nucleic Acids Res.* **16**:3105–3106.

Saiki, R. K., U. B. Gyllensten, and H. A. Erlich. 1988a. The polymerase chain reaction. In *Gene technology: a practical approach* (ed. K. Davies), p. 141–152. IRL Press, Washington, D.C.

Saiki, R. K., D. H. Gelfand, S. Stoffel, S. J. Scharf, R. Higuchi, G. T. Horn, K. B. Mullis, and H. A. Erlich. 1988b. Primer-directed enzymatic amplification of DNA with a thermostable DNA polymerase. *Science* **239**:487–491.

Smith, I., V. Furst, and J. Holton. 1989. Hazards of Sulphur. *Nature (London)* **337**::696.

Torgersen, H., D. Blaas, and T. Skern. 1989. Low cost apparatus for primer-directed DNA amplification using *Thermus aquaticus*-DNA polymerase. *Analytical Biochem.* **176**:33–35.

White, T. J., N. A. Arnheim, and H. A. Erlich. 1989. The polymerase chain reaction. *Trends in Genetics* **5**:185–189.

BASIC EQUIPMENT AND SUPPLIES

Roberta Madej and Stephen Scharf

The purpose of this section is to provide the researcher with information on the basic equipment and supplies needed to perform PCR. The materials listed are those used in setting up a PCR from DNA, running the PCR, and detecting the PCR product on an agarose gel. The tables do not list items needed to extract and prepare DNA for PCR, nor do they list reagents for the several uses of the PCR product. For efficiency and contamination minimization, some of the items should be dedicated to PCR experimentation only; these items are marked with an asterisk (*).

It is particularly important, when large numbers of samples are being amplified, that certain measures be taken to minimize the risk of carry-over of amplified DNA from sample to sample or from sample to reagent stocks (see Chapter 17). It is suggested that a specific area, preferably a laminar flow hood, be dedicated to the preparation of the PCRs. No amplified product should ever be brought into this hood. A set of pipettors, racks, tubes, etc., should be dedicated to the preparation hood and used only to mix the reactions. It is advisable to have one microcentrifuge in the preparation area so that unamplified DNA (sample) and amplified DNA (product) are never centrifuged in the same machine.

Table 1

Equipment for Polymerase Chain Reaction

Item	Use in PCR	Product Information
DNA Thermal Cycler	Automates PCR	Perkin-Elmer Cetus 761 Main Ave., Norwalk, Conn. 06859
		M. J. Research (for 96-well Kendall Square U-bottom plate) Box 363 Cambridge, Mass. 02142
		Ericomp 10055 Barnes Canyon Rd. Suite G San Diego, Calif. 92121
Microcentrifuge with rcf at least $10,000 \times g^*$	Concentrate reaction at bottom of tube	Eppendorf 20901-051/Brinkmann 5415 VWR Scientific Inc. P.O. Box 7900 San Francisco, Calif. 94120
Positive displacement pipettors to deliver $1-20\ \mu l$ and $20-200\ \mu l^*$	Mixing reaction constituents	Microman M-25, M-50, M-250 Rainin Instrument Co. Inc. 1715 64th St., Emeryville, Calif. 94608
Pipettors, adjustable to deliver $1-1000\ \mu l^*$	Prepare reagent stock solutions/ master mix	Gilson P-20, P-200, P-1000 Rainin Instrument Co.
Freezer to $-20°C$	Store reagents	Major lab suppliers
Balance, accurate to 0.1 mg	Reagent preparation	Major lab suppliers
pH meter	Reagent preparation	Major lab suppliers
Eppendorf repeating pipettor*	Dispense reagents for PCR when several are set up	Eppendorf # P5063-20 American Scientific Products 1430 Waukegan Road McGraw Park, Ill. 60085-6787
Laminar Flow Hood* with UV option	Minimize risk of carry-over when preparing reactions	Baker c/o Tegal Scientific P.O. Box 5905, Concord, Calif. 94524 Major lab suppliers
Agarose submarine gel apparatus minigel size	For analytical gel	Bio-Rad 170-4307 Bio-Rad Laboratories 32nd & Griffin Ave. Richmond, CA 94908-9989 Hoefer, BRL, Aqueboque

Table 1

Continued

Item	Use in PCR	Product Information
Electrophoresis power supply (to 100 mA, 500 V)	For analytical gel	BioRad 200/2.0, BioRad Laboratories Hoeffer, Major lab suppliers
UV Trans-illuminator	For analytical gel	American Scientific Products
Pipettors, adjustable to deliver 1–20 μl	To load analytical gel	Gilson P-20 Rainin Instrument Co.

Table 2

Supplies and Reagents for Polymerase Chain Reaction

Item	Use in PCR	Product Information
Cloned Recombinant *Taq* DNA Polymerase	Catalyzes primer extension	Perkin-Elmer Cetus (Can be purchased in GeneAMP kit with dNTPs, *Taq* buffer, and controls or alone.)
		United States Biochemical Corp. P.O. Box 22400 Cleveland, Ohio 44122
Native *Taq* DNA Polymerase	Catalyzes primer extension	Perkin-Elmer Cetus
		Amersham Corp. 2636 South Clearbrook Dr. Arlington Heights, Ill. 60005
		Beckmann Instruments, Inc. 2500 Harbor Blvd. Fullerton, Calif. 92634
Oligonucleotide primers*	To initiate enzymatic extension on DNA template	Synthetic Genetics 10455 Roselle St. San Diego, Calif. 92121
dNTPs (dATP, dCTP, dGTP, dTTP) HPLC grade*	Deoxynucleotide triphosphates for enzymatic extension	USB 14244, 14279, 14314, 22324 US Biochemical Corporation P.O. Box 22400 Cleveland, Ohio 44122

(*Continues*)

Table 2 (*Continued*)

Item	Use in PCR	Product Information
Microcentrifuge tubes 0.5 ml sterile, siliconized	To hold PCR	Perkin-Elmer Cetus
Mineral oil, light, sterile* (store in small aliquots)	Prevent evaporation of reactions <100 μl	Sigma M3516 Sigma Chemical Co. P.O. Box 14508 St. Louis, Missouri 63178 Perkin-Elmer Cetus
Glass-distilled, sterile water* (store in small aliquots)		
Microman pistons/ capillaries*	For Microman pipettors	CP-25, CP-50, CP-250 Rainin Instrument Co.
Pipette tips	For pipettors	RT-96, RT-200 Rainin Instrument Co.
Sterile Combi-tips	For Eppendorf repeator	P-5063-67, -38, -39 American Scientific Products
Gloves, sterile	To minimize DNase and con-tamination of re-action mixes	American Scientific Products
SeaKem Agarose	For PCR product detection gel (1% Agarose, 3% NuSieve)	FMC BioProducts 5 Maple St. Rockland, Maine 04841
NuSieve Agarose	For PCR product detection gel (1% Agarose, 3% NuSieve)	FMC BioProducts
Ethidium Bromide	To detect PCR product on gel	Sigma Chemical Co.
Bromophenol blue	For gel loading solution	Sigma Chemical Co.
Xylene cyanol	For gel loading solution	Sigma Chemical Co.

Glass-distilled water, sterile mineral oil, and buffers should be aliquoted and stored in small amounts, enough for an average single preparation. Once they are brought into the preparation area, the unused portion should be discarded.

The product/purchase information on each item is suggested only. It is provided for clarification of the materials listed as well as for information to the researcher when setting up a PCR section of the laboratory. Substitutions that serve the same purpose may be used.

INDEX

Aerosols, cross-contamination by,
450
Agarose gel electrophoresis
B-cell lymphoma translocation-
specific fragments, 396
hepatitis B virus products, 354
products synthesized from M13 phage
templates, 174
small subunit rRNA coding regions,
331
Allele-specific oligonucleotide probes
for HLA-DQα typing, 125–126,
262–268
hybridization, analysis of hemophilia
A factor VIII gene
polymorphisms, 294–295
procedure, 296
slot or dot blotting, 295–296
tetramethylammonium chloride
wash, 296
Aminocytes, hemophilia A factor VIII,
291–292
5'-Amino oligonucleotides
purification of labeled PCR primers
from, 102–103
synthesis of labeled PCR primers
from, 101–102
Ammonium acetate
inhibitory effect on Taq DNA
polymerase activity, 136
primer removal with, 78
Ammonium chloride, inhibitory effect
on Taq DNA polymerase activity,
136
Amplification
ancient DNA
collection of samples, 160–161
extraction of DNA, 161–162
jumping PCR, 165

maximum length of mitochondrial
DNA segments, 165
procedure, 162–164
apo-B RNA, 241
asymmetric PCR, unequal primer
concentrations for, 77, 82
bacterial DNA, pathogenic, 403
B-cell lymphoma translocation-
specific fragments, 392–393, 395
biotinylated dUTP incorporation
during, 114–116
cDNA, mixed oligonucleotide primed
(MOPAC) procedure
first-strand cDNA synthesis, 51
fractionation of products, 52–53
with Klenow fragment, 51–52
principles, 47–48
with Taq DNA polymerase, 52
techniques, 48–50
co-amplification
competitive templates for
quantitation of mRNA, 60–61
advantages, 67–68
GM-CSF cDNA quantitation with
genomic GM-CSF
competitive template, 64–67
limitations, 68
PCR procedure, 63–64
reagents, 62
reverse transcription, 62–63
interleukin-2 mRNA quantitation
with internal standard AW106
cRNA, 73–74
with 7-deaza-2'-deoxyguanosine
triphosphate mixtures, 58
with degenerate primers
applications
cloning genes related to known
gene families, 45

Amplification (*continued*)
    detection of novel sequences, 45
    identification and cloning of a
      novel but related gene
      segment (herpesvirus gB
      gene), 42–43
    design of, 40–41
    thermal cycling, 41–42
  DNA flanking sequences by inverse
    PCR
    circularization, 221–222
    DNA digestions, 221
    PCR procedure, 222
  estimated costs, 452–453
  genital human papillomavirus DNA
    sequences
    L1 consensus primers for, 356–359
    PCR procedure, 359–361
  genomic DNA
    buffers, 16
    cycling procedures, 17
      automated, 19
      manual, 18–19
    deoxynucleotide triphosphates, 16
    primer selection, 15–16
    sample preparation, 14–15
    Taq DNA polymerase
      concentrations, 16
    with transcript sequencing
      (GAWTS)
      advantages, 197
      dideoxy sequencing with reverse
        transcriptase, 200–202
      disadvantages, 197–198
      first-strand cDNA synthesis, 199
      PCR procedure, 199–200
      transcription with phage
        polymerase, 200
  hemophilia A factor VIII gene
    polymorphisms, 289–291
    from aminocytes, 291
    from blood, 292
    from chorionic villi, 291
    gel analysis, 292
    oligonucleotide primers and probes
      for, table, 290
  hepatitis B virus DNA, 353–354
  HLA class II gene, sequence of
    oligonucleotide PCR primers for,
    264
  HLA-DQα, procedure, 267–268

  human immunodeficiency virus
    DNA, 342–343
  human mitochondrial D loop
    DNA extraction, 420
    PCR procedure, 421–422
    primers for, 418–419
  human T-cell lymphoma/leukemia
    viruses, 332–333
    carry-over-related false positives, 326
    primers and probes for, 328–329
    procedural safeguards, 326–327
    solutions and reagents for, 330–331
  multiplex, dystrophin gene
    development for other genome
      regions, 279–280
    problems and solutions, 277–279
    procedure, 273–274
    reagents, 273
  novel viral sequences with degenerate
    oligonucleotide primers, 379–381
    preparation of genomic DNA, 381
    primer design, 379–380
    procedure, 381–382
    products
      cloning, 383–384
      gel electrophoresis, 382–383
      recovery of PCR fragments by
        DEAE cellulose, 384
  from paraffin-embedded tissue, 156
  primers, *see* Primers
  products, *see* Products
  RACE protocol
    controls, 36
    fidelity of 5′-end, 37
    full-length cDNA construction, 37
    gene-specific primers, 35
    gene-specific primers for, 35
    genomic DNA and, 34
    materials, 29–31
    nonspecific, 35
    products
      analysis, 32
      artifactual, 33
      blunt cloning, 32
      cloning, 36–37
      directional cloning, 32–33
      truncated, 33
    reverse transcription, 34–35
      truncation of cDNA ends during,
      33
    tailing, 35

truncation
cDNA ends during reverse
transcription, 33
cDNAs during amplification, 34
truncation of cDNAs during, 34
ras-specific sequences in genomic
DNA, 387–388
rDNA sequences from M13 phage, 172
retinoblastoma gene, 411–412
RNA, combined cDNA–PCR
methodology
amount of cytoplasmic RNA
sufficient for, 26–27
analysis of products, 24
cycles, 23–24, 26
deoxynucleotide triphosphates, 22,26
first-strand cDNA sequences, 24–25
magnesium concentration, 26
primers
amount of, 26
choice of, 27
method, 24–25
procedure, 23–24
reagents, 22–23
reverse transcriptase
reaction, 23
source and type of, 25
RNA, transcription-based system
(TAS), 245–247
isolation of nucleic acids from
HIV-1-infected cultured
lymphocytes, 248–249
outline, 246
procedure, 249–250
reagents, 248
specificity enhancement with bead-
based sandwich hybridization
system, 250–251
RNA, with transcript sequencing
(RAWTS)
advantages, 197
dideoxy sequencing with reverse
transcriptase, 200–202
disadvantages, 197–198
first-strand cDNA synthesis, 199
PCR procedure, 199–200
transcription with phage
polymerase, 200
rRNA genes, fungal, 315–320
procedure, 320–321
reagents, 320

rRNA genes, mitochondrial or
chloroplast
errors in amplification process,
312–313
modifications of PCR methods,
309–310
preparation of genomic DNA for
PCRs, 308
primers for, 308–310
speed of, 312
samples for denaturing gradient gel
electrophoresis, 213–215
single-copy-gene fragments, cloning
blunt-end cloning, 88–89
primer design, 85
reaction conditions, 86–87
sticky-end cloning, 87–88
standard protocol, 4–5
Anion-exchange column
chromatography, horseradish
peroxidase–oligomer conjugate, 108
Annealing, primers, 7
nonspecific, false positives due to,
231
Apodachlya mycelium, PCR-amplified
rDNA from, 283
Apolipoprotein-B
mRNA post-translational
modification
cDNA synthesis, 240–241
discriminating hybridization of
sequence-specific
oligonucleotides, 241–242
PCR amplification, 241
production of mRNA, 240
quantitation, 242
relationship between single gene and
its two protein products, 238
Aqualinderella mycelium, PCR-
amplified rDNA from, 283
Artifacts
detection of homologous
recombinants, 229–234
jumping PCR, 165
non-templated, 135
primer dimer, 9, 134–135
RACE protocol, 33
Asymmetric PCR
production of ssDNA
of cloned inserts from bacterial
plasmids, 80

Asymmetric PCR (*continued*)
  of cloned inserts from lambda gt11
    phage plaques, 80
  of cloned inserts from M13 phage
    plaques, 77–79
  cycles required, 82
  deoxynuycleotide triphosphate
    concentrations, 78, 82
  from genomic DNA, 80
  sequencing of single-stranded
    products, 194
  direct sequencing without
    purification, 194–195
  purification, 194

Bacterial pathogens, detection, 399–402
  cell and DNA recovery, 402–403
  dot and slot blots, 404
  hybridization, 404
  PCR amplification, 403
  radiolabeled oligonucleotide probes
    for, 403–404
  reverse hybridization, 404–405
  Southern blots, 404
Bacterial plasmids, single-stranded DNA
    of cloned inserts, production by
    asymmetric PCR, 80
Bacteriophages
  lambda *CI*857 *ind*1 *Sam*7,
    amplification, 432
  lambda gt11
    cDNA library, screening
      materials, 254
      procedure, 254–256
      variations, 256–257
    production of ssDNA of cloned
      inserts by asymmetric PCR, 80
  M13
    amplification of rDNA sequences,
      172
    production of ssDNA of cloned
      inserts by asymmetric PCR,
      77–79
    space promoter mutations, *in vitro*
      transcription assay of PCR
      products synthesized from, 171
B-cell lymphoma, translocation-specific
    fragment
  contamination, 398
  detection of amplified products, 396
  PCR procedure, 395

primers for, 393–395
  standardization of procedures, 397
  template DNA, 393
Bead-based sandwich hybridization,
    detection of TAS-amplified HIV-1
    sequences, 251
Bead-based sandwich hybridization
    system, 250–251
Biotin–psoralen labels, 122
Biotinylated primers
  purification by HPLC and PAGE,
    102–103
  synthesis from 5'-amino
    oligonucleotides, 101–102
  synthesis scheme, 100
Biotinylated probes, vector-free
  incorporation of biotinylated dUTP,
    114–116
  reagents, 114
  sensitivity, 116
  specificity of labeling, 116
Blood
  DNA specimens for CMV detection,
    370
  hemophilia A factor VIII gene DNA
    amplification with *Taq* DNA
    polymerase, 292
  peripheral blood mononuclear cells
    DNA extraction, for hepatitis B
      virus detection, 352
    DNA extraction/preparation, 148,
      352
  white cells, DNA sample preparation
    from, 148
  whole blood, DNA sample preparation
    from, 148
Bovine serum albumin, enzyme
    stabilization with, 7
Buffers
  for cDNA—PCR RNA amplification,
    22, 24
  for genomic PCRs, 16
  recommended, 6

Carrier detection, hemophilia A with
    *Bc1*I polymorphism, 292–293
Chloroplast genomes, nuclear rRNA
    genes, amplification, 308–310
Chorionic villi, hemophilia A factor VIII
    gene DNA amplification with *Taq*
    DNA polymerase, 291

Chromosomes
    chromosome 18 major breakpoint
        region, 392–395
    single, haplotype analysis from,
        301–304
    Y-specific sequences
        amplification, 294
        fetal sexing with, 289–292
Chromosome walking, inverse PCR
    amplification of end-specific
        fragments, 224–225
    circularization, 224
    recovery of yeast artificial
        chromosomes, 223
    restriction fragments, 223–224
ChronTrol CD4 timer/controller
    description, 430
    programming, 431
    sample program, 431–432
Circularization, DNA molecules,
    inverse PCR, 221–222, 224
Cloning
    cDNA with degenerate primers
        (MOPAC procedure)
        first-strand cDNA synthesis, 51
        fractionation of products,
            52–53
        with Klenow fragment, 51–52
        principles, 47–48
        with *Taq* DNA polymerase, 52
        techniques, 48–50
    genes related to known gene families,
        degenerate primers for, 45
    RACE products
        blunt, 32
        directional, 32–33
    single-copy gene fragments from
        genomic DNA, 84–85
        blunt-end cloning, 88–89
        primer design, 85
        reaction constituents, 86–87
        sticky-end cloning
            ligation, 88
            restriction endonuyclease
                digestion, 87–88
    viral sequence products,
        383–384
CMV, *see* Human cytomegalovirus
Competitive PCR, for mRNA
    quantitation, 60–61
    advantages of, 67–68

GM-CSF cDNA with a genomic
    GM-CSF competitive template,
        64–67
    limitations, 68
    procedure, 63–64
    reagents, 62
    reverse transcription, 62–63
Complementary DNA
    cloning with degenerate primers
        (MOPAC procedure)
        first-strand cDNA synthesis, 51
        fractionation of products, 52–53
        with Klenow fragment, 51–52
        principles, 47–48
        with *Taq* DNA polymerase, 52
        techniques, 48–50
    combined methodology for RNA
        sequences
        analysis of amplification products,
            24
        cytoplasmic RNA sufficient for,
            26–27
        deoxynucleotide triphosphates, 22,
            26
        first-strand synthesis, 24–25
        magnesium concentration, 26
        PCR amplification, 23–24
        PCR cycles, 23–24, 26
        primers
            amount of, 26
            choice of, 27
            method, 24–25
            reagents, 22–23
        reverse transcriptase
            reaction, 23
        source and type of, 25
    first-strand synthesis
        cDNA-PCR combined methodology
            for RNA amplification, 24–25
        GAWTS/RAWTS procedure, 199
        MOPAC procedure, 51
        for PCR amplification for RNA
            analysis, 240–241
    first-strand tailing, RACE protocol, 35
    GM-CSF, quantitation with genomic
        GM-CSF competitive template,
            64–67
    lambda gt11 libraries, screening
        materials, 254
        procedure, 254–256
        variations, 256–257

Complementary DNA (*continued*)
  RACE protocol
    amplifications
      cDNA 3' ends, 31
      cDNA 5' ends, 31–32
      gene-specific primers for, 35
      genomic DNA and, 34
      nonspecific, 35
      truncation of cDNAs during, 34
    controls, 36
    fidelity of 5'-end, 37
    full-length cDNA construction, 37
    gene-specific primers, 35
    materials, 29–31
    products
      analysis, 32
      artifactual, 33
      blunt cloning, 32
      cloning, 36–37
      directional cloning, 32–33
      truncated, 33
    reverse transcription, 34–35
    tailing, 35
    truncation
      cDNA ends during reverse
        transcription, 33
      cDNAs during amplification, 34
  retinoblastoma gene, reverse
    transcriptase reaction, 411
Complementary RNA, synthetic AW106
  internal standard, 70–71
  preparation, 72
  quantitation of interleukin-2 mRNA
    with, 73–74
Contamination, 450
  via aerosols, 450
  B-cell lymphoma translocation-
    specific fragments, 396
  laboratory space allocation and, 450
  from micropipettes, 451
  minimization of product carry-over
    aliquot reagents, 143
    during HIV PCRs, 338, 340
    during HTLV PCRs, 326
    laboratory techniques, 143–144
    physical separation of
      amplifications, 142–143
    positive displacement pipettes, 143
    selection of controls, 144–145
  during multiplex DNA amplification,
    277–278

samples by pipetting devices, 143
  sources of, 145, 451–452
Controls, selection of, 144–145
Costs, estimated, for PCR and direct
  sequencing, 452–453
Cross-contamination, *see*
  Contamination
Cycles
  amplification of HPV L1 product from
    a paraffin sample, 156
  in asymmetric PCR, 82
  for cDNA–PCR RNA amplification,
    23–24, 26
  for genomic PCRs, 17
    automated procedure, 19
    manual procedure, 18–19
    optimum number, 8–9
    plateau effect, 10
    retinoblastoma gene amplification,
      412–413
  thermal cycling of degenerate PCR
    primers, 41–42
Cycling ovens, low-cost air-driven,
  435–437
  construction, 437–440
  setting parameters on temperature
    controller, 440–441
  supply sources, 441

DEAE cellulose, recovery of PCR
  fragments with, 384
7-Deaza-2'-deoxyguanosine
  triphosphate, PCR with,
  54–55
  DNA containing, restriction enzyme
    digestion, 57
  mixtures, 55, 58
  procedure, 55–56
  products, ethidium staining and
    radioactivity, 56
Denaturation, times and temperatures, 8
Denaturing gradient gel electrophoresis
  amplified retinoblastoma gene
    products, 408
  DNA polymorphism identification
    with
    advantages, 207, 217
    gel preparation, 215
    parallel gel approach, 208
    parallel procedure, 215–216

perpendicular gel approach,
208–213
perpendicular procedure, 216–217
preparation of PCR-amplified
samples for, 213–215
Deoxynucleotide triphosphates,
concentrations, 5–6
in cDNA–PCR RNA amplification, 26
for genomic PCRs, 16
magnesium ion concentrations and,
136
Detergents, nonionic
DNA solubilization with, 151
enzyme stabilization with, 7
Dictyuchus mycelium, PCR-amplified
rDNA from, 283
N,N-Dimethylformamide, inhibitory
effect on *Taq* DNA polymerase
activity, 136–137
Dimethyl sulfoxide
effects on amplification, 6–7
inhibitory effect on *Taq* DNA
polymerase activity, 136–137
Diploid cells, single, haplotype analysis
artifacts, 301, 303
detection of PCR products, 305–306
lysis, 305
method, 301–304
PCRs, 305
Direct sequencing
estimated costs, 452–453
human mitochondrial D loop, 422,
424–425
with phage promoters
genomic amplification with
transcript sequencing
advantages, 197
disadvantages, 197–198
RNA amplification with transcript
sequencing
advantages, 197
dideoxy sequencing with reverse
transcriptase, 200–202
disadvantages, 197–198
first-strand cDNA synthesis, 199
PCR procedure, 199–200
transcription with phage
polymerase, 200
rRNA genes (fungal), 315–320
procedure, 320–321
reagents, 320

single-stranded products of
asymmetric PCR, 194–185
Discriminant hybridization
apo-B RNA, 241–242
hemophilia A factor VIII gene
polymorphisms, 294–295
procedure, 296
slot or dot blotting, 295–296
tetramethylammonium chloride
wash, 296
D loop, *see* Human mitochondrial
D loop
DMSO, *see* Dimethyl sulfoxide
DNA
ancient
amplification, 162–164
extraction, 161–162
jumping PCR, 165
maximum length of mitochondrial
DNA segments, 165
sample collection, 160–161
for B-cell lymphoma translocation-
specific fragment amplication, 393
deletions at Duchenne muscular
dystrophy locus, multiplex DNA
amplification, 275
extraction
for human mitochondrial D loop
amplification and sequencing,
420
from liver biopsy, 353
from peripheral blood mononuclear
cells, 148, 352
from serum, 352
flanking sequences, inverse PCR
amplification
circularization, 221–222
DNA digestions, 221
PCR procedure, 222
genomic
amplification
buffers, 16
cycling procedures, 17
automated, 19
manual, 18–19
deoxynucleotide triphosphates, 16
β-globin gene (human), 16–17
primer selection, 15–16
sample preparation, 14–15
Taq DNA polymerase
concentrations, 16

DNA (*continued*)
  cloning of single-copy gene
    fragments, 84–85
    blunt-end cloning, 88–89
    primer design, 85
    reaction constituents, 86–87
    sticky-end cloning, 87–88
  polymorphisms, identification by
    denaturing gradient gel
    electrophoresis
    gel preparation, 215
    parallel gel approach, 208
    parallel procedure, 215–216
    perpendicular gel approach, 208
    perpendicular procedure,
      216–217
    preparation of PCR-amplified
      samples, 213–215
  ras-specific sequences, 387–388
  single-stranded DNA production by
    asymmetric PCR, 80
  HLA class II, 261–267
    allele-specific oligonucleotide probe
      hybridization, 268
    detection, 268–269
    DQα-PCR amplification, 267–268
    hybridization, 269–270
    preparation of dot blots, 268
    problems with, 270–271
    rehybridization, 269
    reverse dot blots, 269
  isolation
    by detergent/protease methods, 151
    from fungi, 284–285
    reagents for, 147
    from single spores (*Neurospora
      tetrasperma*), 286–287
  linking fragments, generation by
    inverse PCR, 226
  oligonucleotide ligation assay for
    closely related sequences, 92–94
    design and labeling of
      oligonucleotides, 96
    modifications to, 97–98
    reactions, 96–97
    reagents, 96–97
  ^{32}P-end-labeled fragment for DNase I
    footprinting
    end-labeled primer, 185–186
    extraction and precipitation of PCR,
      187

  materials, 184–185
  PCR reaction, 186–187
  polymorphisms, denaturing gradient
    gel electrophoresis
    advantages, 207, 217
    gel preparation, 215
    parallel gel approach, 208
    parallel procedure, 215–216
    perpendicular gel approach,
      208–213
    perpendicular procedure, 216–217
    preparation of PCR-amplified
      samples for, 213–215
  sample preparation
    from clinical swabs, 147–148
    from white cells or whole blood, 148
  sequencing, *see* Sequencing
  single molecules, haplotype analysis
    of, 301–304
DNase I footprinting, ^{32}P-end-labeled
  fragment for
  DNA starting material, 184–185
  end-labeled primer, 185–186
  extraction and precipitation, 187
  PCR reaction, 186–187
Dot blot hybridization
  genital human papillomavirus
    products, 362–364
  with labeled oligonucleotide probes,
    263, 265, 268, 295–296
  reverse dot blot procedure, 267,
    269
  PCR-amplified pathogenic bacteria
    DNA, 404
Drosophila melanogaster, yeast
  artificial chromosome library,
  chromosome walking in,
  224–226
Duchenne muscular dystrophy locus,
  multiplex PCR
  amplification procedure, 273–274
  development for other genome
    regions, 279–280
  primer sets, table, 276
  problems and solutions
    amount of DNA, 277
    contaminations, 277–278
    specificity, 278–279
  reagents, 273
dUTP, biotinylated, incorporation
  during amplification, 114–116

E. coli DNA polymerase I, endogenous
    DNA in, 138
Endogenous DNA, in *E. coli* DNA
    polymerase I and *Taq* DNA
    polymerase preparations, 138
Enteroviruses
    PCR procedure, 375
    PCR product analysis with minigel
        electrophoresis, 375–376
    primer pair and probes for, 374
    results, analysis, 375–376
    reverse transcription, 375
    RNA extraction, 373–374
    specimen preparation, 373
Enzyme concentrations, 5
Equipment and supplies
    air-driven cycling oven, 435–441
    inexpensive simple method for
        thermocyling PCRs, *see*
        Thermocyling PCRs
    inexpensive thermocycling PCR,
        429–433
    modified Histokinettes, 442–445
    for PCRs, table, 456–458
Estimated costs, for PCR and direct
    sequencing, 452–453
Ethanol, inhibitory effect on *Taq* DNA
    polymerase activity, 137
Ethidium bromide gel electrophoresis,
    viral sequence products, 382–383
Ethidium bromide staining
    B-cell lymphoma translocation-
        specific fragments, 396
    products of PCR with 7-deaza-2'-
        deoxyguanosine triphosphate, 56
Extension, primers, 7–8

False positives, *see* Products, carry-over
    minimization
Fidelity
    5'-end, RACE protocol, 37
    optimization, 10–11
    of PCR products, 10–11
    Taq DNA polymerase, 131–134
Fluorescent-labeled primers
    purification, 102–103
    synthesis, 101–102
Formamide, inhibitory effect on *Taq*
    DNA polymerase activity,
    136–137
Fractionation, MOPAC product, 52–53

Fungal DNA, isolation, 284–285
Fungal rRNA genes, amplification and
    direct sequencing, 315–320
    procedure, 320–321
    reagents, 320

GAWTS, *see* Genomic amplification
    with transcript sequencing
Gelatin
    in buffers for genomic PCRs, 16
    enzyme stabilization with, 7
Gel electrophoresis
    agarose
        B-cell lymphoma translocation-
            specific fragments, 396
        HBV products, 354
        products synthesized from M13
            phage templates, 174
        small subunit rRNA coding regions,
            331
    denaturing gradient, *see* Denaturing
        gradient gel electrophoresis
    ethidium bromide, viral sequence
        products, 382–383
    minigel, enterovirus PCR product,
        375–376
    polyacrylamide gels
        B-cell lymphoma translocation-
            specific fragments, 396
        HPV products, 362
    polyacrylamide-urea, sequencing with
        Taq DNA polymerase on M13-
        based single-stranded and
        asymmetric PCR-derived
        templates, 190
Gene fragments, single-copy, cloning,
    84–85
    blunt-end cloning, 88–89
    primer design, 85
    reaction constituents, 86–87
    sticky-end cloning
        ligation, 88
        restriction endonuclease digestion,
            87–88
Genital human papillomaviruses
    dot blot hybridization, 362–364
    gel electrophoresis, 362
    L1 consensus primers for, 356–359
    L1 product, thermocycling conditions
        for amplification from a paraffin
        sample, 156

Genital human papillomaviruses
(*continued*)
PCR amplification procedure,
359–361
reagents and materials for product
analysis, 361–362
restriction enzyme analysis, 364–365
results, analysis and interpretation,
366
Southern blot hybridization, 362
Genomic amplification with transcript
sequencing (GAWTS)
dideoxy sequencing with reverse
transcriptase, 100–202
first-strand cDNA synthesis, 199
PCR with phage promoters, 199–200
transcription with phage polymerase,
200
Genomic DNA
amplification
buffers, 16
cycling procedures, 17
automated, 19
manual, 18–19
deoxynucleotide triphosphates, 16
β-globin gene (human), 16–17
primer selection, 15–16
sample preparation, 14–15
specific exons with modified
Histokinette, 445–446
Taq DNA polymerase
concentrations, 16
cloned, probes for PCR-amplified
B-cell lymphoma translocation-
specific fragments, 396
cloning of single-copy gene fragments,
84–85
blunt-end cloning, 88–89
primer design, 85
reaction constituents, 86–87
sticky-end cloning
ligation, 88
restriction endonuclease
digestion, 87–88
PCR-assisted oligonucleotide ligation
assay, 92–94
design and labeling of
oligonucleotides, 96
modifications to, 97–98
reactions, 96–97
reagents, 96–97

polymorphisms, identification by
denaturing gradient gel
electrophoresis
gel preparation, 215
parallel gel approach, 208
parallel procedure, 215–216
perpendicular gel approach, 208
perpendicular procedure, 216–217
preparation of PCR-amplified
samples, 213–215
preparation, for amplification of
16S-like rRNA genes, 308
RACE products and, 34
ras-specific sequences, amplification,
387–388
single-stranded DNA production by
asymmetric PCR, 80
β-Globin gene, PCR for 536-bp fragment
of, 16–17
GM-CSF, *see* Granulocyte macrophage
colony stimulating factor
Granulocyte macrophage colony
stimulating factor, cDNA
quantitation with genomic GM-CSF
competitive template, 64–67

Haplotype analysis, from single sperm
or diploid cells
artifacts, 301, 303
detection of PCR products, 305–306
lysis of sperm and diploid cells, 305
method, 301–304
PCRs, 305
preparation of sperm, 304–305
Hemophilia A factor VIII gene
polymorphisms
analysis by hybridization of allele-
specific oligonucleotides,
294–295
procedure, 296
slot or dot blotting, 295–296
tetramethylammonium chloride
wash, 296
carrier detection with *Bc1*I
polymorphisms, 292–293
DNA amplification with *Taq* DNA
polymerase, 289–292
from aminocytes, 291–292
from blood, 292
from chorionic villi, 291
gel analysis, 292

fetal sexing via Y-specific sequence
    amplification, 294
oligonucleotide primers and probes
    for, table, 290
prenatal diagnosis
    with BclI polymorphisms, 293–294
    strategies for, 296–298
Hepatitis B virus
    DNA amplification, 353–354
    DNA extraction
        from liver biopsies, 353
        from peripheral blood mononuclear
            cells, 148, 352
        from serum, 352
    oligonucleotide locations in genome,
        351
    PCR-product analysis
        by agarose gel electrophoresis, 354
        by Southern blot and hybridization,
            354–355
    primers for, 349
    production of biotinylated probe, 115
    reagents, 352
    sequence of synthetic
        oligonucleotides, 350
Herpesvirus gB gene, degenerate primers
    for detection of new sequence
    related to, 42–43
High-performance liquid
    chromatography
    HRP-oligomer conjugates, 108–110
    labeled primers from 5'-amino
        oligonucleotides, 102–103
    5'-tritylmercapto oligonuycleotide,
        104–106
Histokinettes
    amplification of
        gene fragments from reverse-
            transcribed mRNA with, 446
        specific exons from genomic DNA
            with, 445–446
    modification, 442–444
    modification procedure, 444–445
    site-directed mutagenesis with, 446
HIVs, see Human immunodeficiency
    viruses
HLA-DQα DNA typing
    with allele-specific oligonucleotide
        probes, 125–126
    restriction fragment length
        polymorphism approach, 262

sequence-based analysis, 262–267
    allele-specific oligonucleotide probe
        hybridization, 268
    detection, 268–269
    DNA sequence and location of
        allele-specific oligonucleotide
        probes, 266
    DQα-PCR amplification, 267–268
    hybridization, 269–270
    preparation of dot blots, 268
    problems with, 270–271
    protein sequences in, 265
    rehybridization, 269
    reverse dot blots, 269
Homologous recombinants
    artifacts
        false negatives, 229–231
        false positives due to nonspecific
            primer annealing, 231
        false positives due to polymerase
            halt-mediated linkage of
            primers, 231–234
    follow-up, 235
    PCR procedure, 235
    preparation of DNA, 234–235
    sample collection, 234
    screening strategy, 229
Horseradish peroxidase
    labeled probes, from 5'-mercapto
        oligomers
        purification of 5'-tritylmercapto
            oligonucleotide, 104–106
        removal of trityl group from
            oligonucleotide, 106–107
        synthesis, 103–104
    maleimido, preparation, 107–108
    oligomer conjugates
        purification, 108–110
        stoichiometry and quantitation, 110
HPVs, see Genital human
    papillomaviruses
HTLVs, see Human T-cell lymphoma/
    leukemia viruses
Human cytomegalovirus
    hybridization and washing conditions,
        369–370
    primers and probes for, 68–369
    reagents, 369
    specimen isolation and purification,
        370
    thermal profile, 369

Human immunodeficiency virus-1
  detection of transcription-based
      amplification-amplified
      sequences by bead-based
      sandwich hybridization, 251
  isolation of total nucleic acids from
      infected cells, 248–249
  RNA transcription-based
      amplification system, 249–250
Human immunodeficiency viruses
  applications of PCRs, 345–346
  cross-contamination of samples, 338,
      340
  DNA amplification, 342–343
  DNA extraction, 341–342
  effect of primer/template mismatches
      on amplification, 132–133
  interpretation of results, 344–345
  oligomer hybridization of amplified
      products, 343–344
  primers and probes for, table, 339
  product carry-over, 338, 340
  reagents, 340–341
  sample preparation, 341
Human mitochondrial D loop
  amplification
      DNA extraction, 420
      PCR procedure, 421–422
      primers for, 418–419
  determination of material
      relationships via, 424–425
  direct sequencing, 424–425
  population genetics of, 422–423
Human papillomaviruses, see Genital
      human papillomaviruses
Human T-cell lymphoma/leukemia
      viruses, detection
  carry-over related false-positive
      results, 326
  hybridization
      liquid, procedure, 333–334
      solutions and reagents, 330–331
      spot blot method, 334–335
  PCRs
      procedure, 332–333
      solutions and reagents, 330–331
  primers and probes for, table, 328–329
  procedural safeguards, 326–327
Hybridizations
  bead-based sandwich system, 250–251
  discriminant

apo-B RNA, 241–242
  hemophilia A factor VIII gene
      polymorphisms, 294–295
  procedure, 296
  slot or dot blotting, 295–296
  tetramethylammonium chloride
      wash, 296
dot blot
  HPV products, 362–364
  with labeled oligonucleotide probes,
      263, 265, 268, 295–296
  reverse dot blot procedure, 267, 269
  liquid, HTLV DNA, 333–334
  with nonradioactive labeled probes,
      122–124
oligomer, detection of amplified HIV
      products by, 343–344
pathogenic bacterial DNA, 404
Southern blot
  HBV products, 354–355
  HPV products, 362
spot blot, HTLV DNA, 334–335

Instrumentation, ChronTrol CD4 timer/
      controller, 430–432
Interleukin-2 mRNA, quantitation in
      Jurkat cells with internal standard
      AW106 cRNA, 73–74
Inverse PCR
  for chromosome walking within yeast
      artificial chromosomes, 222–223
      amplification of end-specific
          fragments, 224–225
      circularization, 224
      recovery of yeast artificial
          chromosomes, 223
      restriction fragments, 223–224
  circularization, 221–222
  DNA digestions, 221
  for generation of linking fragments,
      226
  PCR reaction, 222
  for site-directed mutagenesis, 226–227
Isopropanol, primer removal with, 78

Jumping PCR, 165

Klenow fragment
  MOPAC procedure with, 51–52
  3' termini repair for blunt-end
      cloning, 89

Laboratory organization, 447–449
  equipment for automation, 448–449
  prevention of cross-contamination,
    see Contamination
Laboratory techniques
  carry-over-related HIV false positives,
    338, 340
  carry-over-related HTLV false
    positives, 326
  to minimize PCR-product carry-over,
    143–144, 450–452
Lambda phage plaques, single-stranded
    DNA of cloned inserts, production
    by asymmetric PCR, 80
Liquid hybridization, HTLV DNA,
    333–334
Liver biopsies, DNA extraction, for HBV
    detection, 353
Lymphomas, B-cell, see B-cell
    lymphomas

Magnesium ion
  in cDNA–PCR RNA amplification,
    26
  deoxynucleotide triphosphate
    concentration and, 136
  standard, 6
  Taq DNA polymerase requirements,
    136
Major breakpoint region, chromosome
    18, 392–395
5'-Mercapto oligomers
  purification of 5'-tritylmercapto
    oligonucleotide, 104–106
  removal of trityl group from
    oligonucleotide, 106–107
  synthesis of horseradish peroxidase-
    labeled probes from, 103–104
Messenger RNA
  apo-B, post-translational modification
    discriminating hybridization of
      sequence-specific
      oligonucleotides, 241–242
    PCR amplification, 241
  quantitation
    competitive PCR, 60–61
      advantages of, 67–68
      GM-CSF cDNA with a genomic
        GM-CSF competitive
        template, 64–67
      limitations, 68

procedure, 63–64
      reagents, 62
      reverse transcription, 62–63
    by PCR with synthetic AW106
      cRNA internal standard,
        70–71
      gel electrophoresis, 73
      interleukin-2 quantitation in
        Jurkat cells, 73–74
      PCR procedure, 72–73
      plasmid pAW106 structure, 71
      quantitation procedure,
        73–74
      reagents, 72
      reverse transcriptase reaction,
        72
      RNA preparation, 72
    reversed-transcribed, amplification
      gene fragments with modified
      Histokinette, 446
Micropipettes, contamination from,
    451
Minigel electrophoresis, enterovirus
    PCR product, 375–376
Misincorporation, 10–11
Mismatches, 11
Mitochondrial genomes, nuclear rRNA
    genes, amplification, 308–310
Mixed oligonucleotides primed
    amplification of cDNA (MOPAC)
  first-strand cDNA synthesis, 51
  fractionation of products, 52–53
  with Klenow fragment, 51–52
  principles, 47–48
  with Taq DNA polymerase, 52
  techniques, 48–50
MOPAC procedure, see Mixed
    oligonucleotides primed
    amplification of cDNA
M13 phage, see Bacteriophages
Multiplex PCR, Duchenne muscular
    dystrophy locus
  amplification procedure, 273–274
  development for other genome
    regions, 279–280
  primer sets, table, 276
  problems and solutions
    amount of DNA, 277
    contaminations, 277–278
    specificity, 278–279
    reagents, 273

Mutagenesis
  by combining two PCR products with
    overlapping sequence, 177–180
  misincorporation by *Taq* DNA
    polymerase, 182–183
  primary PCRs, 180–181
  removal of excess primers,
    181–182
  secondary PCR, 182
  saturation, 22-bp spacer promoter
    (yeast), 170–172
  site-directed
    inverse PCR for, 226–227
    with modified Histokinette, 446
  Taq DNA polymerase, 134
Mutants, spacer promoter (yeast), *in
  vitro* transcription assay, 175–176

Neurospora tetrasperma single spores,
  PCR-amplified rDNA from,
  286–187
Nonisotopically labeled primers
  fluorescein-labeled
    purification, 102–103
    synthesis, 101–102
  purification, 102–103
  synthesis from 5'-amino
    oligonucleotides, 101–102
Nonisotopically labeled probes
  dot-blot-based format
    biotin–psoralen labels, 122
    color development, 124–125
    HLA-DQα typing with, 125–126
    horseradish peroxidase labels, 122
    hybridization with, 122–124
    instruments and supplies, 120–121
    membrane preparation, 121–122
    reagents, 120
    troubleshooting
      faint signals, 126
      high background color on
        membrane, 126–127
      high background signals, 126
      negative control reaction with
        positive signal, 127
  horseradish peroxidase-labeled,
    synthesis from 5'-mercapto
    oligomers, 103–104
  horseradish peroxidase–oligomer
    conjugates

  purification, 108–110
  stoichiometry and quantitation,
    110–111
  preparation of maleimido horseradish
    peroxidase, 107–108
  purification of 5'-tritylmercapto
    oligonucleotide, 104–106
  removal of trityl group from
    oligonucleotides, 106–107

Oligomers
  5'-biotinylated
    purification, 102–103
    synthesis, 101–102
  fluorescent-labeled primers
    purification, 102–103
    synthesis, 101–102
  hybridization, detection of amplified
    HIV products, 343–344
  radioactively labeled, screening of *in
    vitro* amplified *ras*-specific
    sequences, 386, 389
  5'-thiolated, conjugates
    preparation of maleimido
      horseradish peroxidase,
      107–108
    purification, 108–110
    stoichiometry and quantitation,
      110
Oligonucleotide ligation assay,
    PCR-assisted, 92–94
  design and labeling of
    oligonucleotides, 96
  modifications to, 97–98
  reactions, 96–97
  reagents, 96–97
Oligonucleotide primers
  degenerate, amplification of novel
    viral sequences, 378–381
    amplification procedure, 381–382
    cloning of amplified products,
      383–384
    design of primers, 379–380
    gel electrophoresis of amplified
      products, 382–383
    preparation of genomic DNA, 381
    recovery of PCR fragments by DEAE
      cellulose, 384
  for hemophilia A factor VIII gene
    polymorphisms, table, 290

Oligonucleotide probes
allele-specific
for hemophilia A factor VIII gene
polymorphisms, 290
procedure, 296
slot or dot blotting, 295–296
tetramethylammonium chloride
wash, 296
for HLA-DQα typing, 125–126,
262–268
for B-cell lymphoma translocation-
specific fragments, 396
for pathogenic bacteria, 403–404
Optimization
bovine serum albumin, 7
buffers, 6–7
cycle number, 8–9
denaturation time and temperature, 8
deoxynucleotide triphosphates, 5–6
DMSO use, 6–7
enzyme concentration, 5
fidelity, 10–11
gelatin, 7
magnesium concentration, 6
nonionic detergents, 7
plateau effect, 10
potassium chloride, 6–7
primers
annealing, 7
concentration, 9
design of efficient, 9–10
extension, 7–8
standard amplification protocol, 4–5
Tris–HCl, 6
Ovens, cycling, 435–437
construction, 437–440
supply sources, 441
temperature controller, 440–441

Paraffin-embedded tissues, sample
preparation from
deparaffinizing sections, 155
PCR with, 156
proteinase digestion, 155–156
reagents, 154
tissue section preparation, 154
Peripheral blood mononuclear cells
DNA extraction, for hepatitis B virus
detection, 352
DNA extraction/preparation, 148, 352

Phage promoters, direct sequencing
with, 199–200
Phylogenetic relationships, analysis by
amplification and direct sequencing
of rRNA genes, 315–320
Pipetting devices, cross-contamination
of samples by, 143
Plasmids
bacterial, production of ssDNA of
cloned inserts by asymmetric
PCR, 80
pAW106, structure, 71
Plateau effect, 10
Platelet glycoprotein Ib, sequence
amplification from HEL cell cDNA
library, 256
Point mutations, detection in *ras* gene
filter preparation, 388–389
oligonucleotide hybridization, 389
PCR procedure, 388
prehybridization, 389
principle of method, 387–388
washing procedures, 390
Polyacrylamide-urea gel electrophoresis,
sequencing with *Taq* DNA
polymerase on M13-based single-
stranded and asymmetric
PCR-derived templates, 190
Polymerase halt-mediated linkage of
primers, 231–232
Polymorphisms
genomic DNA, identification by
denaturing gradient gel
electrophoresis
gel preparation, 215
parallel gel approach, 208
parallel procedure, 215–216
perpendicular gel approach,
208–213
perpendicular procedure, 216–217
preparation of PCR-amplified
samples, 213–215
hemophilia A factor VIII gene, DNB
amplification with *Taq* DNA
polymerase, 289–291
Population genetics, human
mitochondrial D loop, 422–423
Potassium chloride
in buffers for genomic PCRs, 16
effect on primer annealing, 6

Potassium chloride (*continued*)
inhibitory effect on *Taq* DNA
polymerase activity, 136
Prenatal diagnosis, hemophilia
with *Bcl*I polymorphism, 293–294
strategies for, 296–298
Preservation, samples, 149–150
Primer–dimer products, 134–135
Primers
annealing, 7
nonspecific, false positives due, 231
for B-cell lymphoma translocation-
specific fragment, 393–395
biotinylated
purification by HPLC and PAGE,
102–103
synthesis from 5'-amino
oligonucleotides, 101–102
synthesis scheme, 100
cDNA–PCR RNA sequence
amplification
amount of, 26
choice of, 27
method, 24–25
for cloning single-copy gene
fragments, 85
concentrations, 9
degenerate, for cDNA cloning
(MOPAC procedure)
first-strand cDNA synthesis, 51
fractionation of products, 52–53
with Klenow fragment, 51–52
principles, 47–48
with *Taq* DNA polymerase, 52
techniques, 48–50
degenerate, for DNA amplification
applications
cloning genes related to known
gene families, 45
detection of novel sequences, 45
identification and cloning of novel
but related gene segments
(herpesvirus gB gene),
42–43
design, 40–41
thermal cycling, 41–42
degenerate oligonucleotide, for novel
viral sequences, 378–381
cloning of amplified products,
383–384
design of primers, 379–380

gel electrophoresis of amplified
products, 382–383
preparation of genomic DNA, 381
procedure, 381–382
recovery of PCR fragments by DEAE
cellulose, 384
for Duchenne muscular dystrophy
gene multiplex amplification,
276–277
efficient, design of, 9–10
5'-end add-on sequences, 85
for enterovirus PCR amplification, 374
extension, 7–8
fluorescent-labeled
purification, 102–103
synthesis, 101–102
for genomic PCRs, 15–16
for hepatitis B detection, 349
for human cytomegalovirus detection,
368–369
for human immunodeficiency viruses,
table, 339
for human mitochondrial D loop
amplification and sequencing,
428–429
for human *ras* gene amplification, 387
for human T-cell lymphoma/leukemia
virus amplification/detection,
328–329
mismatched, *Taq* DNA polymerase,
131–134
oligonucleotide, for hemophilia A
factor VIII gene polymorphisms,
290
for ^{32}P-end-labeled DNA fragment,
185–186
polymerase halt-mediated linkage of,
231–232
primer–dimer products, 134–135
RACE protocol, 30, 35
removal
with ammonium acetate or
isopropanol, 78
procedure, 181
for retinoblastoma gene, 408–409, 413
sequential extension during
amplification of ancient DNA,
165
for 16S-like rRNA genes, 308–310
unequal concentrations, asymmetric
PCR with, 77

Probes
  allele-specific oligonucleotide, for
    HLA-DQα DNA typing,
    262–268
  cloned genomic DNA, for PCR-
    amplified B-cell lymphoma
    translocation-specific fragments,
    396
  for enterovirus PCR amplification, 374
  horseradish peroxidase-labeled, from
    5'-mercapto oligomers
    purification of 5'-tritylmercapto
      oligonucleotide, 104–106
    removal of trityl group from
      oligonucleotide, 106–107
    synthesis, 103–104
  for human cytomegalovirus detection,
    368–369
  for human immunodeficiency virus
    detection, table, 339
  for human T-cell lymphoma/leukemia
    virus amplification/detection,
    328–329
  nonisotopically labeled, dot-blot-
    based format
    biotin–psoralen labels, 122
    color development, 124–125
    horseradish peroxidase labels, 122
    hybridization with, 122–124
    instruments and supplies, 120–121
    reagents, 120
    troubleshooting, 126–127
  oligonucleotide
    for B-cell lymphoma translocation-
      specific fragments, 396
    for hemophilia A factor VIII gene
      polymorphisms, 290
    for pathogenic bacteria, 403–404
  vector-free biotinylated
    incorporation of biotinylated dUTP,
      114–116
    reagents, 114
    sensitivity, 116
    specificity of labeling, 116
Products
  carry-over minimization
    aliquot reagents, 143
    during HIV PCRs, 338, 340
    laboratory techniques, 143–144
    physical separation of
      amplifications, 142–143

    positive displacement pipettes, 143
    selection of controls, 144–145
  detection
    B-cell lymphoma translocation-
      specific fragment, 396
    pathogenic bacterial DNA, 403–404
    single DNA molecules, PCR-
      amplified, 305–306
  enterovirus PCR, minigel
    electrophoresis, 375–376
  genital human papillomavirus
    dot blot hybridization, 362–364
    gel electrophoresis, 362
    reagents and materials for analysis
      of, 361–362
    restriction enzyme analysis,
      364–365
    Southern blot hybridization, 362
  hepatitis B virus DNA
    agarose gel electrophoresis, 354
    Southern blot and hybridization,
      354–355
  nonisotopic detection
    allele-specific oligonucleotide
      probes for HLA-DQα typing,
      125–126
    biotin–psoralen labels, 122
    color development, 124–125
    horseradish peroxidase labels, 122
    hybridization with with probes,
      122–124
    instruments and supplies, 120–121
    reagents, 120
    troubleshooting, 126–127
  with overlapping sequence,
    combination of, 177–180
  primer–dimer, 134–135
  single-stranded asymmetric PCR-
    derived, sequencing
    direct sequencing without
      purification, 194–195
    purification for, 194
  viral sequences, novel
    cloning, 383–384
    gel electrophoresis, 382–383
Proteinase digestion, paraffin-embedded
  samples, 155–156
Proteinase K, DNA/RNA release with,
  151
Pythium mycelium, PCR-amplified
  rDNA from, 283

Quantitative PCR
competitive templates, 60–67
mRNA with synthetic AW106 cRNA
internal standard, 71–74

RACE protocol, *see* Rapid amplification
of cDNA ends protocol
Radioactivity, products of PCR with
7-deaza-2'-deoxyguanosine
triphosphate, 56
Rapid amplification of cDNA ends
(RACE)
amplifications
gene-specific primers for, 35
genomic DNA and, 34
nonspecific, 35
truncation of cDNAs during, 34
controls, 36
fidelity of 5'-end, 37
full-length cDNA construction, 37
gene-specific primers, 35
materials, 29–31
products
analysis, 32
artifactual, 33
blunt cloning, 32
cloning, 36–37
directional cloning, 32–33
truncated, 33
reverse transcription, 34–35
truncation of cDNA ends during, 33
tailing, 35
truncation
cDNA ends during reverse
transcription, 33
cDNAs during amplification, 34
ras gene, point mutations
filter preparation, 388–389
oligonucleotide hybridization,
389–390
PCR procedure, 388
prehybridization, 389
primers for, 387
principle, 387–388
washing procedures, 390
RAWTS, *see* RNA amplification with
transcript sequencing
Reagents
aliquotted, 143
for DNA and RNA isolation, 147
for PCRs, table, 457–458

Recombinant PCR
add-on mutagenesis, 180
5'-add-on sequences, 178
base substitution, 180
deletions, 178, 180
insertions, 180
misincorporation by *Taq* DNA
polymerase, 182–183
primary PCRs, 180–181
process of, 178–180
removal of excess primer, 181–182
secondary PCR, 182
Recombinants, homologous, *see*
Homologous recombinants
Restriction endonuclease cleavage maps,
M13mp10::sp, 171
Restriction endonucleases, for inverse
PCR, 221
Restriction enzyme digestion
7-deaza-2'-deoxyguanosine
triphosphate-containing DNA, 57
genital human papillomavirus
products, 364–365
Retinoblastoma gene
enzymes and buffers, 409–410
PCR procedure, 411–412
PCR temperature profile, 412
primers for PCR, 408–409, 413
reverse transcriptase reaction for
cDNA, 411
Reverse dot blot hybridization, with
labeled oligonucleotide probes, 267,
269
Reverse transcriptase
cDNA synthesis from retinoblastoma
RNA, 411
in competitive PCR for mRNA
quantitation, 62–63
dideoxy sequencing with, GAWTS/
RAWTS procedure, 200–202
first-strand cDNA synthesis, for PCR
amplification for RNA analysis,
240–241
for RNA amplification with cDNA–
PCR methodologies, 22–23, 25
Reverse transcription
enteroviruses, 375
in RACE protocol, 34–35
cDNA 3'ends, 31
cDNA 5'ends, 31–32
truncation of cDNA ends during, 33

Ribosomal DNA, isolation
  from fungi, 283
  from single ascospores (*Neurospora
    tetrasperma*), 283
Ribosomal RNA
  fungal genes, amplification and direct
    sequencing, 315–320
    procedure, 320–321
    reagents, 320
  mitochondrial or chloroplast
    genomes, amplification,
    307–308
    errors in amplification process,
    312–313
    modifications to PCR methods,
    309–310
    preparation of genomic DNA for
    PCRs, 308
    primers for PCR amplification,
    308–310
    speed of amplification, 312
  from M13 phage, PCR amplification,
    172
RNA
  amplification with transcript
    sequencing (RAWTS)
    dideoxy sequencing with reverse
     transcriptase, 100–202
    first-strand cDNA synthesis, 199
    PCR with phage promoters,
     199–200
    transcription with phage
     polymerase, 200
  analysis, PCR amplification for, 241
  extraction, enteroviruses, 373–374
  isolation, reagents for, 147
  sample preparation
    from blood cells, 148–149
    from viral pellets, 149
  sequence amplification, combined
    cDNA–PCR methodology
    amount of cytoplasmic RNA
     sufficient for, 26–27
    analysis of products, 24
    cycles, 23–24, 26
    deoxynucleotide triphosphates, 22,
     26
    first-strand cDNA synthesis,
     24–25
    magnesium concentration, 26
    PCR procedure, 23–24

primers
  amount of, 26
  choice of, 27
  reagents, 22–23
  reverse transcriptase
    reaction, 23
    source and type of, 25
  transcription-based amplification
    system, 245–247
    detection of HIV-1 in cultured
     lymphocytes
     bead-based sandwich
      hybridization, 250–251
     isolation of toal nucleic acids,
      248–249
    procedure, 249–250
    reagents, 248
    specificity enhancement with bead-
     based sandwich hybridization
     system, 250
RNA : DNA hybrid RNase protection
  assays, amplified retinoblastoma
  gene products, 408

Sample contamination, *see*
  Contamination
Sample preparation
  ancient DNA
    collection, 160–161
    extraction, 161–162
  DNA
    from clinical swabs, 147–148
    from white cells or whole blood,
     148
    genomic DNA, 14–15
     for denaturing gradient gel
      electrophoresis, 213–215
    human immunodeficiency virus
     DNA, 341
    from paraffin-embedded tissues
     deparaffinizing sections, 155
     PCR with, 156
     proteinase digestion, 155–156
     reagents, 154
     tissue section preparation, 154
  RNA
    from blood cells, 148–149
    from viral pellets, 149
Sample preservation, 149–150
Saturation mutagenesis, 22-bp spacer
  promoter (yeast), 170–172

Screening, lambda gt11 cDNA libraries
 materials, 254
 procedure, 254–256
 variations, 256–257
Sequencing
 direct
  human mitochondrial D loop, 422,
   424–425
  with phage promoters
   genomic amplification with
    transcript sequencing
    advantages, 197
    dideoxy sequencing with
     reverse transcriptase,
     200–202
    disadvantages, 197–198
    first-strand cDNA synthesis,
     199
    PCR procedure, 199–200
    transcription with phage
     polymerase, 200
   RNA amplification with
    transcript sequencing
    advantages, 197
    dideoxy sequencing with
     reverse transcriptase,
     200–202
    disadvantages, 197–198
    first-strand cDNA synthesis,
     199
    PCR procedure, 199–200
    transcription with phage
     polymerase, 200
  rRNA genes (fungal), 315–320
   procedure, 320–321
   reagents, 320
   single-stranded products of
    asymmetric PCR, 194–195
  rRNA genes (mitochondrial,
   chloroplast), 312
  with Taq DNA polymerase
   annealing template and primer,
    192–193
   labeling reaction, 193
   reagents, 191–192
   single-stranded products of
    asymmetric PCR
    direct sequencing without
     purification, 194–195
    purification, 194
   termination reactions, 193–194

Serum, DNA extraction for HBV
 detection, 352
Shuffle clones, 90
Site-directed mutagenesis, with
 modified Histokinette, 446
Slot blotting
 bacterial pathogens, 404
 hemophilia A factor VIII gene
  polymorphisms, 295–296
Sodium chloride, inhibitory effect on
 Taq DNA polymerase activity, 136
Sodium dodecyl sulfate, inhibitory
 effect on Taq DNA polymerase
 activity, 136–137
Solvents, Taq DNA polymerase,
 136–138
Southern blot hybridization
 genital human papillomavirus
  products, 362
 hepatitis B virus products, 354–355
 pathogenic bacteria, 404
Specificity, Duchenne muscular
 dystrophy PCRs, 278–279
Sperm, single, haplotype analysis
 artifacts, 301, 303
 detection of PCR products,
  305–306
 lysis, 305
 method, 301–304
 PCRs, 305
 preparation, 304–305
Spores, single, rDNA PCR amplification,
 286–187
Spot blot hybridization, HTLV DNA,
 334–335
Supplies, for PCRs, table, 457–458
Swabs, DNA sample preparation from,
 147–148

Tailing, first-strand cDNAs, RACE
 protocol, 35
Taq DNA polymerase
 addition of single nontemplate-
  directed nucleotide to blunt-
  ended duplex DNA fragment, 135
 bovine serum albumin effects, 7
 concentrations
  for genomic PCRs, 16
  recommended range, 5
  for RNA amplification, 23

denaturation time and temperature, 8
dimethyl sulfoxide effects, 6
DNA sequencing with
    annealing template and primer,
        192–193
    autoradiograph, 190
    labeling reaction, 193
    reagents, 191–192
    single-stranded products of
        asymmetric PCR
        direct sequencing without
            purification, 194–195
        purification, 194
    termination reactions,
        193–194
endogenous DNA in, 138
estimated cost of, 453
extension of mismatched primers,
    131–134
fidelity, 131–134
gelatin effects, 7
half-life at elevated temperatures,
    table, 131
hemophilia A factor VIII gene DNA
    amplification, 289–291
    from aminocytes, 291–292
    from blood, 292
    from chorionic villi, 291
    gel analysis, 292
inhibitors, 136–138
ionic requirements, 136–138
KCl effects, 6
magnesium ion concentration, 6
misincorporation by, 182–183
MOPAC with, 52
mutagenesis, 134
nonionic detergent effects, 7
primer annealing, 7
properties, 134–135
solvents, 136–138
temperature optimum, 130–131
template requirements, 134–135
Taq I restriction endonuclease, cleavage
    of 7-deaza-2'-deoxyguanosine
    triphosphate-amplified DNA, 58
TAS, *see* Transcription-based
    amplification system
Temperature optimum, *Taq* DNA
    polymerase, 130–131
Temperature probes, construction,
    431

Temperatures
    denaturation, 8
    primer extension, 7–8
Templates
    Taq DNA polymerase requirements,
        134–135
    yeast spacer promoter mutants, *in
        vitro* transcription assay,
        175–176
Tetramethylammonium chloride wash,
    296
Tetramethylbenzidine chromogen, 124
Thermocycling PCRs, 429–433
    amplification of bacteriophage lambda
        *CI*857 *ind*1 *Sam*7, 432
    ChronTrol CD4 timer/controller
        description, 430
        programming, 431
        sample program, 431–432
    flow meter and adjustment valve,
        430
    heater, 430
    solinoid valve, 430
    supply sources, 434
    temperature probes, 431
Thermophilic organisms,
    characterization, 129–130
Thermus aquaticus, isolation and
    characterization, 130–131
Thraustotheca mycelium, PCR-
    amplified rDNA from, 283
Times
    denaturation, 8
    primer extension, 7–8
Tissue, DNA specimens for CMV
    detection, 370
Total coliform bacteria, PCR-gene probe
    methods for, 400–401
Transcription-based amplification
    system, 245–247
    detection of HIV-1 in cultured
        lymphocytes with
        bead-based sandwich hybridization,
            250–251
        isolation of toal nucleic acids,
            248–249
    procedure, 249–250
    reagents, 248
    specificity enhancement with bead-
        based sandwich hybridization
        system, 250

Transcription *in vitro*, PCR products synthesized from M13 phage carrying promoter mutations, 173–175

Tris–HCl
in buffer for genomic PCRs, 16
standard, 6

5′-Tritylmercapto oligonucleotide, purification, 104–106

Truncated products, RACE protocol, 33

Truncation
cDNA ends during amplification, 34
cDNA ends during reverse transcription, 33

Urea, inhibitory effect on *Taq* DNA polymerase activity, 137

Urine, DNA specimens for CMV detection, 370

Viral pellets, RNA sample preparation from, 149

Von Willebrand factor, sequence amplification from endothelial cell cDNA library, 255–256

Water, bacterial pathogens, *see* Bacterial pathogens

White blood cells, DNA sample preparation from, 148

Whole blood, DNA sample preparation from, 148

Y chromosome sequences, hemophilia A factor VIII gene
DNA amplification with *Taq* DNA polymerase, 289–292
fetal sexing with, 294

Yeast 35S rRNA spacer promoter mutants, *in vitro* transcription assay, 175–176